Reading Architectural Working Drawings

Residential and Light Construction

Edward J. Muller

Philip A. Grau III

Upper Saddle River, New Jersey
Columbus, Ohio

Library of Congress Cataloging in Publication Data

Muller, Edward John
  Reading architectural working drawings: residential and light construction/Edward J. Muller, Philip A. Grau III. —6th ed.
    p. cm.
  Includes index.
  ISBN 0-13-111468-9 (pbk)
    1. Architecture—Designs and plans—Working Drawings. I. Title.

NA2713 .M85 2004
720'.28'4—dc22

2003053697

Editor in Chief: Stephen Helba
Executive Editor: Ed Francis
Assistant Editor: Linda Cupp
Production Editor: Holly Shufeldt
Production Coordination: Carlisle Publishers Services
Design Coordinator: Diane Ernsberger
Cover Designer: Mark Shumaker
Cover Art: Corbis
Production Manager: Matt Ottenweller
Marketing Manager: Mark Marsden

This book was set in Times by Carlisle Communications, Ltd. It was printed and bound by Banta Book Group. The cover was printed by Phoenix Color Corp.

Copyright © 2004, 2000 by Pearson Education, Inc., Upper Saddle River, New Jersey 07458.
Pearson Prentice Hall. All rights reserved. Printed in the United States of America. This publication is protected by Copyright and permission should be obtained from the publisher prior to any prohibited reproduction, storage in a retrieval system, or transmission in any form or by any means, electronic, mechanical, photocopying, recording, or likewise. For information regarding permission(s), write to: Rights and Permissions Department.

Earlier editions, entitled *Reading Architectural Working Drawings: Basics, Residential, and Light Construction,* © 1996, 1988 by Prentice-Hall, Inc.

Pearson Prentice Hall™ is a trademark of Pearson Education, Inc.
Pearson® is a registered trademark of Pearson plc
Prentice Hall® is a registered trademark of Pearson Education, Inc.

Pearson Education Ltd.
Pearson Education Singapore Pte. Ltd.
Pearson Education Canada, Ltd.
Pearson Education—Japan

Pearson Education Australia Pty. Limited
Pearson Education North Asia Ltd.
Pearson Educación de Mexico, S.A. de C.V.
Pearson Education Malaysia Pte. Ltd.

ISBN 0-13-111468-9

PREFACE

Reading Architectural Working Drawings, Sixth Edition, is intended to be an introduction to architectural working drawings for architectural and civil technology students, as well as for tradespeople, contractors, and the vast number of other people who must read and interpret the information found in construction documents including working drawings and specifications.

This useful text covers the basics of architectural graphics, reading symbols, conventions, terms of the trade, and reading residential and other light-frame construction drawings. In addition, material on freehand sketching is presented so that students and others can further improve their communication skills when dealing with the technical information found in construction documents. A set of working drawings of a popular style of home by Design Traditions of Atlanta, along with related worksheets, is included.

This new edition has been revised in a number of areas. All material graphic symbols have been updated to comply with the Uniform Drawing System and the United States National CAD Standard. The section on architectural abbreviations has been updated to comply with the UDS and the National CAD Standard. It has also been reorganized with the associations and organizations listed separately. The sections on computers and drawing printing have been updated to reflect current practices.

ACKNOWLEDGMENTS

I would like to thank the Construction Specifications Institute and *BUILDER* magazine for their permission to use the updated MasterFormat™ and Figure 1–14 of the Punch List/Strata System, respectively. I would like to acknowledge a special debt to the late Edward J. Muller, the original author, who initially put together this excellent and unique combination of construction information and drawing reading assignments. I plan to keep this solid foundation while continuing to update the text body and drawings as changes in construction and construction documentation occur in the future.

Philip A. Grau III
NCARB, AIA, CSI

CONTENTS

1. **Introduction**   1

   A. Early Working Drawings   1

   B. Present Use of Construction Drawings   4

   C. Working Drawings as a Means of Communication   5

   D. Developing Visualization Ability   7

   E. Regional Variations of Drawings   7

   F. Construction Procedure: Residential and Light Commercial   9

   G. Types of Construction Drawings   12

   H. How Various Types of Prints Are Made   15

   I. Computer-Aided Design and Drafting (CADD)   17

   J. How to Get the Most from This Workbook   23

2. **Background Principles**   25

   A. Principles of Technical Projection   25

   B. Architectural Applications   28

   C. Sections   30

   D. Auxiliary Views   36

   E. Pictorial Drawings   36

   F. The Meaning of Lines on Drawings   39

   G. Review of Construction Mathematics   41

H. Symbols Used on Working Drawings   51

I. Glossary of Construction Terms   58

J. Abbreviations Used in Architectural Drawings   65

K. Reading Scales and Dimensions   73

L. Reading Modular Dimensions   79

M. Metric Dimensions   83

N. Pencil Sketching   86

O. Exercises in Basic Principles   100

3. Residential Working Drawings   135

A. Survey Plats and Plot Plans (Site Plans)   135

B. Foundation (Basement) Plans   140

C. Floor Plans   156

D. Elevations   174

E. Sections and Details   182

F. Interior Elevations   195

G. Heating and Cooling Plans (HVAC)   199

H. Plumbing Plans   202

I. Electrical Plans   203

J. Framing Plans   203

4. Specifications   209

A. Their Purpose   209

B. How Specifications Relate to Drawings   210

C. How Information Is Arranged in Specifications   210

D. The CSI Format   212

E. How to Find Information Quickly in Specifications   219

5. Reading the Working Drawings of a Small Church 265

    A. Start with a General Concept of the Building Structure   268

    B. Reading the Small Church Drawings   269

Appendix   281

Index   295

Architectural Working Drawings

(These architectural plans are located after the index)

Chapter 1

Introduction

A. EARLY WORKING DRAWINGS

Architectural working drawings, although probably very crude in appearance before the invention of paper, have nevertheless been in use almost as long as recorded history. Temples and dwellings occupied people's interest long before the Machine Age, which, because of the discovery of iron and steel, brought with it the need for engineering drawings. Architectural drawings, however, preceded any of the other types of technical drawings by many years and have continued to play an important role in human activity to the present time.

The earliest evidence available showing the use of a working drawing for the construction of a building is now found in the Louvre in Paris (see Figure 1–1). It is a floor plan of a temple shown on the statue of Gudea. The statue, which dates from about 2130 B.C., is a well-preserved example of early Chaldean art. The sculptor evidently thought enough of the Chaldean engineer Gudea to preserve not only his likeness but also evidence of his skill by including this ancient temple layout. Although records are not clear, the plan is probably the temple of Ningirsu, located in the area later known as Babylon. It is remarkable how this plan, engraved on a tablet of stone, resembles those in use today although "drawn" thousands of years before paper was invented.

Many years after early Babylonian civilizations, construction drawings presumably still consisted merely of a plan laid out directly on the ground. Eventually, single-line drawings of the plan made on paper or other convenient material came into use; but for a long time only the plan drawing was made. Even as late as several hundred years ago, the only drawing generally required for residential buildings was a simple plan. Later, as buildings became more ornate and complex, appearances of specific features like arches or decorations had to be formulated beforehand and drawings of them were required. Then entire exterior views were needed and eventually scaled the same as the plan so that measurements could be taken off one view for the other and direct projections could be made. Architects have retained this practice of direct graphic projection between views to the present day.

The development of drawings took place in various civilizations in different times throughout history (see Figure 1–2), making it difficult to give construction drawings as we know them a definite time of origin.

We see, then, that the need for drawings existed even in early civilizations when building concepts were limited by the materials available. Through the years, new materials, tools, and scientific knowledge, to say nothing of the aspirations for esthetic beauty, have made our buildings more complex and therefore more dependent on carefully executed working drawings. If architecture itself has reflected a measure of our civilization in terms of shelter and the need for artistic expression, then surely the working drawings that provide

FIGURE 1–1

Statue of Gudea, from about 2130 B.C., showing the oldest working drawing in existence.

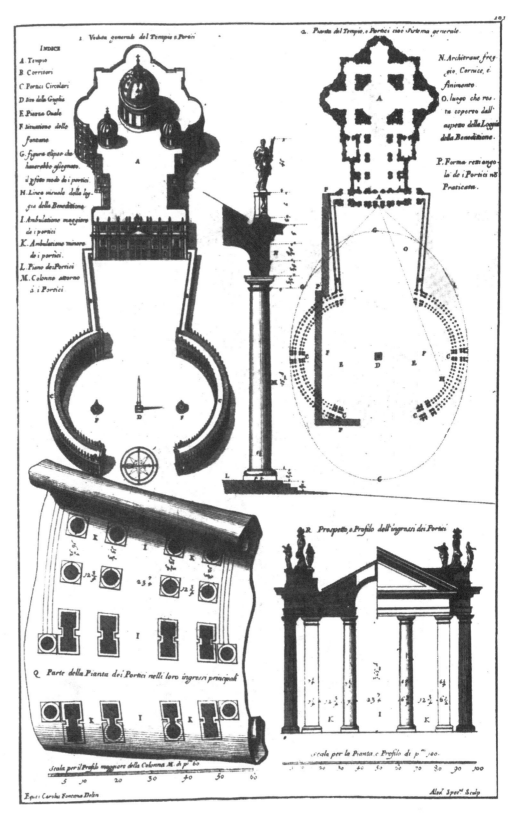

FIGURE 1–2

Working drawing from the Vatican, cir. 1500.

the communication among creative designers, engineers, and tradespeople should also reflect a similar measure.

Modern buildings, as well as the older monuments still in evidence throughout the world, testify to the creativity of architects and builders and to their graphic ability for representing complex ideas on drawings.

B. PRESENT USE OF CONSTRUCTION DRAWINGS

Today, construction occupies 8.3 million Americans in one capacity or another, according to the U.S. Department of Labor. This represents 8.1 percent of the country's private industry nonagricultural payroll. The continually increasing demand for buildings of all types indicates that the number of people involved in their construction will steadily increase in the future.

The construction industry engages the services not only of architects, engineers, and draftspeople, who prepare the drawings, but also of numerous skilled tradespeople, such as carpenters, masons, plumbers, and electricians, who actually do the construction (Figure 1–3). Also, administrators, material suppliers, supervisors, foremen, inspectors, contractors, and others are directly involved in a supporting role and must communicate with many of the rest. The magnitude of the industry and the interdependence of the many services involved make a common language for accurate communication essential.

FIGURE 1–3

Construction is one of our major industries and requires accurate communication between designers and builders.

C. WORKING DRAWINGS AS A MEANS OF COMMUNICATION

The pencil has been referred to as the world's most potent instrument, for it gives reality to human thoughts and ideas. Mental pictures have no value unless they can be drawn graphically. In technology, particularly, ideas must be put on paper before they can be of any use to anyone. Drawings thus become the "windows" through which we see things created in someone's mind, even though the creative effort is often taken for granted. The drawings must have a consistent method for depicting information in order to be useful, and the builder must be able to understand the designer's intention. No drawing would be useful unless the user could interpret the intent of the draftsperson. That is why a working drawing must have consistent methods for showing a building and still retain some latitude to allow for individual variations. Basic technical projection concepts, which have been in use for many years, are of course the foundation for this graphic representation, and we shall consider these concepts more fully in the next chapter. Consistent symbols representing complex features of construction are also used (see Figures 2–25, 2–26, and 2–27) and must be understood by those reading a drawing. Other consistencies, such as abbreviations, methods of showing dimensions, and placements of specific information on drawings, will also be noted as the student becomes more familiar with the various working drawings.

We must remember that drawings are prepared with the intention of being read by those somewhat experienced in construction work. Often minor "standardized" features of construction, for the sake of economy in preparing drawings, have been entirely omitted. Yet this omission must be understood by the reader. For example, you would seldom see on a floor plan the actual layout of studs used in wood-frame construction or the nails used to fasten them together; yet it is taken for granted that standard 16″ or 24″ spacing of studs will be used to build the wall and sufficient nails will be used to fasten them together. Unless aware of these omissions, the novice might find the drawing incomplete or difficult to understand. Sometimes, however, you see excessive omissions of important information; these omissions weaken the working drawings and often result in costly mistakes by workers on the job. Critical details are completely shown on well-developed drawings that give all information clearly. Skimpy details and poorly thought-out drawings contribute to construction mistakes. Such mistakes can be costly to contractors and often result in inferior quality. The preparation of working drawings, then, requires knowing the precise directions needed by the various trades to complete their work correctly, as well as the unimportant information that can be omitted.

Many trades must of necessity work closely together if the job is to progress satisfactorily. Therefore, information for one trade cannot always be isolated from other trade information on drawings. Much of it is interrelated. The complete picture is the usual intent of a set of drawings. Thus, when reading any part of a set, the total concept must be kept in mind. Often, bits of information from one drawing must be verified on other parts of the set. But the total impression should leave little chance for misinterpretation.

Drawings represent not only a graphic language but also a way of technical thinking. The concepts of a building must be created in terms of the conventional methods of drawing. Only those who can draw are able to conceive the building and its structure. The discipline required in drawing is reflected in predetermining shapes and construction and is thus a primary tool in design. But the interpretation of a drawing need not require the ability to draw it. For example, one need not be a writer to be able to enjoy a book. The ability to interpret drawings comes with learning the consistent conventions of the language, as previously mentioned, and with deliberate application of these conventions to the reading of various types of working drawings.

The word *plan* has several meanings in connection with construction work. We often hear it used in a general sense, "a set of plans" to build a house, for instance. This layperson's usage commonly refers to the complete set of drawings used for the construction. However, in architectural terminology a *plan* is a graphic diagram showing a building's layout on a horizontal plane, such as is shown in Figure 1–4, as well as in other figures in this manual. A *Roof Plan* would be the *top view* of a building, whereas a *Floor Plan* would be the view of a horizontal cut through the structure revealing the layout of the walls and the

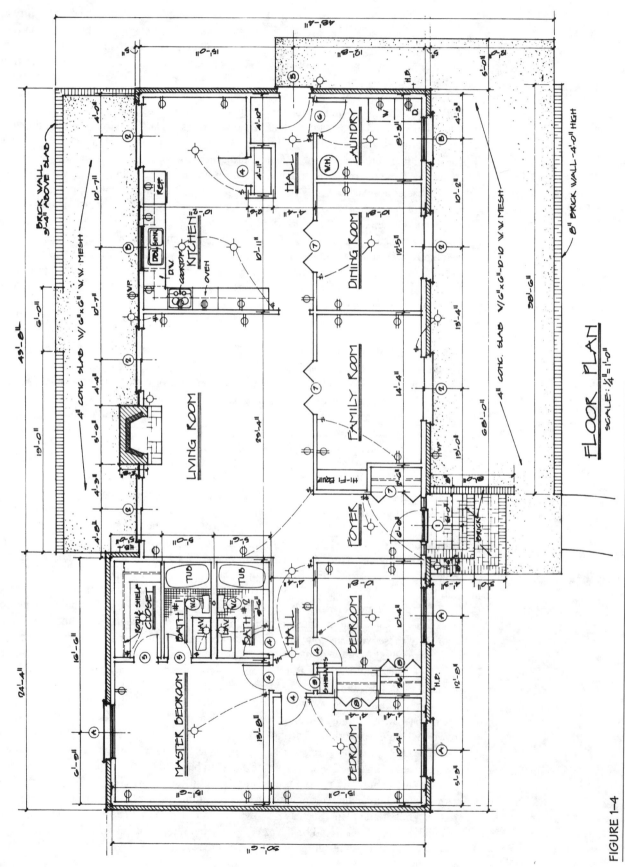

FIGURE 1–4

A floor plan shows the layout of a building.

room arrangement. Plans such as Basement Plans, Foundation Plans, and Floor-Framing Plans would also be horizontal section views through corresponding levels of the building necessary for revealing information specific to those areas. Remember that a plan refers only to *horizontal* views or layouts. Moreover, the word *plan* might be used as a method of procedure or as a plan of action. We plan things when we think them out prior to doing them. One plans a building when actually making the drawings for it. Surely a floor plan, for instance, represents considerable planning before the drawing can be made.

In observing the various plans throughout this manual, notice that floor plans show much more information than mere wall and room arrangement. Symbols are combined on the plans to show type and location of doors, windows, electrical outlets, plumbing fixtures, heating equipment, fireplaces, stairs, flues, and cabinets, as well as other necessary features. Furthermore, symbols indicate the actual material intended, such as wood, concrete, and steel. Numerical dimensions also indicate the size and placement of features. Because the inclusion of so much information on a floor plan might become confusing, it is sometimes necessary to provide separate plans for specific trade information, for instance, an Electrical Plan or a Heating Plan, in addition to the regular plan. These separate plans simplify the sifting of specific trade information on complex drawings.

The complete instructions for the construction of a building consist of two major parts: a *set of working drawings* and a *set of written specifications* to accompany the drawings. Both are important in conveying complete information. Together with the agreement and conditions of the contract, these instructions constitute what are known as the *Contract Documents*.

In summary, the following suggestions will help you to understand the language of working drawings in the shortest possible time:

1. Learn the terminology currently used in the trade.
2. Learn the meanings of various lines, symbols, and abbreviations found on drawings.
3. Learn the principles of technical projection.
4. Learn the specific application of these principles to architectural drawings.
5. Learn conventional architectural dimensioning methods, architectural scales, and the arithmetic involved.
6. Learn the basic construction techniques commonly found in modern buildings.
7. Practice reading various types of drawings so that precise information can be quickly obtained.

D. DEVELOPING VISUALIZATION ABILITY

Reading working drawings requires an innate *visualization* ability. In looking over a drawing, you must be able to see in your mind the actual full-size building and the structural features within it. You must be aware of the three-dimensional volumes represented, even though their representations are shown on the flat planes of the paper. While searching for information on the drawings, imagine yourself within the building, walking about as you move from one symbol to another. This is a necessary state of mind and is not difficult to attain. The designer who created the building possessed this visualization when he or she made the drawing; you must similarly utilize it in interpreting the ideas. Each symbol or line represents an actual feature. *Therefore, you must train your mind to visualize the features and be continually conscious of the reality of the building if you are to understand the drawing clearly.*

E. REGIONAL VARIATIONS OF DRAWINGS

Slight variations on working drawings are found in different regions of the country. Minor symbol variations or practices adopted in local areas occasionally appear and are used for no reason other than habit or the whim of the drafters. Generally, these variations are self-explanatory; if they do have major significance, legends are included to explain them.

FIGURE 1–5

Buildings vary according to local climate.

Other variations that must be kept in mind concern *local climate* and weather conditions existing in different areas (Figure 1–5). Construction, of course, must be modified to cope with these conditions. For example, in cold climates, footings must be placed deep in the ground below the frost line to prevent upheavals or fracturing of the foundations, whereas in warmer climates, the footings are usually placed only far enough below grade to rest upon solid, undisturbed soil. In areas plagued by hurricanes, earthquakes, unstable subsoils, and so on, more steel may be necessary (in the concrete foundations) than elsewhere. In some areas where *subsoils* are unreliable, wood or concrete pilings must be driven deep into the ground to provide a stable base for foundations. The amount of dampness in a given area often dictates the choice of materials and the type of construction. In addition, the local *availability* of building materials is reflected in the working drawings. To avoid shipping costs, it is usually more economical to use locally obtained materials. However, if cost is not an object, materials are commonly transported to nearly all areas of the country to satisfy modern construction.

Moreover, one finds certain *styles* of architecture more popular in one area than in another; these tastes are reflected in the appearance of working drawings in different areas.

Local building *codes* usually dictate variations in construction. To protect both the owner and the public, restrictions must be made on both the choice of materials and the type of construction. Fire hazards, public health hazards, and zoning regulations prohibiting inferior construction adjacent to quality construction—all place restrictions on the buildings as well as the drawings, and compliance with all governing codes is mandatory. Many of the standards spelled out in the codes have been reached through years of experience with local conditions.

If the student reading a drawing realizes the need for these variations, there will be little difficulty in understanding the variations encountered.

F. CONSTRUCTION PROCEDURE: RESIDENTIAL AND LIGHT COMMERCIAL

The broad field of construction is commonly divided into residential, light commercial, heavy commercial, and industrial types of buildings. *Residential* and *light commercial* buildings are for the most part wood-framed with some masonry and little steel or concrete construction, whereas heavy commercial and industrial buildings utilize mainly steel, concrete, and masonry materials in their structures. The construction procedure for residential and light commercial buildings involves different techniques, even different types of tradespeople, than heavy commercial construction, and we will concern ourselves mainly with them in this basic manual.

In the last decade the boom in the home-building industry has spawned procedures that make the purchase of homes simpler for the individual buyer. Contacting a real estate firm is usually the first step that a family takes to purchase a home. Only a small percentage of homes are now built in the traditional way by securing a lot, hiring an architect to draw the plans, and contracting a builder to construct the house. This procedure is still used, however, for the most expensive and individualized homes that are built in exclusive residential areas. More commonly, the buyer is shown a finished house by the real estate agent in a residential development that has been constructed by a home-builder as a speculative venture.

Typically, the residential building process starts in the office of a *developer*. This is an entrepreneur who usually specializes in residential development.

A market survey to determine the economic feasibility of the demand for homes in a particular area is usually the first step that the developer takes. Also, a preliminary engineering study must be made to determine the possible drainage problems, the availability and capacity of utilities, and the number of lots the property would yield. With an estimate of the cost of improvements that must be made, the developer can then determine the approximate *cost per lot* in the subdivision. If the results of these studies are favorable, the tract of land is purchased (or options to purchase are obtained) to develop the proposed subdivision. Many times it becomes necessary to have the land *rezoned* so it will conform to the type of homes planned within the development.

Final engineering drawings (see Figure 1–6) are then made. These show street and lot layout, necessary drainage structures, road profiles, utility connections, and curb and gutter locations. Local government approval is usually required before construction can be started. Periodic inspections are also made by government officials to ensure compliance with local codes and ordinances, since streets are usually deeded to (and maintenance is usually assumed by) the local government. The cost of this construction is borne by the developer and often is financed by local lending institutions.

Most developers then sell the lots to a number of selected home-builders who agree to construct houses of a similar style and price range in the subdivision. While the developer may aid in promoting the development, the responsibility for the sale of the houses rests with the home-builders. Many times a real estate firm may assume the marketing responsibility for an entire subdivision, but the buyer purchases the home from the builder. To protect all builders who are working in the subdivision, the developer generally requires that each builder construct houses that are within the quality and price range predetermined by the market survey. Since the quality of the finished home is usually dependent on the capability of the builder, the successful developer is careful in the selection of each builder working in the subdivision.

Construction drawings used by home-builders are generally obtained from a *home plan service*. Most of the plan services provide plan books that show elevations and schematic floor plans. From these, various styles are selected. The construction drawings usually include the floor plan(s), foundation plan, all elevations, and some of the details. The number of details needed depends on the type of builder constructing the house. Those who build only a few homes each year usually do not require carefully detailed plans, since they are experienced with the traditional building techniques and they personally supervise the entire construction. Builders who construct many homes per year need completely detailed plans, since the supervision will usually be delegated to various supervisors in their employ.

FIGURE 1-6

Site plan of a new development.

Whereas architectural control of the subdivision is retained by the developer, and approval from him or her must be obtained, the style of a home for a particular lot is usually selected by the builder. When a buyer contracts to purchase a home from a builder before construction starts (known as *pre-sale* by builders), the buyer may be allowed to make some changes in the plan or to select the desired floor plan from several designs. Usually, most plan services will make minor changes in the construction drawings for a small additional cost. Many also offer *custom-design* services, in which case the plans are extensively detailed and the additional cost is figured accordingly. This solution is often satisfactory inasmuch as the finished home will meet the requirements of the buyer. These custom-designed drawings usually require the approval of the subdivision developer.

1. MANUFACTURED HOMES

A number of companies throughout the country are now engaged in fabricating complete homes within their plants for delivery in various stages of completion. Even larger frame buildings such as motels, shopping centers, apartments, and condominiums are now panelized in-plant and transported directly to on-site locations. These firms usually have catalogs of their stock designs available for customer selection. Other than choosing a suitable exterior design and floor plan, the home buyer is confronted with few, if any, working drawings. Manufactured housing minimizes labor costs while making a wide variety of home styles and final costs available to the public. Reputable firms now produce manufactured buildings of quality construction.

2. ARCHITECT-DESIGNED HOMES

Higher-quality homes, as previously mentioned, are usually custom designed by architects or home designers who specialize in residential work. After a lot is obtained by the client, whether an isolated lot or one within a subdivision, the services of an architect are secured to generate the working drawings and specifications. Numerous idea-sketches and consultations are required to develop a client's exact needs and wishes. Particular attention is given to utilizing the natural aspects of the property in the design. Although considerable expense is involved in engaging the services of the architect or designer, the result is usually more satisfactory than the previously described procedures, since all the desires of the client are combined before final working drawings are made.

Once the owner has a set of drawings and specifications, a construction loan must be obtained from one of the local lending institutions, unless the owner has sufficient money for the entire construction. Many lenders are authorized to offer FHA (Federal Housing Administration) loans at minimum interest rates if the construction meets FHA standards. The FHA merely backs the security of the loan through the lending institution. After sufficient financing is secured, a reputable contractor is hired to complete the project. Often the architect will assist in searching for a reliable contractor. The set of drawings and specifications becomes part of the contract documents, and full compliance with its intent is agreed upon.

Payment for the construction is usually made to the contractor in prearranged installments as major phases of the construction are satisfactorily completed, thus ensuring compliance with the agreement. After completion of the entire project, the temporary construction loan is converted into a permanent-type loan between the lending institution and the owner, and a monthly repayment schedule covering principal, interest, and other minor fees is established.

G. TYPES OF CONSTRUCTION DRAWINGS

For identification, drawings associated with construction work may be divided into several categories closely associated with their purposes: Preliminary Drawings, Presentation Drawings, Working Drawings, and Shop Drawings.

1. PRELIMINARY DRAWINGS

As the name implies, preliminary drawings are the beginning drawings prepared by the architect during the promotional stage of the building development. They form a convenient basis for communication between the designer and the owner during this period. The drawings are not meant to be used for the construction but merely for exploration of original concepts, functional studies, material selection, preliminary cost estimates, and preliminary approval by civil authorities, and as a basis for preparation of the final working drawings.

A large portion of the design work is reflected in the preliminary drawings. Although drawn at smaller scales than working drawings, they include many of the structural, mechanical, and electrical concepts coordinated with the architectural features to form a skeleton outline in graphic form. Generally, only a Site Plan, a Floor Plan, one or two Elevations, and a Typical-Wall Section are included.

Only a few sets of prints are made of the preliminaries, and corrections and revisions are recorded on them. After the working drawings are prepared from these preliminary studies, the latter are of no further use.

2. PRESENTATION DRAWINGS

Presentation drawings usually consist of perspective views of a tentatively planned building (see Figures 1–7 and 1–8). Occasionally, we even see elevation views carefully landscaped with shadows and color. Their purpose is to show the building attractively in its natural setting, mainly for promotional purposes. In general, a building must be *sold* before it reaches the working drawing stage. So presentation drawings are actually *selling tools.* Architectural artists are often employed to prepare colorful and artistic drawings for this purpose, with the esthetic characteristics of the building as the predominant consideration. They are sometimes called Display Drawings. If a project justifies it, a three-dimensional architectural model may be made for a similar purpose.

3. WORKING DRAWINGS

The term *working drawings* is general in that it includes all the drawings needed by the various tradespeople to complete a building project (Figure 1–9). These drawings are the technical directions in graphic form, showing the size, quantity, location, and relationships of the building's components. Working drawings are prepared by the architect or engineer in detailed form on transparent paper or film so that a sufficient number of prints can be made from them. The time and effort expended to produce them comprise a major part of the architectural service. Ordinarily, a set consists of Site Plans, Floor Plans, Elevations, Framing Plans, Sections, and Details.

The following are the major functions of working drawings:

1. *Instrument for material take-offs:* Labor, material, and equipment estimates are made from working drawings prior to construction.
2. *Instructions for construction:* Working drawings show specific sizes, location, and how all materials are put together.
3. *Means for granting a building permit:* Before construction can begin, the local building authority must review the working drawings to see that they meet the requirements of public safety in terms of structural soundness, fire, and other hazards. A building permit is issued to the builder only after approval of the drawings.

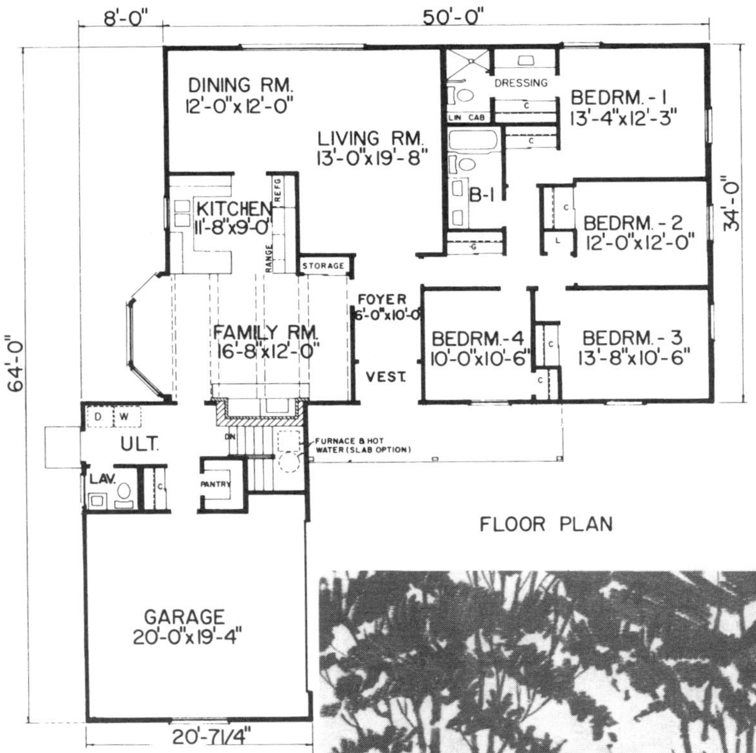

THE RANCH COLONIAL

FLOOR PLAN

FIGURE 1-7
Presentation drawing of a home.

FIGURE 1-8

Presentation drawing of a roadside rest station.

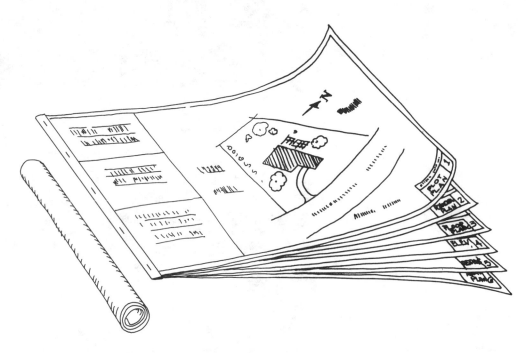

FIGURE 1-9

Working drawings are used to construct a building and are the major part of architectural service.

4. *Instrument for competitive bidding:* In a free-enterprise system, working drawings allow potential contractors a uniform guide for preparing bids, thereby providing the owner the most economical cost of construction.
5. *Means of coordination among the various trades:* Working drawings are the basis of agreement and harmony between material suppliers and specialized trades working together.
6. *A permanent record for future remodeling or expansion or for legal use in the advent of a dispute:* Working drawings eliminate remeasurements in case of future reconstruction. Drawings must be furnished in case of legal disputes. Building failures might possibly occur by natural, unavoidable causes, design errors, or neglect, but the drawings and design calculations are used as evidence and should be available for the life of the building.
7. *Basis for agreement between owner and tenant:* In leasing all or portions of a building, the owner must use the working drawings in the contract agreement.

8. *A complement to the specifications:* In obtaining the written information from the specifications, contractors need the working drawings for interpreting the information. Information from one source is incomplete without the other.

Working drawings for each job are arranged in a logical sequence and numbered to form a set. The set for a small job such as a house usually has the following sequence:

1. Plot (site) Plan
2. Foundation Plan (basement plan)
3. Floor Plan(s)
4. Elevations
5. Details/Sections
6. Interior Elevations
7. Mechanical Equipment (heating, etc.)

Sets of working drawings for larger buildings are usually numbered in a similar sequence; however, because of the greater number of drawings required, a slightly different numbering system is used (see Figures 5–6 to 5–20). Bulky sets require careful numbering; hence an index is shown on the cover sheet to facilitate finding the needed drawing.

It will be noticed, in reading various working drawings, that some sets are more detailed and thorough than others. From an economic standpoint, it is evident that the more complete sets must justify their greater expense in preparation. They are actually worth no more to the owner than they save on construction costs and returns from the building. If a building is constructed with a few simple drawings, much of the planning has to be done by the builder either before or during the construction process; but it must be done. This procedure may be possible for an uncomplicated building. Some buildings have been erected with only one or two drawings, some even with only a sketch on the back of an envelope. But for a more complex structure, it would be difficult to plan mentally all the details and sizes necessary in the building without very complete drawings. Costly changes and mistakes might very well occur—factors that could have been avoided had better drawings been used. It is much cheaper to erase a wall on a drawing, for example, than to tear it down on the job. Complete and well-developed working drawings demand a more precise type of construction in every detail of the building, whereas drawings with fewer details allow greater latitude for decisions by the builder and usually result in more mistakes.

4. SHOP DRAWINGS

These are technical drawings prepared by various suppliers participating in the construction; they are used mainly by their mechanics. On many jobs the architect or designer must rely on these specialists to furnish precise information about the components. For example, if complex cabinetwork is required, it must be built to exact sizes and specifications. A shop drawing is needed to ensure that it will fit into the structure and the structure will accommodate it. Approval of the shop drawings usually precedes the actual fabrication of the component. With shop drawings the architect or designer is able to also check the quality of other components that subcontractors propose to furnish.

H. HOW VARIOUS TYPES OF PRINTS ARE MADE

Architectural working drawings are generally made on vellum, film, or lightweight bond paper that can be economically reproduced. Drawings have little value unless they can be satisfactorily reproduced. Sets of prints (often referred to as blueline prints) must be furnished to contract bidders, estimators, subcontractors, workers, and others who are concerned with the construction of the proposed building; the original drawings are retained

FIGURE 1–10

Floor model diazo machine (GAF Corp.).

FIGURE 1–11

The two steps in making a diazo print.

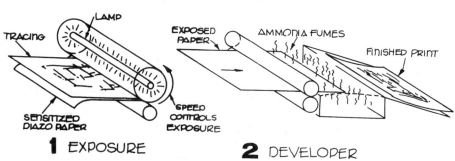

by the architectural or engineering office. Several types of reproduction equipment are used for making prints; each type requires the original tracing to have opaque and distinct linework and lettering in order to produce legible copies. Some architectural offices have their own print-making equipment; others send their tracings or computer files to local printing companies which charge a nominal square-foot fee for their service.

1. DIAZO PRINTS

Commonly known as Ozalid prints, diazo prints are produced on paper coated on one side with light-sensitive diazo chemicals. The sensitive paper, with the tracing above it in direct contact, is fed into a print machine (Figure 1–10) and is exposed to ultraviolet light for a controlled amount of time (Figure 1–11). The pencil or ink lines of the tracing prevent exposure of the chemical directly below, while the translucent areas are completely exposed. After the exposure, only the sensitized paper is again fed into a dry developer, utilizing ammonia vapors that turn the background of the print white and the lines blue, black, or sepia, depending on which type of paper has been selected. Sepia prints, which are reproducible, are available in both paper and Mylar and can be erased or modified directly on the sepia rather than on the original. Dry development neither shrinks the paper nor changes the scale of the drawings. The white background of the prints also allows changes or notes to be added, if necessary, directly on the paper with ordinary pencils.

2. PLAIN-PAPER COPIES

Using plain-paper machines prints can be made from opaque originals, whereas diazos require translucent originals. Plain-paper copiers are mainly used in offices where changes and modifications are needed during the development stages or where the originals need to be put on vellum or film, from which volume prints are made for distribution using the less expensive diazo process. This method is especially useful in printing CADD plots done on bond paper. Plain-paper copiers are also effective in making copies from shop-worn originals onto durable polyester film or vellum media. Some copiers can also enlarge or reduce the size of the original drawing.

3. PHOTOGRAPHIC REPRODUCTION

Various types of photographic reproduction are available today. Photostats can be made from an original drawing by use of a large, specially designed camera that produces enlargements or reductions from the original work. This direct print process delivers a negative with white lines and a dark background. When a photostat is made of the print, a positive image with dark lines and a light background results. A number of high-quality reproduction methods using a film negative made from the original drawing are available. Projection prints from a photographic negative can be reproduced on matte paper, glossy paper, vellum, and Mylar (Cronoflex™). These excellent-quality prints can be enlarged or reduced in scale with accuracy and are very durable. Microfilm negatives can be made from original drawings using a microfilm processing camera that prepares film ready for use within a few seconds. The negative is usually mounted on a standard-size aperture or data card and is systematically filed for future retrieval. A microfilm enlarger reader-printer with a display screen is used to review the images as well as reproduce print copies of various sizes from the negative. Microfilming is an excellent means of storing drawings, thus eliminating the need to retain cumbersome original tracings.

4. MULTICOLOR OFFSET PRINTS

Some prints are now being made in several colors (usually blue and red) via offset printing. Generally the original drawings are made in smaller scales and sizes for further economy. The various colors are used to show new work in relation to existing construction or to distinguish highly complex mechanical or electrical systems in new projects. The chief advantage in using colors is ease in reading the drawings; fewer mistakes result and less time is required in interpreting them.

I. COMPUTER-AIDED DESIGN AND DRAFTING (CADD)

1. INTRODUCTION

Computers were introduced to the architecture and building professions as automated drafting machines. Their purpose was to make the most tedious and expensive part of traditional practice more efficient. Figures 1–12 and 1–13 show examples of a computer-generated floor plan and exterior elevation. Today the role of the computer in architectural firms goes far beyond drafting into the realm of information technology (IT).

The ability of members of a design team to turn 3D computer-generated concept drawings into computer files that can be used to create construction documents has become more important than the ability to just draft. This is a major shift in architects' perception of the role of computers and technology. Never before in the history of architecture and engineering has the underlying technology been changing as fast as it is today. Architects and builders are reinterpreting the computer as a tool for developing, processing, and communicating information about buildings. Included in this

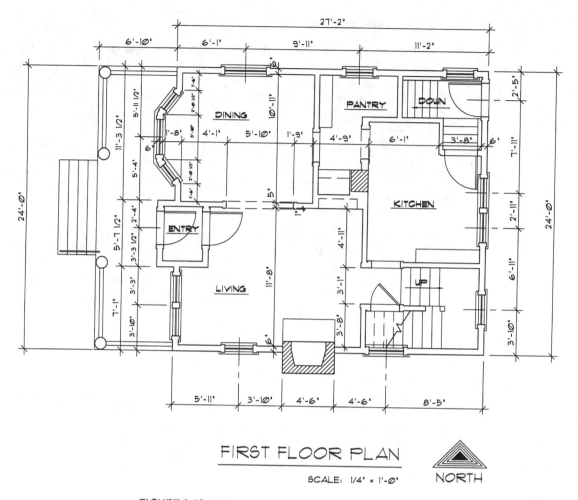

FIGURE 1–12

A floor plan of a small residence drawn with the use of a computer.

process is the development of animation, virtual reality scenes, interactive facilities management models, sun studies, real-time cost analyses, and, last but not least, working drawings.

Computer-assisted design-based software forms the basis for highly interactive computer-based presentations in which the designer and client explore variations together and experiment with all aspects of a project. It further enhances design by providing better communication of information among design professionals. A CADD drawing becomes an information resource readily accessible to all professionals involved in the design and construction process.

Project managers benefit from systems that track how fees are being spent. Electronic mail and file exchange improve communications within the firm and with clients and consultants. Project data can be saved and reused for marketing and on subsequent projects. Construction administration is increasingly automated. Well-managed firms tie all these functions together in a unified information infrastructure with the goal of increasing the amount and changing the nature of information available about a proposed or existing building.

Some of the tools being used to accomplish these goals are:

- Model-centric design instead of drawing-oriented design practices.
- Object-oriented software that facilitates the exchange of information between project participants irrespective of the design system being used.

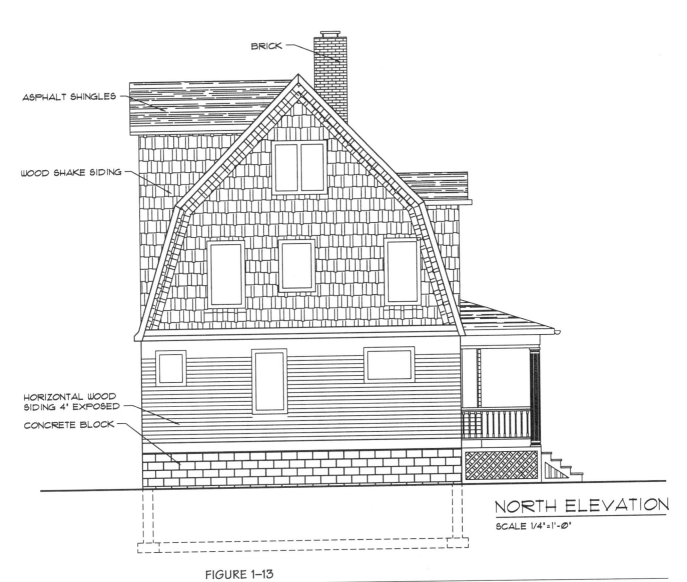

FIGURE 1–13

An elevation of the house shown in Figure 1–12 drawn with the use of a computer.

- ∞ Photorealistic visualization as both a sales and a planning tool.
- ∞ Digital cameras for acquiring site data and information about current facilities.
- ∞ Use of the Internet to obtain and disseminate information including the use of project Web sites that ensure that everyone has the latest data.
- ∞ High-performance computers that can quickly handle large projects.

2. OPERATING SYSTEMS

An operating system (OS) is the software between the applications that a person uses and a computer's hardware such as memory, disk drives, and input/output devices. It is the foundation upon which all other programs are built. The performance and usability of software applications are determined by the performance and usability of the operating system.

Architectural software programs run on a variety of operating systems including, Windows 98, Windows 2000, Windows XP–Home and Professional, Windows NT, and MacOS. Today most architects use Windows-compatible computers, Macintosh computers, or a combination of both. Some important features of current operating systems include multitasking, 32-bit memory addressing, and integrated networking.

Multitasking refers to the ability of a computer to run more than one program at the same time. The term is misleading, however. A multitasking system switches back and forth quickly between different programs. This gives the illusion that two or more things are happening at the same time.

32-bit memory addressing means that the operating system can access up to four gigabytes (four billion bytes) of memory at any given time.

Integrated networking includes capabilities for built-in networking, e-mail, and access to on-line services.

3. SOFTWARE

The types of software applications used by most architects have expanded dramatically. In addition to CADD, software programs used by architects include word processing, spreadsheets, databases, contract management, project scheduling, desktop publishing, multimedia presentation, digital image editing, electronic mail, and on-line services. Three-dimensional model building, visualization, animation, and virtual reality tools are becoming part of the design and communication process for many firms.

Most software programs are organized into small modules that perform a specialized function on a set of data stored in files outside the program. This program structure is followed by most of the software currently used by architects. Because most software programs arrange data in a very specialized way, it is very difficult to share information between different programs. It is important, however, to architects, engineers, and their clients that their electronic drawings are compatible so that they can check and consult with one another. The DXF and IGES standards have been developed to allow a drawing created in one package to be displayed in another, but they are not always accurate.

CADD is best understood as a design information management system and should be well integrated as a general-purpose tool with other software applications. In many architectural firms the documents that describe the total building process are being developed using a variety of smaller, general-purpose software applications that share information with each other, not a single monolithic CADD program.

The basic idea of object-oriented software is to combine software and data into the same object. Applications allow information from other applications to be combined into compound documents. Users can drag and drop objects from one document into another and create a link between the two.

Over time, traditional CADD software packages have become systems supplying all the functions needed within a self-contained environment. Whereas the traditional CADD file was an all-inclusive monster document, the new, object-oriented software will be a compact set of files with intelligent links. The linked technologies will make it possible to experiment with changes wherever it is most convenient to do so. Making changes in the design module will update the spreadsheet, database, and word processing document. Making changes in the spreadsheet or database program will update the drawing and the word processor. The new software will make it easier for architects and builders to balance all considerations when planning and designing buildings.

4. HARDWARE RECOMMENDATIONS

The best way to decide what kind of computer you should purchase is to talk to other people who have recently purchased hardware. Current high-end PC systems are using Intel Pentium 4 2.2 GHz, Mac PowerPC G4 800MHz, or AMD Athlon XP 2200 processors. Systems should have a minimum of 512 MB RAM for CADD and 3D work. Graphics cards should have 3-D accelerated (AGP) graphics with at least 32 MB of video RAM. Hard drives should be 60 GB or larger with seek times under 13 ms. DVD-ROM drives should have minimum transfer rates of 6× for DVD and 32× for CD. If your computer is not networked, buy a tape backup unit or a recordable CD-ROM drive. You should have a 56k v. 90 fax-modem if not a cable connection. Finally, a 15-inch flat panel or a 19-inch monitor with 0.28-mm dot pitch is a minimum for CADD work.

5. PERIPHERAL DEVICES

Over the last 15 years, a variety of plotting technologies have been introduced. Pen, thermal, electrostatic, inkjet, and electrophotographic (LED) plotters have been introduced with great fanfare. Today only the LED and inkjet plotters continue to hold significant market share. The resolution of most printers and plotters (except for pen/pencil plotters) is typically measured in dots per inch (dpi).

a. Inkjet Plotter

Inkjet plotters range in price from $2000 to $7500 and are available from companies such as Hewlett Packard, CalComp, Encad, and Summagraphics.

Inkjet has become the choice for many firms. Inkjet plotters use standard ink cartridges, the kind used in inexpensive desktop inkjet printers. They provide very good print quality with little or no operator intervention. All that is required is the loading of a new ink cartridge when the old one runs out. Most inkjet plotters are capable of creating a D-size, draft-quality plot in less than 3 minutes.

Inkjet plotters are available in both monochrome and color models. Higher-end color models are suitable for printing photorealistic images when loaded with specially coated media, while lower-end color inkjet units are more suitable for spot color on CADD drawings.

A few limitations are associated with inkjet plotting. One is that inkjet plotting involves spraying ink on paper, and filling large black areas of paper is costly. Also, the ink used in inkjet plotters is water based, which creates questions about the permanence of the image. Some of the newer inkjet cartridges use inks containing a polymer-based pigment, and these inks are more resistant to fading and more archival than earlier inks. Media manufacturers offer specially coated inkjet media, which are designed to be stable for long-term storage.

Monochrome D-size plotters have reached retail prices of $2000 or less. The Hewlett Packard DesignJet 230 monochrome sheet-fed plotter often sells in this range.

b. LED Plotter

The high end of the plotter market is dominated by electrophotographic (LED) plotters. They are available from Xerox, CalComp, Océ and JDL. LED plotters have a resolution of 400 dpi and are designed for unattended operation. They are configured for roll-feed media, automatic cutting, and automatic data format recognition.

Top-end LED plotters are generally available as part of hybrid digital copying systems combining both a plotter and a scanner. Xerox and Océ offer units that can copy or plot equally well. These units can scan a number of drawings and then create multiple collated sets without having to rescan the drawings. LED printed sets are fully collated, have no errors, and are of original quality. Large-format LED plotters are economical when used for high-volume production.

c. Desktop Printers

The preferred resolution for desktop, small-format printers is 600 dpi; 300 dpi is a minimum. One of the most important recent developments in printing technology for architects is the affordable 600-dpi, 11 17-inch black-and-white laser printer. The 11 × 17-inch format, at 600 dpi, is suitable for a wide range of small projects and is excellent for reduced-size printing of large projects.

6. THE INTERNET

Imagine potential clients looking up your office in a directory and finding a history of your work, including drawings and photos. Also imagine that the viewer can quickly look up clients' testimonials and credentials of your staff and consultants, find a map to your office, and receive an invitation to call or correspond by e-mail to review information and services that might be of value to them. A home page on the World Wide Web offers this. It provides a chance for prospective clients to see what you have to offer, to see what sets you apart from others, and to see what you offer that most closely matches their needs.

The Internet is a term used to describe the interconnection of worldwide computer networks that took place in 1995. Operating on the Internet are a variety of computer services such as e-mail, UseNet newsgroups, and the World Wide Web. The Internet is a massive global network connecting literally millions of computers.

The Internet ends traditional limitations of state and international borders. This is especially exciting for those whose specialized services have had limited local markets or those with small practices who could not see themselves offering services in other locales. Clients, consultants, and employees are accessible from all over the world. Virtually every possible potential design client is reachable on the Internet. Some are accessible directly through their own Web sites, while others are available through their membership in professional and trade associations.

The Internet has become a major source of information. Research data from many sources, satellite images, and reference information from standards organizations like ASTM are all available. There is a growing presence of design resources on the Internet as well. A large number of vendors have current product information available at their Web sites. The Internet can be a global library and a universal communications medium.

Internet and Web technologies are rapidly becoming key tools for collaboration between design professionals and their clients and builders. Progressive design firms are already using the Internet. E-mail is displacing faxes as the medium of preference in many firms; the global e-mail capabilities of the Internet make it possible for project managers to operate "virtual offices" anywhere. Some firms are upgrading their brochures using 3D, sound, animation, and walk-throughs in their Web pages. Firms search for new personnel by reviewing portfolios and résumés on the Internet. Many students have already created personal portfolio Web sites. Architects and their consultants work together on data maintained on Web servers. On-line "project management information groups" house drawings and meeting minutes and allow project managers to view project information wherever and whenever they wish. Drawings and specifications are sent back and forth between architects, engineers, clients, associates, consultants, contractors, and job-site representatives.

7. FUTURE DEVELOPMENTS

Converting from manual drafting to computer-aided design and drafting was a small step compared to the potential of today's technology. The next major jump will be the linking of digital data and remote construction activities by affordable telecommunications. There is already a new way to automate the processing of construction punch lists. Punch List software from Strata Systems (Figure 1–14) was written and designed for use with 3Com's PalmPilot handheld computer. This software allows you to define punch list items on your desktop computer and transfer the information to the PalmPilot, which is then carried into the field. Field personnel record punch list item status on the PalmPilot and can then transfer the information back to the desktop computer in the office from the field. Once this information has been sent to the desktop, the Punch List software can automatically fax reminder lists and schedules to subcontractors. Tomorrow's industry will be an interactive environment of virtual buildings constructed from binary bits, design and construction teams connected by keyboards and video eyeballs, and a new set of uniform drawing standards to help organize the entire design–build process.

FIGURE 1–14

Punch List software by Strata Systems and PalmPilot by 3Com. (Reprinted from the July 1998 issue of *Builder* magazine © Hanley-Wood, Inc.)

J. HOW TO GET THE MOST FROM THIS WORKBOOK

1. IN FORMAL CLASSWORK

The learning areas in this workbook have been organized in the correct sequence to lead you from the most elementary background needed in learning to read architectural working drawings to the more involved points necessary to interpret more complex drawings of larger buildings. Unless you have had prior experience in the background areas, it is best to progress through the workbook in the same sequence that the material is given. It is important, of course, that the text material be studied before the exercise sheets are attempted.

In some situations the instructor may require that all exercise sheets be removed from the book during the first formal meeting of a class. You will notice that they can be torn easily from the binding. Then, after class discussion of each basic subject, the appropriate exercise sheet should be returned to you for completion. Some instructors, on the other hand, may require that you retain the sheets in the workbook until the appropriate time for their use.

Reading the text material assumes that you have also carefully studied the drawings that accompany it. Acute observation of the drawings will methodically show you the conventional methods of representing construction information in current practice. If slight variations are encountered in the appearance of various symbols, don't let it confuse you; architectural drawings traditionally vary somewhat in appearance.

At the very outset, spend sufficient time in learning construction terms and their abbreviations. They should not be stumbling blocks when your mind is occupied in searching for information on a drawing. When you encounter a term that is not entirely clear, take the time to look it up immediately in the glossary at the end of Chapter 2 (beginning on page 58) so that you will remember it. If further explanation is necessary, possibly of a pictorial nature, look through the various illustrations and drawings in the book to find corroborating identification of the term. It may be helpful to locate the glossary with a small tab for easy reference.

The instructor may also find it helpful to supplement the drawings with others done by local architects and draftspeople to strengthen important points and to introduce the variations that may exist in local practice.

When doing the exercise sheets, be sure you are aware of the correct set of drawings that some are keyed to, and be sure the set is easily available so that little time is lost. All exercises requiring working drawings to work from will have the drawings identified at the top of the page. Follow the directions as given so that each answer is in the form requested. Usually, only short answers are needed. It is also good practice to check all answers when the sheet is finished to ensure that no careless mistakes have been made. As we will mention again, answers must often be verified from several places on drawings when giving positive information.

It may be advisable to eliminate the use of some of the basic-work exercise sheets, depending on the objectives of the course and your background ability. For example, if sufficient background in construction arithmetic has been previously attained, only a review test may be necessary to see if further work in that area is needed. If a satisfactory grade is achieved, your time may be better spent in other areas.

When reading the larger drawings as you complete the exercises, it will be helpful to work on a sufficiently large desk or table so that the drawings can be laid out flat and you can turn from one to the other easily. Remember that the drawings in this workbook had to be reduced in size from their originals and therefore cannot be scaled directly. It is preferable to leave the sets of drawings bound together as they appear in the book, rather than tear them from the binding, so that they will always be in the correct order. Exercise sheets, on the other hand, should be torn from the workbook before they are completed. The arithmetic sheets may require the use of additional scratchpaper for calculation.

2. IN SELF-STUDY AT HOME

If you are using this workbook for self-instruction at home, you will find that being your own instructor can be rewarding; and if the proper procedure is followed, there is no reason why the same benefits of formal classroom work cannot be obtained.

First, as we have already mentioned, read over the text material one chapter at a time and acquaint yourself with the drawings that supplement it. Then tear out one worksheet at a time and complete it without referring to the following information—similar to the procedure in an actual classroom situation. To make the worksheets effective, you must do the sheets individually, as though they were given as a test under the supervision of an instructor. When completed, each sheet should be graded from the text itself; the result will be an indication of your progress. If a set of drawings in the book is to be used with a worksheet, be sure that you *use only that designated material.*

When you have studied the entire workbook, learned the terms and symbols involved, and completed the worksheets in the correct manner, you will have obtained a comprehensive and useful background in reading architectural working drawings.

Chapter 2

BACKGROUND PRINCIPLES

A. PRINCIPLES OF TECHNICAL PROJECTION

For many years drafters and engineers have used a system known as ***orthographic projection*** for accurately representing three-dimensional objects graphically on paper. Recently, the term *multiview drawing* has come into general use, indicating that more than one view is used to illustrate an object, but the terms are synonymous. Technical drawings have a standardized method of development in their representation so that interpretation will be simple and universally understood. A pictorial drawing, of course, would be the most realistic in appearance, and architects frequently utilize this type of drawing when concerned only with the visual characteristics of a proposed building (see Figure 2–7). However, when working drawings must show accurate graphic descriptions, true shapes, correct relationships, and true dimensions, pictorial drawings are inadequate and orthographic views must be used instead.

"Ortho" means *straight* or *at right angles* and gives us a clue as to the positions of the views in relation to the object being drawn. "Projection" indicates that relating views are made by direct projection and therefore have common relationships. Anyone reading working drawings must understand the principle of orthographic development.

The easiest way to understand how orthographic views are developed is to think of a simple house enclosed within transparent planes—for example, the "glass box" shown in Figure 2–1. If we were standing in front of the Front Plane of the box at an infinite distance away, we would see the front side of the house through the glass plane. Now, if the visible lines and features of the house were brought to the glass plane with parallel projectors as shown, a true representation of the front of the house would be obtained; it is known as an orthographic view. Notice that the planes of the box are parallel to the walls of the house and that receded features, such as the ridges of the roof, are also brought to the plane to form the image. In forming the right-side view, the visible features of the right side of the house would also be brought to the right plane with parallel projectors to form the right-side view. Views of an object from either side, top, back, or bottom can be formed in this manner (Figures 2–2 and 2–3). Notice that hidden features are shown with broken lines. Usually, small objects such as machine parts require three views—a front, a right side, and a top—to show all features clearly. When the glass planes of the box are opened as shown in Figure 2–1C, the views are then on a single plane, as it would be necessary to show them on a sheet of drawing paper.

On engineering drawings, the views have a definite relationship so that features from one view can be projected to the other and clarification of features can be obtained by

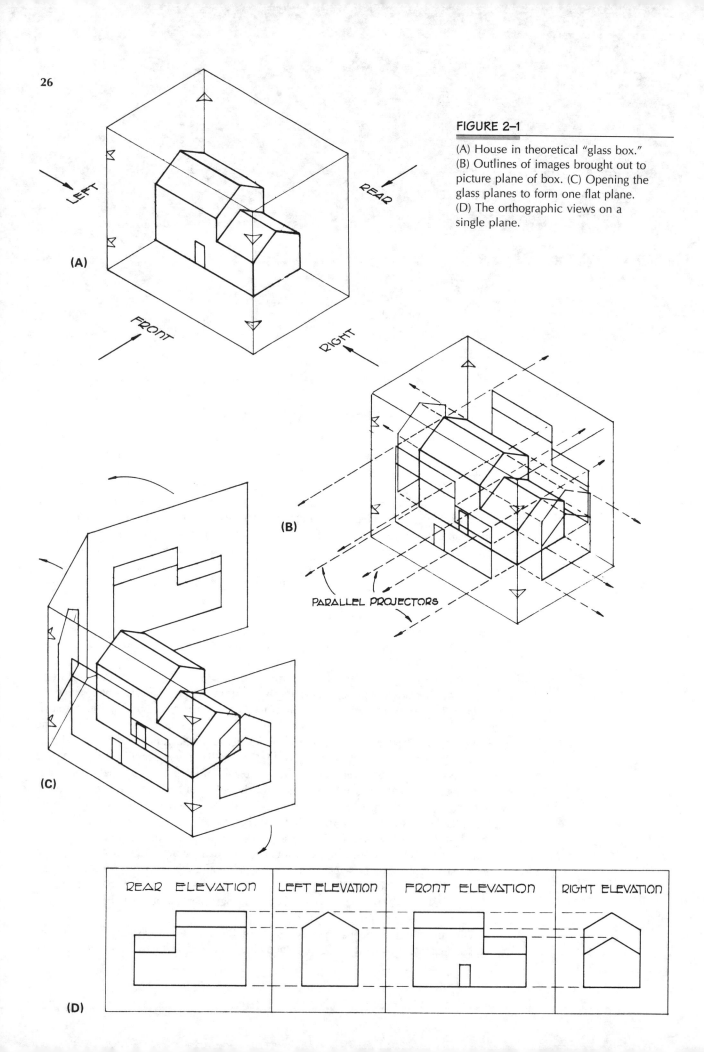

FIGURE 2-1

(A) House in theoretical "glass box." (B) Outlines of images brought out to picture plane of box. (C) Opening the glass planes to form one flat plane. (D) The orthographic views on a single plane.

FIGURE 2-2

A concrete block and the three orthographic views of the same block.

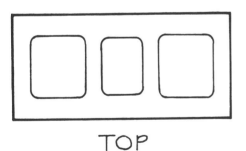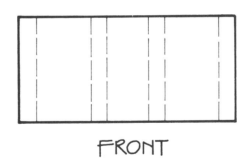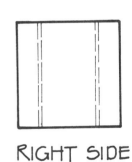

FIGURE 2-3

Orthographic drawing of a concrete block.

cross-observation of more than one view. The top view is placed directly above the front view; the right-side view is placed directly to the right of the front; the left-side view is placed to the left of the front, and so on.

On architectural drawings of large buildings, this relationship would be difficult to portray because of the size of the views; nevertheless, the method for developing either type of drawing is the same.

The student must understand the conventional method of relating views of small objects on engineering drawings, since these multiviews are occasionally found on architectural working drawings to show smaller, usually mechanical, components.

B. ARCHITECTURAL APPLICATIONS

Views of a building projected to vertical planes are known as **Elevation Views** on architectural drawings. The view from the front of the building is called the Front Elevation, the view from the right side the Right Elevation, and so on (see Figure 2–4). If a building has a definite north-south orientation, the elevations are sometimes labeled according to the compass directions. Thus, if a side of the building faces south, it would be called the South Elevation; the side facing east would then be called the East Elevation, and so on, as shown in Figure 2–5. The building would also have to be oriented with a *North Arrow* on the plan to make this method of labeling satisfactory (Figure 2–6). When a building has one side facing, say, the northeast, the labeling could be misleading to the reader and is usually avoided in preference to the former method.

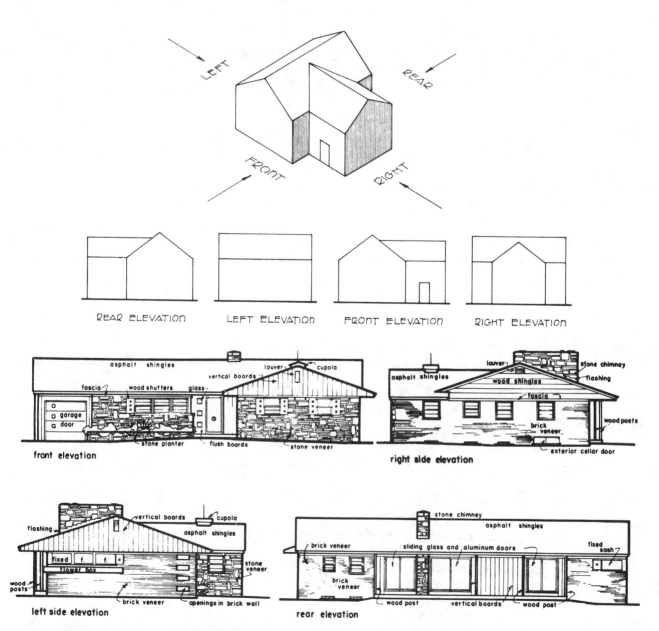

FIGURE 2–4

Elevation views of a house labeled in relation to the front.

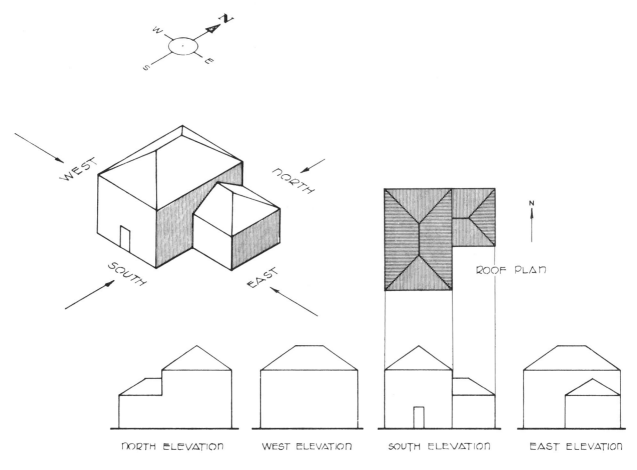

FIGURE 2-5

Elevation views of a house labeled in relation to their compass directions.

FIGURE 2-6

Various types of north-point arrows used on drawings.

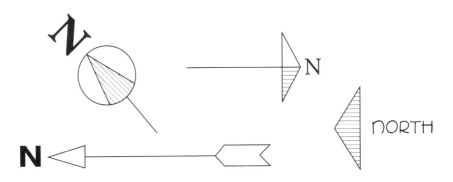

Interior views of buildings, such as kitchen and bath walls, are also known as Elevations as long as they show vertical surfaces.

As mentioned previously, because of their size, architectural elevations are not conventionally placed in relating positions in sets of drawings. Often they must be placed on separate sheets so that they can be drawn large enough for easy interpretation. Each view is then labeled for correct identification. Also, elevation views are placed to show their natural appearance; that is, vertical surfaces are represented with vertical lines and horizontal surfaces with horizontal lines. Seldom would one find an elevation in a turned position or on end. However, if views are small enough, they are placed in their related positions so far as possible in order that convenient projections can be made during drawing; the right elevation is placed to the right of the front, the left elevation to the left of the front, and so on.

Another practice found on architectural drawings is the elimination of hidden lines on major views. Broken lines represent hidden surfaces on orthographic views. But if all the hidden surfaces found within a building were shown on the elevations, the maze of broken lines would merely be confusing. Except for a special reason, such as the outline of a foundation below ground or the finish floor level within the building, hidden lines are avoided on elevation views.

Still another convention on construction drawings is the use of the word *Plan* for all views shown on a horizontal plane. A top view of a building would be called a **Roof Plan.** The layout of a building, commonly called the **Floor Plan,** is actually a section view of the building with a horizontal cutting plane placed between the floor and ceiling of that particular floor level (see Figure 2–7). A floor plan is not the bottom view of a building but, rather, a view from above showing the layout of the walls and arrangement of rooms, with the top of the building removed. If we realize the purpose of the drawing, we can understand why this convention is necessary. It allows the plan to include interior layout information as well as other important information in the simplest manner. Even features that may be above the cutting plane are sometimes placed on the plan for convenience (see Figure 2–8). Therefore, a plan view does not always represent the building strictly through a single plane; occasionally, it may include features on other offset planes in order to make the plan more informative. Remember that plan views show horizontal layouts, whereas elevations show heights and vertical surfaces. As we shall see later, heights can also be shown on Section Details.

C. SECTIONS

It is common practice to show much of the detailed technical information about the construction with isolated or individual section views, called **details.** Detailed information about various members and their construction is revealed by theoretically cutting features through the structure and showing the cut members with section views (see Figure 2–13). The plans and elevations (shown in smaller scales because of their size) then allow the reader to position and relate these details within the structure. Along with the plans and elevations, complete sets of working drawings contain various isolated sections to show workers much of the specific information.

A section is obtained by theoretically cutting through an object, or combination of members, with the material in front removed. To orient the section on a relating view, the plane of the cut is shown by a heavy, broken *Cutting-Plane Line* (see Figure 2–18). Arrows on the ends of cutting-plane lines indicate the direction of observation when the section is drawn. Normally, identification letters are also used to relate the line to its proper view. You will notice that a section does not have any meaning unless it can be oriented properly (Figure 2–9).

Cut surfaces on sections are emphasized by the addition of *material symbols* (see Figure 2–25 for architectural materials symbols) that reveal the material and help in identifying adjacent members. The darker tones produce contrast from the other linework and help the reader to see each member and what material it is to be.

Various types of sections are used; the majority are partial sections of walls, floors, foundations, and so on, where important information must be exposed (Figure 2–10). Larger sections might be views entirely through the building, such as a *transverse* section, a vertical section through the narrow width of the building (Figure 2–11), or a *longitudinal* section, which is also vertical but through the length of the building (Figure 2–12). Horizontal sections (other than regular floor plans) may also be found on drawings, but they are not as common as vertical section views through walls.

We often see the word *typical* used in labeling sections on drawings—for example, a Typical Wall Detail or a Typical Sill Detail (Figure 2–13). This means that the section applies to similar situations throughout the building and there is no need to repeat the drawing. If a building has uniform construction within all its walls, only one Typical Wall Detail

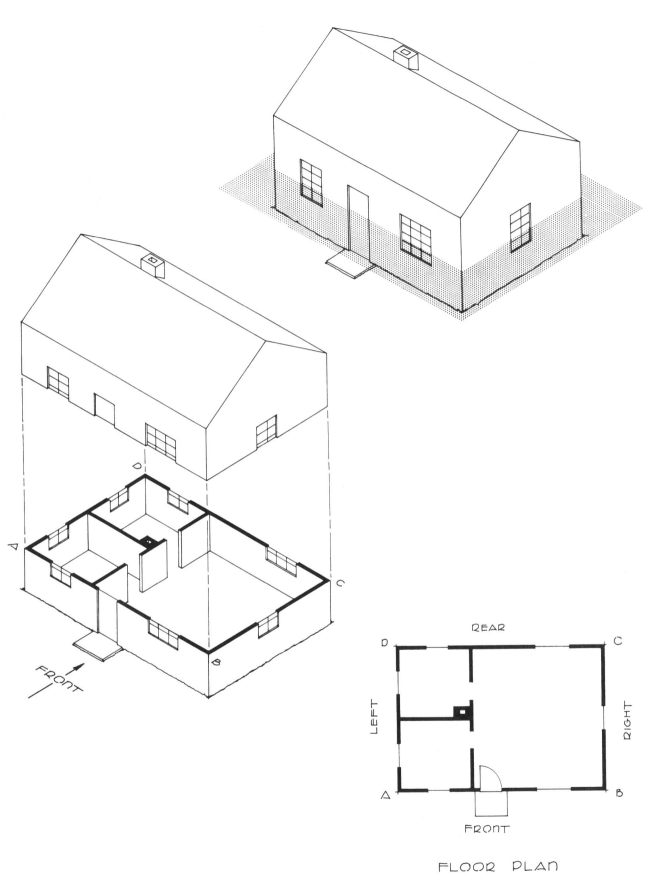

FIGURE 2-7

The floor plan is actually a horizontal section with the cutting plane through windows and doors.

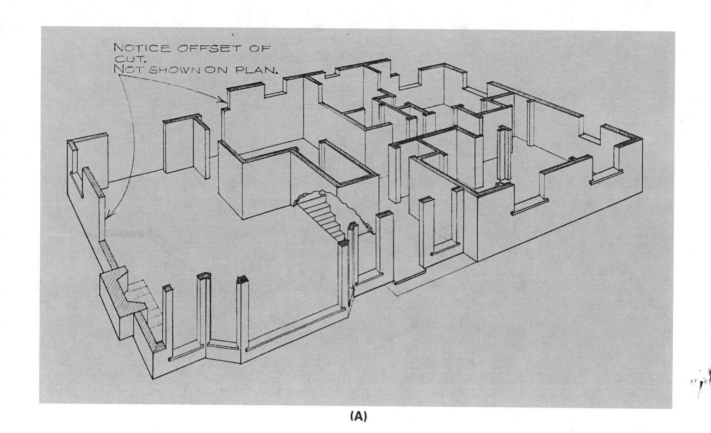

FIGURE 2-8

Pictorial plan of a house showing offsets in the cutting plane for the purpose of including important features. Notice in the lower plan that the offsets are not shown.

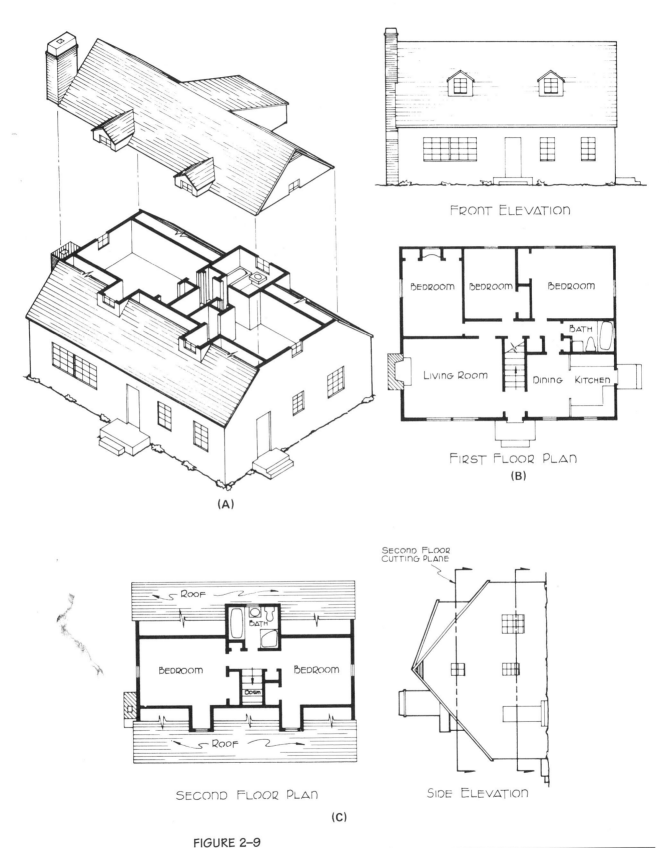

FIGURE 2-9

(A) A one-and-a-half story house with the upper portion removed. (B) Relationship of the first-floor plan and the front elevation of the same house. (C) Partial second-floor plan and the relating side elevation. (Cutting planes for full plans are not conventionally shown.)

FIGURE 2–10

Horizontal section.

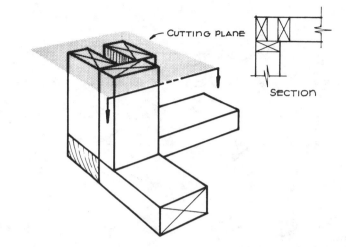

FIGURE 2–11

Transverse section.

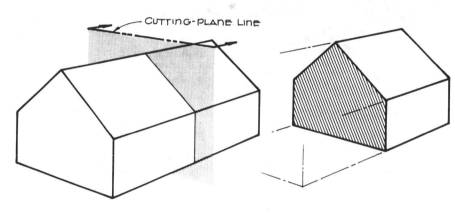

FIGURE 2–12

Longitudinal section.

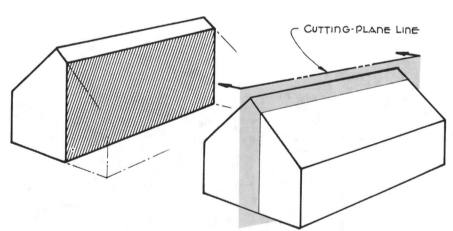

would need to be shown. In this case, no cutting-plane line is used, for the detail refers to a uniform construction throughout.

Sections are conventionally used to show information about stairs, fireplaces, foundations, beam connections, wall construction, windows, doorjambs, concrete reinforcement, and so on. Details having the smallest parts or members are usually shown in the largest scale for better readability. Notice also that often long lengths in sections are broken and much of the unnecessary member is omitted to allow the section to be drawn at a larger scale. It is also not uncommon to find various sections relating to a feature clustered around other drawings of that feature and labeled in a consistent manner for easy reference. Experience in reading different drawings by different architects shows that slightly different

FIGURE 2-13

Typical wall section.

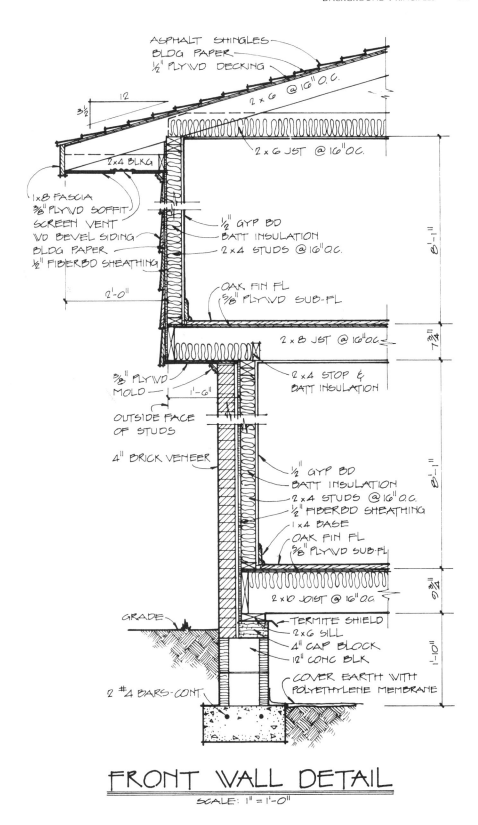

mannerisms are used to combine and relate detail sections on drawings. Typical variations are shown throughout the drawings in this workbook.

Notice that fasteners such as bolts, rivets, screws, shafts, and rods that fall in the plane of the section are not cut. This practice is conventional on working drawings of all types.

D. AUXILIARY VIEWS

Occasionally, we find surfaces or features on drawings that are oblique to the principal planes of projection and still are shown in true shape, such as the house in Figure 2–14, which shows one wing at other than right angles to the rest of the house. On the major elevations, this wing would appear distorted. Other features in modern construction are also designed at various angles to the principal planes of projection. To show these features or surfaces in true shape for accurate description, *Auxiliary Views* are used. An auxiliary view is similar to an orthographic view except that it is projected to a plane *parallel to the auxiliary surface* rather than to the customary orthographic planes. The view then shows the true shape of the oblique feature and also eliminates much of the time-consuming projection for the drafter. Notice that a conventional break line is used to show the termination of the auxiliary portion of the view when only a partial view is required. When reading a drawing with auxiliary views, usually the labeling and the logical association applied to relative elevations or plans will quickly orient the auxiliary view to the rest of the drawings. In Figure 2–14 the true shape of the diagonal wing is better described in the auxiliary elevation than in the major elevation

E. PICTORIAL DRAWINGS

On some working drawings, pictorial views of features will be used to reveal information that would be difficult to show with orthographic views; other situations may require a pictorial drawing merely to supplement a major view. One of the commonly used pictorial drawings is called an *Isometric Drawing* (see Figure 2–15). This type of drawing is drawn with its receding sides at 30 degrees from the horizontal, and its corners are drawn vertically (Figure 2–16). The lines on the three mutual planes are drawn parallel with the use of conventional 30–60 degree drafting triangles. Lines on the three axes are measured true length to scale. An isometric conveniently exposes three sides of an object in a tipped manner so that important information is easily visible; hidden lines are seldom needed. The height, width, and depth of an object can then be seen in one composite view for easy com-

FIGURE 2–14

Use of auxiliary views.

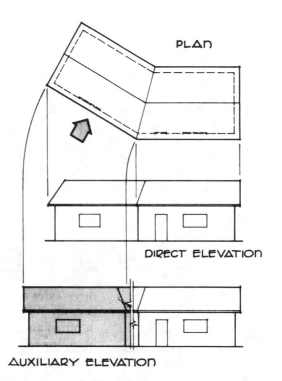

FIGURE 2-15

The appearance of an object drawn by different methods.

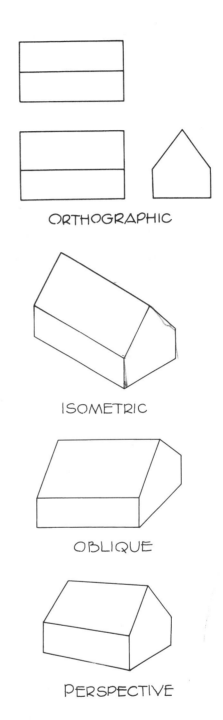

prehension. Even pictorial isometric sections are employed in working drawings occasionally to show a detail more clearly.

Another pictorial-type drawing sometimes found on drawings is the *Oblique Drawing*. It is similar to the isometric except that one side (usually the important side) is placed on the frontal plane, and the receding top and edge are shown with parallel oblique lines. Three sides are shown, as with the isometric drawing, and the three sides have parallel lines representing true lengths on the object. The front surface appears similar to an orthographic view (as though we were looking frontally at that surface), while the oblique lines provide the appearance of depth. Various angles for the receding lines are used, although the 45-degree angle from the horizontal is the most popular (see Figure 2–17).

FIGURE 2–16

The 30-degree axes of an isometric drawing.

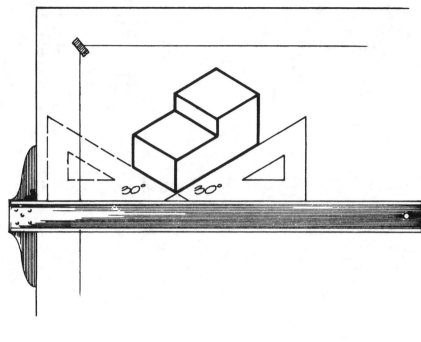

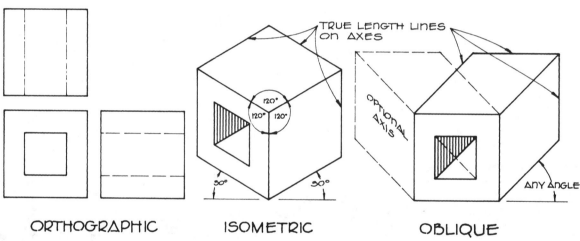

FIGURE 2–17

Characteristics of isometric and oblique pictorial drawings.

Dimensions for either type of pictorial drawing are conveniently shown on their respective planes.

Compared to the pictorial drawings just mentioned, a *Perspective Drawing* represents an object in the most realistic manner. In fact, the representation is similar to a photo taken with a camera or even the image we see with our own eyes (see Figure 2–15). However, this type of drawing is time-consuming to draw and hence is usually unsatisfactory for showing technical information on working drawings. It can be useful on architectural presentations and construction display drawings; students dealing with working drawings should be able to recognize its characteristics.

Briefly, to appear realistic, a perspective drawing is drawn so that all horizontal features receding from the observer appear to converge at a distant point. If the object is positioned in an angular manner of observation, the receding lines will appear to vanish at two distant points. Also, these points of convergence, called *vanishing points,* fall on the horizon line, which is theoretically the same level as the observer's eye. In drawing perspective views, projectors are used to bring the image to a transparent picture plane. This procedure is similar to the one used in creating orthographic views except that the projectors of a perspective converge as though emanating from a point, similar to an eye looking through the transparent plane instead of being made parallel as in orthographic projection. After observing perspective drawings, one can see that they would have little use on working drawings; however, they are unequaled for showing the layperson the realistic qualities of a building.

F. THE MEANING OF LINES ON DRAWINGS

The lines that make up a drawing have various meanings to different people, depending on their knowledge of the graphic language. To the average person, a line drawing merely means a recognizable picture of a physical object. To the person trained in graphics, lines have several meanings. First, some may be *pictorial* lines that outline the picture of an object in a lineal manner; second, others may be *symbolic* lines that represent something more complex than what the lines actually show. Lines may also be *quantitative* in nature, representing quantities and number values, such as shown on charts and graphs or as used in graphic calculations. On working drawings we are concerned mainly with the first two types of lines.

Pictorial lines merely outline the views of orthographic or pictorial drawings. Some are drawn to scale, while others may be made freehand; but their meanings become evident if we keep in mind the purpose of the drawing we are reading.

Symbolic lines used on construction drawings are simplified representations or conventional symbols of frequently used components (see Figure 2–18) and must be learned by the reader so that they are quickly recognizable. When we realize the great reduction in size in which views of a building must be made, we can understand the need for symbols on architectural drawings, not only from the standpoint of reading but also from the standpoint of making the drawing.

Lines themselves are expressive on well-executed drawings: heavier lines are meant to have more importance, whereas fine, thin lines are meant to have less importance. Drawings with all lines of the same intensity are often difficult to interpret and usually monotonous to read.

Listed next are the names and descriptions of the various types of lines used on working drawings. These descriptions relate directly to Figure 2–18.

> OBJECT OR VISIBLE LINES: Prominent lines representing the edges of surfaces or the intersection of two surfaces.
>
> DASHED OR HIDDEN LINES: Medium-weight broken lines made with uniform dashes. They represent hidden surfaces or intersections, or on floor plans they may be used to represent minor features that fall above the plane of the drawing, such as high wall cabinets in a kitchen. Occasionally, on drawings used for remodeling a building, they indicate the position of the old construction. In other cases they are used for relationship clarification or to show alternative positions of a movable component.
>
> CENTER LINES: Fine lines made with alternate dots and dashes. They show symmetry on the axis of circular features and are used to locate centers of windows and door symbols on floor plans (see Figure 3–26). To further distinguish center lines from object lines, center lines are extended slightly beyond the outline of the view they show symmetry for.
>
> DIMENSION LINES: Fine lines with arrowheads at their extremities to show the extent of a given dimension. Numerical dimensions are placed directly above the lines near their center.

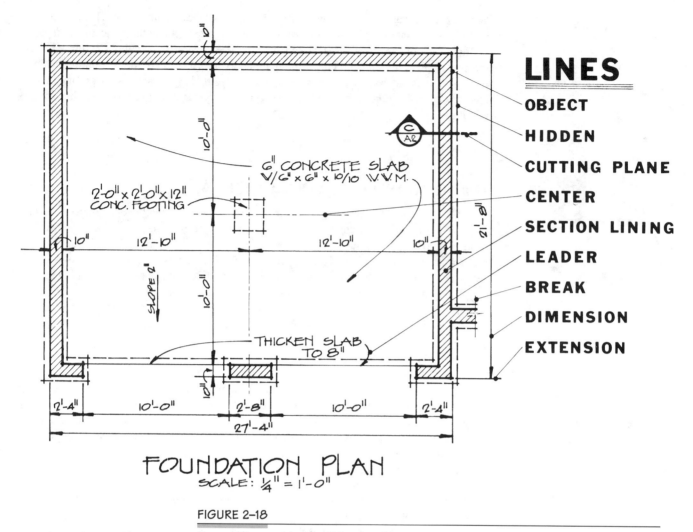

FIGURE 2–18

Types of lines used on working drawings.

EXTENSION LINES: Fine lines that relate the dimension lines to their features. They do not touch the features; instead, they start about $\frac{1}{6}''$ from the feature and extend about $\frac{1}{8}''$ beyond the arrowheads of the dimension line.

LEADERS: Fine lines with an arrow or dot at one end to relate a note or callout to its feature. Generally, they are drawn at an angle from the principal lines on the drawing or in a free-curved manner to distinguish them easily from the object lines.

CUTTING-PLANE LINES: Very prominent broken lines (usually two dots and a dash) used to locate the exact location of the plane of a section. Arrows on their ends indicate the direction in which the section is observed. Circular symbols are sometimes used at their ends to relate the cutting planes to their section views.

BREAK LINES: Fine lines with offsets to show a break or the termination of a partial view. Architects frequently use break lines to eliminate unimportant portions of details, thus allowing the important portions to be made larger. Small break lines are often fine, ragged lines done freehand.

SECTION LINING: Fine linework, usually angular, giving a tone to sectioned surfaces and made with the conventional symbol of the surface being cut (see Figure 2–25 for the various symbol representations found on sectioned surfaces).

G. REVIEW OF CONSTRUCTION MATHEMATICS

The fundamentals of arithmetic and geometry are also the fundamentals of graphics. One cannot be learned without the other. In fact, a facility with numbers is the starting point of any technical subject. Reading construction drawings is no exception, for mathematics is used from the very beginning of design concepts to the final completion of a building. Simple errors even in addition and subtraction frequently result in needless waste on many construction projects. Learn the accurate methods for making simple calculations so that correct habits will be formed and numbers will become a working tool for you.

Nearly everyone in technical work now has access to a calculator, whether an elaborate, expensive one or merely a small hand-held type. Either is useful in dealing with working drawings. However, one must have a facility with the numerical functions before the calculator is effective, especially since most calculators operate in the decimal system rather than with fractions, as is found on the majority of U.S. working drawings. It becomes necessary, therefore, to be able to convert from one system to the other. Some small calculators are now available with modes that allow calculations with fractions and whole numbers, thus saving time for the reader of architectural drawings. The decimal system, of course, is compatible with metric dimensioned drawings. The availability of a calculator still does not eliminate the need for students to have a sound background in the various functions of arithmetic. Review the following calculations, which are usually associated with construction work and drawings, to gain this important background.

1. BASIC ARITHMETIC TERMINOLOGY

ADDITION

   3   Addend
+ 3   Addend
   6   Sum

MULTIPLICATION

   3   Multiplicand
× 3   Multiplier
   9   Product

FRACTIONS

  5/8   Numerator / Denominator
1 ½   Mixed number

SUBTRACTION

   6   Minuend
− 3   Subtrahend
   3   Difference

DIVISION

   9   Dividend
÷ 3   Divisor
   3   Quotient

DECIMALS

5.375 — Whole number, Decimal point, Tenths, Hundredths, Thousandths

SYMBOLS

=	Equals: 4 + 2 = 6	∟	Right angle
>	Greater than: 4 > 3	⊥	Perpendicular
<	Less than: 3 < 4	‖	Parallel
::	As, used between ratios	□	Square
∴	Therefore	1^2	Exponent (squared)
∠	Angle	√	Square root
:	Is to, the ratio of		

2. ADDITION AND SUBTRACTION OF FEET-AND-INCH DIMENSIONS

Since the English system of feet and inches is the conventional method of showing dimensions on working drawings, care must be taken in adding and subtracting combinations of these mixed units. *When dimensions are added,* and they are whole numbers (fractions will be treated later), and the sum of the inch units is less than 12, only simple addition is necessary. For example:

$$3' 7'' + 4' 2'' = \mathbf{7'9''}$$

$$\begin{array}{r} 3' \ 7'' \\ +4' \ 2'' \\ \hline 7' \ 9'' \end{array}$$

But if the sum of the inch units is greater than 12, they must be totaled and then divided by 12 to obtain the greatest number of foot units in the final sum. For example:

$$3' 7'' + 4' 8'' = 8' 3''$$

$$\begin{array}{r} 3' \ 7'' \\ +4' \ 8'' \\ \hline 7' \ 15'' \end{array} \quad (15'' \div 12'' = 1\tfrac{1}{4}' \text{ or } 1' \ 3'')$$

$$1' \ 3'' + 7' = \mathbf{8'3''} \text{ Ans.}$$

Also, as an example of adding a series:

$$12' 1'' + 3' 0'' + 23' 11'' + 5' 6'' = 44' 6''$$

$$\begin{array}{r} 12' \ 1'' \\ 3' \ 0'' \\ 23' \ 11'' \\ + \ 5' \ 6'' \\ \hline 43' \ 18'' \end{array} \quad (18'' \div 12'' = 1\tfrac{1}{2}' \text{ or } 1' \ 6'')$$

$$1' \ 6'' + 43' = \mathbf{44'6''} \text{ Ans.}$$

When dimensions are subtracted, the inch units of the subtrahend must be smaller than the inch of the minuend (Example 1); otherwise, if the inch units in the subtrahend are larger, enough foot units must be converted into inch units before the calculation can be made (see Example 2).

Example 1: $12' 6'' - 3' 4'' = \mathbf{9' 2''}$

$$\begin{array}{r} 12' \ 6'' \\ - \ 3' \ 4'' \\ \hline 9' \ 2'' \end{array}$$

Example 2: $12' 2'' - 3' 4'' = \mathbf{8' 10''}$

$$\begin{array}{r} 11' \ 14'' \\ \cancel{12' \ 2''} \\ - \ 3' \ 4'' \\ \hline 8' \ 10'' \end{array}$$

3. MULTIPLICATION OF FEET-AND-INCH DIMENSIONS

When multiplying two dimensions with mixed units (feet and inches), simply change the feet units to inches by multiplying them by 12, total the inches in each dimension, and then multiply. The product will be in square inches. For example:

$$3' 4'' \times 2' 6'' = \mathbf{1200 \ sq \ in.}$$

$$\begin{array}{ll} 36'' + 4'' = 40'' & \quad 40'' \\ 24'' + 6'' = 30'' & \underline{\times \ 30''} \\ & \quad 1200 \text{ sq in.} \end{array}$$

To convert square inches to square feet, divide the product by 144 (number of square inches in one square foot).

$$1200 \text{ sq in.} \div 144 = \mathbf{8\tfrac{1}{3} \text{ sq ft}}$$

4. CONVERSION OF FRACTIONS TO DECIMALS

To convert a fraction to a decimal number, divide the numerator by the denominator, as shown by the following example:

Change $\tfrac{11}{16}$ to a decimal

$$\begin{array}{r}
0.6875 \\
16\overline{)11.0000} \\
\underline{9\,6} \\
1\,40 \\
\underline{1\,28} \\
120 \\
\underline{112} \\
80 \\
\underline{80} \quad \mathbf{0.6875 \text{ Ans.}}
\end{array}$$

Some of the most frequently used decimal equivalents are

$$\frac{1}{4} = 0.25 \qquad \frac{1}{2} = 0.50 \qquad \frac{1}{8} = 0.125$$
$$\frac{3}{4} = 0.75 \qquad \frac{1}{16} = 0.0625 \qquad \frac{7}{8} = 0.875$$

Other equivalents are given in Table A–4 in the Appendix.

Some fractions have no exact equivalent; for instance, $\tfrac{1}{3} = 0.333\ldots$ and $\tfrac{2}{3} = 0.666\ldots$. When calculations involve these numbers, round off the repeating decimal to agree with the accuracy of the other numbers in the calculation.

5. CONVERSION OF DECIMALS TO FRACTIONS

To convert a decimal to a fraction, first multiply it by 10, or 100, or 1000, and so on, to bring the decimal point to the right of the last digit; then divide the whole number by the same 10, or 100, or 1000, and so on, and reduce as shown by the following examples:

$$0.25 = 0.25 \times \frac{100}{100} = \frac{25.}{100} = \frac{1}{4} \text{ Ans.}$$
$$0.125 = 0.125 \times \frac{1000}{1000} = \frac{125.}{1000} = \frac{1}{8} \text{ Ans.}$$
$$0.875 = 0.875 \times \frac{1000}{1000} = \frac{875.}{1000} = \frac{7}{8} \text{ Ans.}$$

6. CALCULATIONS INVOLVING BOTH DECIMALS AND FRACTIONS

When both decimal numbers and fractions are given for a calculation, change all fractions to their decimal form and then continue the calculation in decimals, as indicated by the following examples:

$$4.37 + 1\tfrac{1}{3} = 4.37 + 1.33 = \mathbf{5.70 \text{ Ans.}}$$

$$8.63 - 1\tfrac{1}{2} = 8.63 - 1.50 = \textbf{7.13 Ans.}$$

$$12.48 \times 2\tfrac{1}{4} = 12.48 \times 2.25 = \textbf{28.08 Ans.}$$

$$18.6 \div \tfrac{3}{8} = 18.60 \div .375 = \textbf{49.6 Ans.}$$

7. CONVERSION OF FEET AND INCHES TO DECIMALS

Dimensions in both feet and inches are frequently handled more easily in decimal form when calculating, especially if many numbers are involved. If each foot of our measurement were divided into 10 inches, the conversion would be relatively simple; with 12 inches in each foot, however, it becomes a little more complex, yet not difficult. One inch equals one-twelfth of a foot; then, if we write the inches of a dimension as the numerator of a fraction and 12 as the denominator, $1'' = \tfrac{1}{12}'$, $2'' = \tfrac{2}{12}'$, $3'' = \tfrac{3}{12}'$, and so on, the conversion is easily made, as previously shown. For example, convert 5′ 3″ into decimal form.

$$5'\ 3'' = 5\tfrac{1}{4}',$$
$$5 + 0.25 = 5.25$$
$$5'\ 3'' = \textbf{5.25}'\ \textbf{Ans.}$$

$$1.00 \div 4.0 = 0.25$$

$$\begin{array}{r} 0.25 \\ 4\overline{)1.00} \\ \underline{8} \\ 20 \\ 20 \end{array}$$

Similarly, the following dimensions can be converted into decimals: 4′ 1″ = 4.083′, 13′ 6″ = 13.5′, 6′ 9″ = 6.75′, 12′ 4″ = 12.333′, and so on. Therefore, *to change inches to decimal parts of a foot, write the inches as twelfths and convert the fraction to a decimal by dividing the numerator by the denominator.*

If the dimension has feet, inches, and fractional parts of an inch, such as $8'\tfrac{7}{12}''$, convert the $7\tfrac{1}{2}$ to an equivalent fraction by multiplying by a convenient number (in this case 2).

$$2 \times \frac{7\tfrac{1}{2}}{12} = \frac{15}{24}$$

Then change to the decimal as usual.

$$\begin{array}{r} 0.625 \\ 24\overline{)15.000} \\ \underline{14.4} \\ 60 \\ \underline{48} \\ 120 \\ 120 \end{array}$$

$$8' + 0.625' = \textbf{8.625}'\ \textbf{Ans.}$$

8. CALCULATING BOARD FEET

Boards and lumber are priced and sold at lumber dealers on the basis of *board feet*. One board foot is 1″ thick, 12″ wide, and 1′ long (see Figure 2–19). Prices are quoted per thousand board feet ($/M). This measurement is the *nominal size* of the lumber (before it is dressed); dressed lumber is slightly smaller than the nominal size because of the waste (Figure 2–20). For example, the dressed thickness of 1″ boards is $\tfrac{3}{4}''$, of 2″ × 4″ lumber is $1\tfrac{1}{2}'' \times 3\tfrac{1}{2}''$, of 2″ × 6″ lumber is $1\tfrac{1}{2}''\ 5\tfrac{1}{2}''$, and so on. All lumber less than 1″ thick is con-

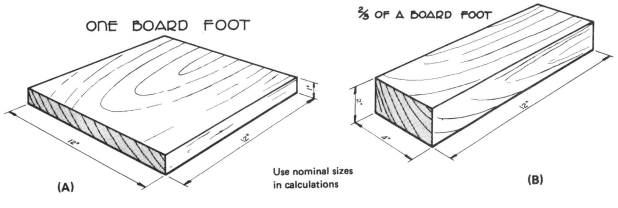

FIGURE 2-19

(A) Use nominal sizes in board-feet calculations: thickness in inches × width in feet × length in feet = board feet. (B): 2 × 4/12 12/12 = 2 × 1/3 × 1 = 2/3 foot board measure.

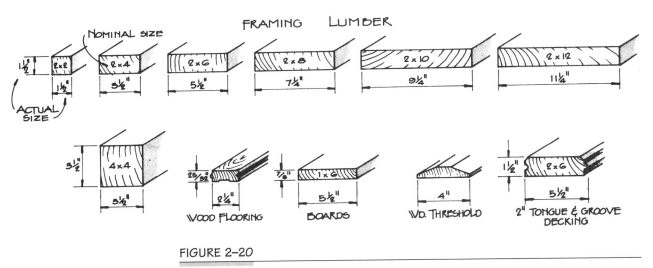

FIGURE 2-20

Typical lumber nominal and actual sizes.

sidered as a full 1″ thick in measuring board feet. Several methods can be used for calculating board feet.

FIRST METHOD: To find the board feet in the quantity of lumber, convert one linear foot to board feet and multiply this factor by the total number of linear feet required. For example, to convert one linear foot of 2″ × 4″ lumber into board feet, multiply the thickness in inches by the width in inches and divide by 12:

$$\frac{2'' \times 4''}{0} = \left(\frac{8}{12}\right)' = \frac{2}{3} \text{ bd ft in 1 lin ft}$$

To use this conversion in a typical problem, suppose that a builder needs 90 pieces of 2″ × 4″ studs (8′ 0″ long). How much will they cost if the price of the lumber is $150/M? (See Figure 2–20.)

TABLE 2–1

Conversion Table of Board Feet in One Linear Foot of Lumber

Nominal Size in Inches	Board Feet per Linear Foot
1 × 4	1/3
1 × 6	1/2
1 × 12	1
2 × 2	1/3
2 × 4	2/3
2 × 6	1
2 × 8	1-1/3
2 × 10	1-2/3
2 × 12	2
4 × 4	1-1/3

$$90 \times 8 \times \frac{2}{3} \times \frac{150}{1000} = \textbf{72.00 Ans.}$$

Use Table 2–1 to find the number of board feet in, say, 150 lin ft of 2″ × 10″ lumber. The conversion factor is $1\frac{2}{3}$ bd ft. Thus:

$$150 \times 1\frac{2}{3} = \textbf{250 bd ft Ans.}$$

SECOND METHOD: To find board feet with this alternative method, multiply thickness in inches by *width in feet* by length in feet. This method eliminates the division by 12, as done previously, but requires the width of the lumber to be converted to feet. For example:

Find the board feet in 5 pieces 2″ × 6″ × 10′ long:

$$5 \times 2 \times \frac{1}{2} \times 10 = \textbf{50 bd ft Ans.}$$

Find the board feet in 24 pieces 2″ × 8″ × 14′ long:

$$24 \times 2 \times \frac{2}{3} \times 14 = \textbf{448 bd ft Ans.}$$

9. RATIO AND PROPORTION

DEFINITION OF RATIO: The relation of one number to another, or the quotient of one quantity divided by another of the same kind, such as is stated by a fraction. Thus the fraction $\frac{8}{4}$ equals 2; the 2 being the "ratio of 8 to 4." Both the fraction $\frac{8}{4}$ and the quotient of 8 ÷ 4 have the same ratio, 8 to 4. In mathematics, the "to" is written (:), so we write the ratio 8:4.

DEFINITION OF PROPORTION: The condition of equality between two ratios, or the relationship between four numbers in which the quotient of the first divided by the second is equal to that of the third divided by the fourth. For example, 8 is to 4 as 6 is to 3, which is written 8:4 :: 6:3 in mathematical terms. The proportion sign (::) has the same meaning as the frequently used equal sign (=).

In a proportion such as 8:4 :: 6:3, the first and last numbers are called *extremes,* and the second and third, or middle numbers, are called *means*. Notice that *the product of the extremes equals the product of the means.*

$$8:4::6:3 \quad 8 \times 3 = 4 \times 6$$
$$24 = 24$$

This is a fundamental rule of proportion, and it is true of any proportion. If any three terms of the proportion are known, the fourth can be found. The product of two divided by the third known number produces the fourth number. In construction work, three known quantities and one unknown find many applications for solving problems; and because of their proportional relationship, the principle is often called the *rule of three.*

In setting up a proportion problem, the number being sought is usually represented by X for convenience. Thus, if the number 4 in the proportion above were unknown, it would be written 8:X :: 6:3. Solving for X, we obtain $8 \times 3 = 24$, $24 = 6X$, $X = 4$.

Another point to remember in setting up a proportion is that it is better to have numbers representing the same things in each ratio. That is, arrange the numbers in each ratio so that dollars are to dollars, hours are to hours, feet are to feet, and so on. In a typical problem, if 100 bd ft of lumber costs $75, how much would 60 bd ft cost? Set the ratio up to say 100 bd ft is to 60 bd ft as $75 is to X dollars, rather than board feet are to dollars as board feet are to X. This will result in an easier analogy of units. The foregoing example is completed as follows:

$$100:60::75:X$$
$$(\text{extremes}) \; 100X = 4500 \; (\text{means})$$
$$X = \frac{4500}{100} = \textbf{45 Ans.}$$

As another example: If a vertical stake in the ground is 5′ high and casts a shadow 6′ long, what is the height of a nearby building that casts a shadow 156′ long? In setting up the proportion, the 6′ shadow is to the 156′ shadow as the 5′ height is to the X′ height. Thus:

$$6:156::5:X$$
$$6X = 780$$
$$X = \frac{780}{6} = \textbf{130 ft Ans.}$$

10. MATHEMATICS OF PLANE FIGURES

A plane figure is a surface on a single plane surrounded by either straight or curved lines, such as a square, rectangle, triangle, or circle. These figures have many interesting characteristics; their study is called *plane geometry*. All figures surrounded by straight lines are known as polygons, and there are many of them (see Figure 2–21). The sum of the lengths of the sides of any figure is its *perimeter*. The plane figures shown in Figure 2–21 and the methods of finding unknown dimensions relative to them are frequently encountered in reading architectural working drawings.

Refer to the formulas in Table 2–2 for solving various area and perimeter problems when encountering plane figures. To review the calculations further, refer to the step-by-step instructions dealing with plane figures in the Appendix.

11. MATHEMATICS OF SOLID FIGURES

A solid figure (Figure 2–22) is one having volume or one having thickness as well as height and length, such as a block of wood, a box, or a tin can. The volumes within buildings have these three-dimensional characteristics and require the use of solid geometry in measuring them and finding their volumes.

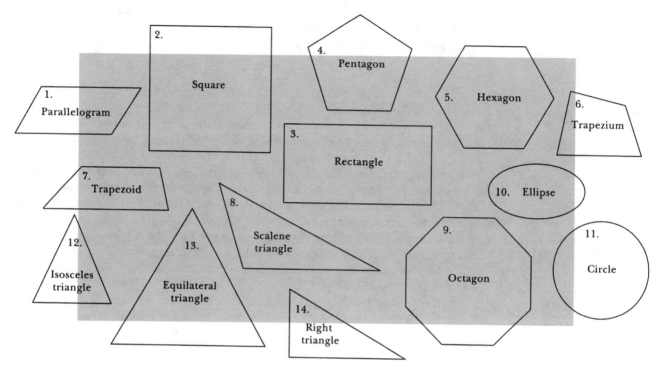

FIGURE 2–21

Plane figures.

Study the solid figures and their formulas shown in Table 2–2. Calculations of volumes are useful in reading architectural working drawings and in estimating. Here again, further step-by-step instructions for calculating the volumes and other features are found in the Appendix in case you need to review.

12. ROOF PITCHES

Roof pitches or slopes are shown on drawings by one of three methods: (1) a slope-ratio diagram (Figure 2–23), (2) a fractional pitch indication, or (3) an angular dimension. The slope-ratio right triangle is made with its hypotenuse parallel to the roof profile and its opposite sides, representing the rise and run slope ratio, drawn vertically and horizontally. The horizontal leg of the triangle is usually measured 1 in. long for convenience and given a numerical value of 12. The rise value shown on the vertical leg is measured off with the $1'' = 1'\,0''$ scale and given a relative value to 12. This ratio of the rise of the roof to its run is stated as 3:12, 7:12, $1\frac{1}{2}$:12, and so on, making it consistent with conventional references associated with roofing materials and construction applications. In Figure 2–23A the 4:12 right triangle symbol means, briefly, that for every 12– of horizontal run, the roof rises 4″. This symbol is found on the majority of drawings showing roof profiles.

The fractional pitch indication is derived from the following formula:

$$\text{Pitch} = \frac{\text{Rise}}{\text{Span}}$$

The span is the total distance between top-plate supports (twice the run). The pitch of the roof in Figure 2–23A would be $\frac{4}{24}$, or $\frac{1}{6}$. Fractional pitches are infrequently found on drawings.

TABLE 2-2

Formulae for Major Geometric Shapes.

AREAS OF PLANE FIGURES

NAME / FORMULA	SHAPE
(A = Area) Parallelogram $A = B \times h$	
Trapezoid $A = \dfrac{B + C}{2} \times h$	
Triangle $A = \dfrac{B \times h}{2}$	
Trapezium (Divide into 2 triangles) A = Sum of the 2 triangles (See above)	
Regular Polygon $A = \dfrac{\text{Sum of sides (s)}}{2}$ $\times$ inside Radius (R)	
Circle $A = $ (1) πR^2 (2) $.7854 \times D^2$ (3) $.0796 \times C^2$	
Sector $A = $ (1) $\dfrac{a^2}{360°} \times \pi R^2$ (2) Length of arc $\times \dfrac{R}{2}$	
Segment A = Area of sector minus triangle (See above)	
Ellipse $A = M \times m \times .7854$	
Parabola $A = B \times \dfrac{2h}{3}$	

VOLUMES OF SOLID FIGURES

NAME / FORMULA	SHAPE
(V = volume) Cube $V = a^3$ (in cubic units)	
Rectangular $V = L \times W \times h$	
Prisms $V(1) = \dfrac{B \times A}{2} \times h$ $V(2) = \dfrac{s \times 6}{2} \times R \times h$ $V = $ Area of end $\times h$	
Cylinder $V = \pi R^2 \times h$	
Cone $V = \dfrac{\pi R^2 \times h}{3}$	
Pyramids $V(1) = L \times W \times \dfrac{h}{3}$ $V(2) = \dfrac{B \times A}{2} \times \dfrac{h}{3}$ $V = $ Area of Base $\times \dfrac{h}{3}$	
Sphere $V = \dfrac{1}{6} \pi D^3$	
Circular Ring (Torus) $V = 2\pi^2 \times Rr^2$ $V = $ Area of section $\times 2\pi R$	

FIGURE 2–22

Triangular prisms in different positions.

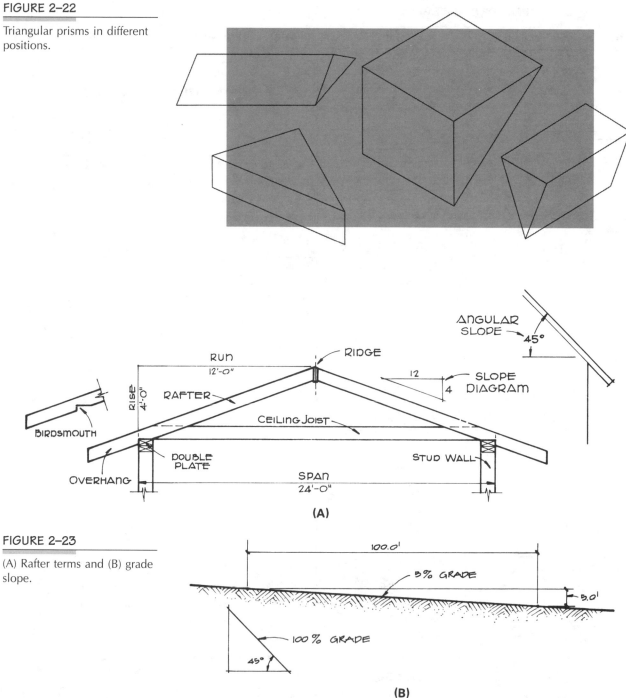

FIGURE 2–23

(A) Rafter terms and (B) grade slope.

Angular slopes are indicated with an arc dimension line, revealing the angular dimension in degrees from the horizontal (Figure 2–23A). Its use is limited mainly to minor construction features.

13. GRADE SLOPE INDICATIONS

The slope of land surface on plot plans and survey maps is indicated as *percentage of grade*. The amount of rise in feet in relation to 100′ of horizontal distance is conveniently shown in percentage. Percentage means the rate per 100. Therefore, if an indication stated a 10 percent grade, it would mean that the land rose 10′ for every 100′ of horizontal distance. Similarly, if the land surface rose $2\frac{1}{2}'$ per 100′, the grade indication

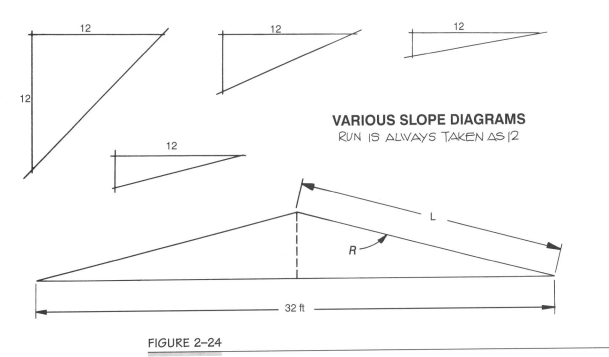

FIGURE 2-24

Slope diagrams and rafter lengths.

would be $2\frac{1}{2}$ percent grade. The percentage-of-grade type of indication is used mainly on civil engineers' drawings.

14. RAFTER LENGTH CALCULATION

Sometimes it becomes necessary to find the length of rafters in gable roofs or the hypotenuse of a right triangle in other construction work. Rafter lengths, actually the hypotenuse of a right triangle, can be calculated either by trigonometry or by the Pythagorean theorem (see A–11 in the Appendix). The latter method is the simpler if you can extract the square root of a number or if you have a square-root table available (see Table A–5 in the Appendix). Slope-ratio diagrams allow you to find the vertical leg (rise) of the triangle if the horizontal leg (run) is given. In Figure 2–24, for example, find the length of the rafter R if the run is 16′ and the slope ratio is 3:12. To find the *rise* of the rafter:

$$3:12::X:16$$
$$12X = 48; \quad X = 4'(\text{rise}); \quad \text{Run} = 16'$$

$$\text{Length of rafter } R = \sqrt{(\text{rise})^2 + (\text{run})^2} = \sqrt{16 + 256}$$

$$= \sqrt{272} = \textbf{16 ft 6 in. Ans.}$$

H. SYMBOLS USED ON WORKING DRAWINGS

Proper use of graphic symbols representing the materials used in buildings is essential to clearly communicate to the building trades what is to be used in the construction of a building. It is equally important to correctly use construction terms and abbreviations. The remainder of the chapter (sections H, I, and J) contains examples of graphic symbols, construction terms, and abbreviations used in architectural drawings. Note that most symbols used to represent materials on elevation views are different than the symbols used to represent those same materials on section views. Abbreviations are used to save space and reduce production time but the drafter must use discretion and abbreviate only those words that have commonly identified abbreviations.

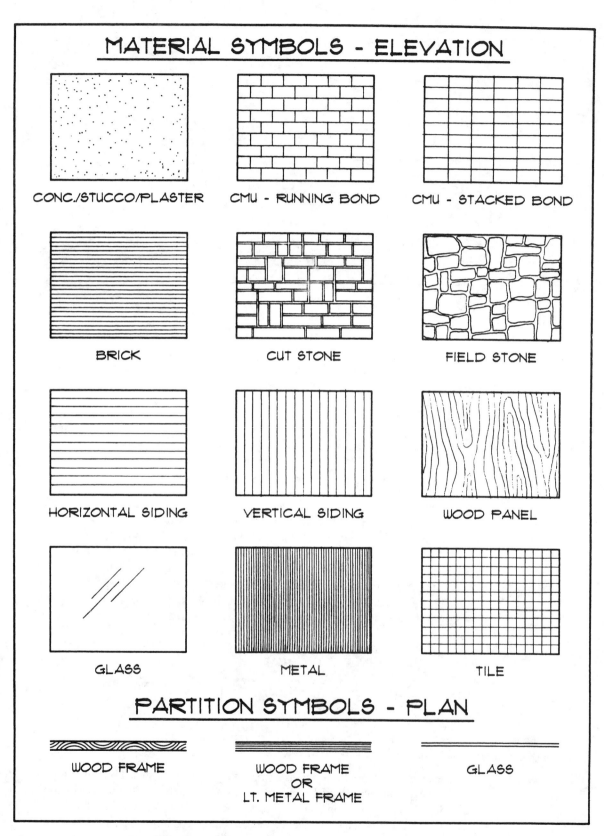

FIGURE 2-25

Architectural materials symbols used on working drawings. Continued on next page.

MATERIAL SYMBOLS – SECTION

EARTHWORK		CONCRETE		WOOD	
EARTH UNDISTURBED		CONCRETE		ROUGH WOOD	
EARTH COMPACTED		CONC. PLANK		WOOD BLOCKING	
ROCK		CMU		FINISH WOOD	
GRAVEL		CMU		LAMINATED WOOD	
SAND		CMU GLAZED		HARD-BOARD	
RIPRAP		ADOBE		PLYWOOD SM. SCALE	

STONE		MASONRY		WOOD	
ASHLAR		BRICK		PLYWOOD LG. SCALE	
CUT		GLAZED BRICK		PARTICLE BOARD	
CAST		FIRE BRICK		ORIENTED STRAND BD	
RUBBLE		GLAZED CLAY TILE		WOOD FIN. ON STUDS	

FIGURE 2-25

Continued.

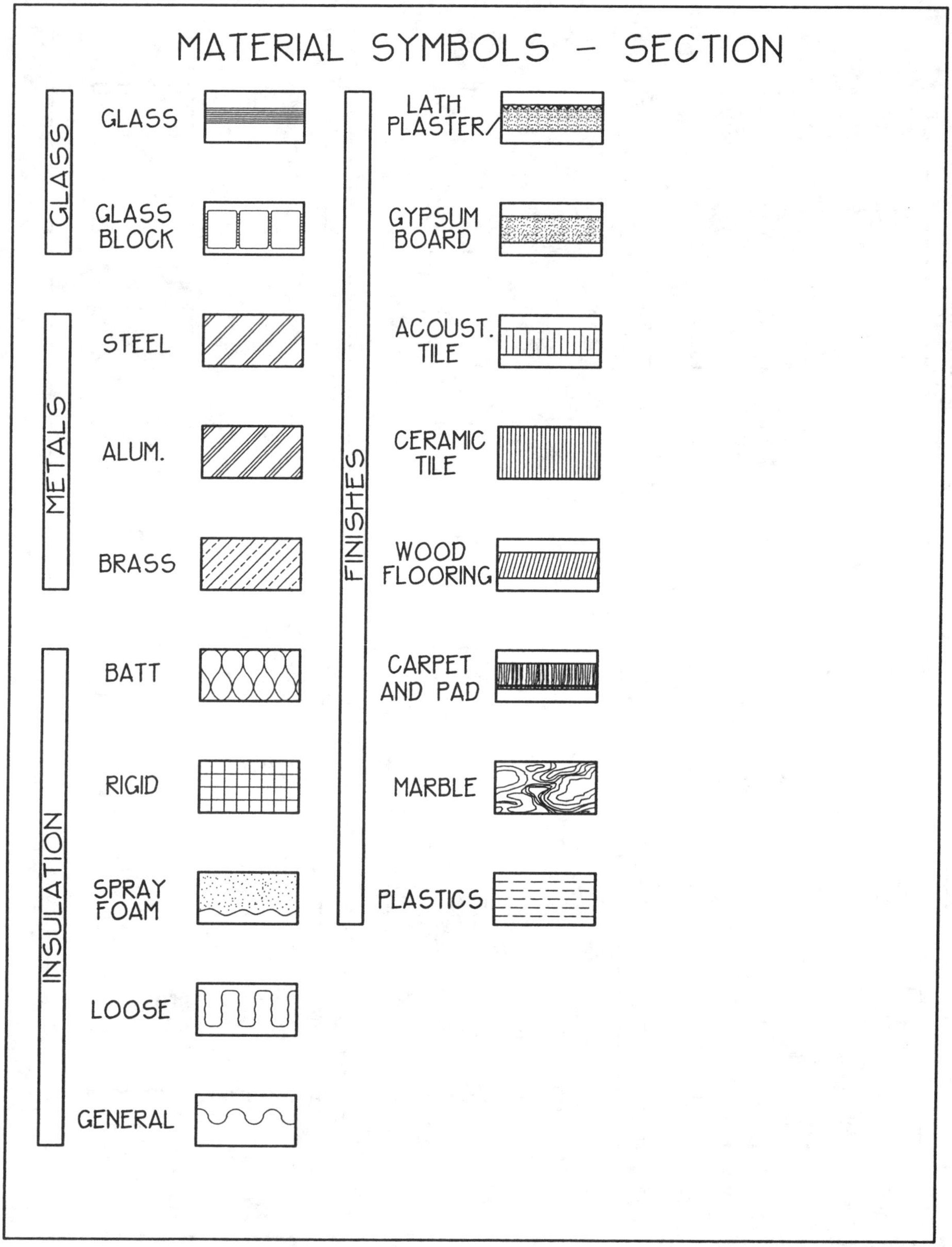

MATERIAL SYMBOLS – SECTION

GLASS		FINISHES	
GLASS		LATH PLASTER	
GLASS BLOCK		GYPSUM BOARD	
METALS		ACOUST. TILE	
STEEL		CERAMIC TILE	
ALUM.		WOOD FLOORING	
BRASS		CARPET AND PAD	
INSULATION		MARBLE	
BATT		PLASTICS	
RIGID			
SPRAY FOAM			
LOOSE			
GENERAL			

FIGURE 2-25

Continued.

GENERAL OUTLETS

- ⌀ CEILING OUTLET
- ⊢O WALL OUTLET
- ⓕ FAN OUTLET
- ⊢O_PS LAMP & PULL SWITCH

CONVENIENCE OUTLETS

- ⊢⊖ DUPLEX OUTLET
- ⊢⊖_WP WATERPROOF OUTLET
- ⊢● RANGE OUTLET
- ⊢⊖_S SWITCH & DUPLEX OUTLET
- ⊢⊖_3 TRIPLEX OUTLET
- ● SPECIAL PURPOSE OUTLET
- ⊙ FLOOR OUTLET
- ⊢⊖_GR DUPLEX OUTLET FOR GROUND

SWITCH OUTLETS

- $ SINGLE POLE SWITCH
- $_2 DOUBLE POLE SWITCH
- $_3 THREE WAY SWITCH
- $_4 FOUR WAY SWITCH

AUXILIARY

- ⊡ DOOR BELL BUTTON
- ▱ BUZZER
- ▱O BELL
- ◀ TELEPHONE
- ◁ INTERCOM
- Ⓜ MOTOR OUTLET
- ▬ LIGHTING PANEL
- O_a,b,c SPECIAL AUXILIARY OUTLETS (SUBSCRIPTS REFER TO NOTES)

SWITCHING ARRANGEMENTS

FROM TWO STATIONS

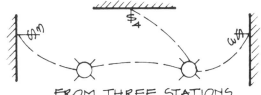
FROM THREE STATIONS

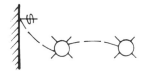
TWO OUTLETS

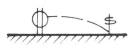
DUPLEX OUTLET

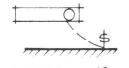
FLOURESCENT FIXTURE

FLOODLITE

FIGURE 2–26

Electrical symbols used on working drawings.

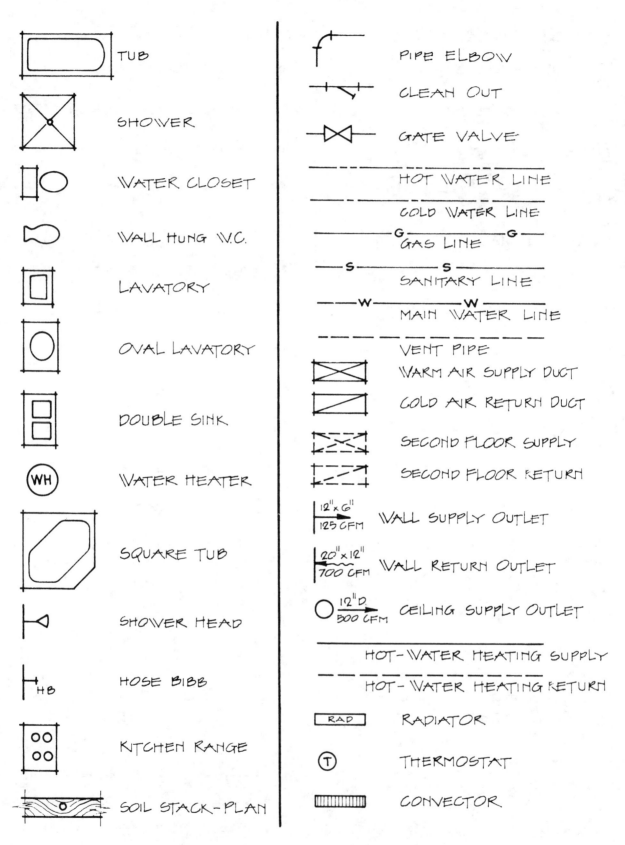

FIGURE 2-27

Plumbing and heating symbols used on working drawings.

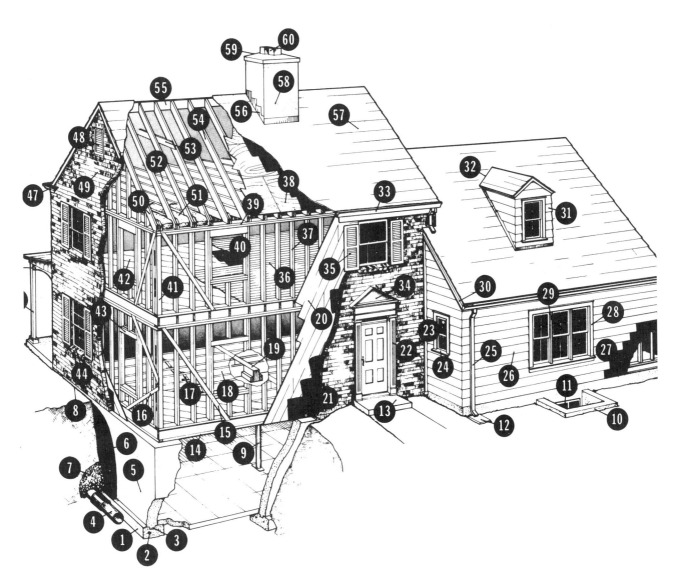

1. Footing
2. Reinforcing rod
3. Keyway
4. Drain tile
5. Foundation wall
6. Waterproofing
7. Gravel fill
8. Grade line
9. Metal column
10. Areaway wall
11. Basement window
12. Splash block
13. Stoop
14. Sill plate
15. Corner brace
16. Knee brace
17. Bridging
18. Floor joist
19. Beam; girder
20. Sheathing
21. Building paper
22. Trim pilaster
23. Double-hung window
24. Window sill
25. Downspout; leader
26. Bevel siding
27. Fiberboard sheathing
28. Window trim
29. Mullion
30. Rake mold
31. Dormer
32. Valley
33. Gutter
34. Pediment door trim
35. Shutter
36. Finish flooring
37. Stud
38. Roof decking
39. Double top plate
40. Flooring paper
41. Corner post
42. Subfloor
43. Lintel; header
44. Brick sill
45. Porch post
46. Porch frieze board
47. Return cornice
48. Louver
49. Brick veneer; gable
50. End rafter
51. Insulation
52. Ceiling joist
53. Collar beam
54. Common rafter
55. Ridge board
56. Flashing
57. Shingles
58. Chimney
59. Cement wash; cap
60. Chimney flues; pots

FIGURE 2-28

Typical residential terms.

GLOSSARY OF CONSTRUCTION TERMS

(Also see Fireplace Terms, page 191; Stair Terms, page 193)

A

Abut: To join the ends of construction members.

Adobe construction: Construction using sun-dried units of adobe soil for walls; usually found in southwestern United States.

A-frame: Structural system utilizing members that when fastened together resemble the letter **A**.

Aggregate: Gravel (coarse) or sand (fine) used in concrete mixes.

Anchor bolt: A threaded bolt anchored in the masonry foundation to fasten a wood sill plate.

Apron: Inside window trim placed under the stool and against the wall (Figure 3–51).

Arch: A curved structure that supports itself and spans an opening.

Areaway: Recessed area below grade around the foundation to allow light and ventilation into a basement window (Figure 3–43F).

Arris: Sharp edge formed by two surfaces; usually on moldings.

Asbestos board: Fire-resistant sheet made from asbestos fiber and portland cement.

Ashlar: Masonry utilizing cut, squared stone.

Asphalt: Dark, thick by-product of petrocarbon that is used for roof shingles and road surfaces when mixed with mineral particles.

Astragal: T-profiled molding usually used between meeting doors or casement windows (Figure 2–29).

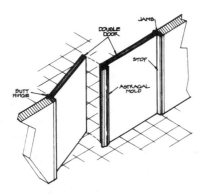

FIGURE 2–29

Astragal mold on double doors.

Atrium: An open court within a building.

Attic: Interior part of gable house that is directly under the roof.

Awning: Rooflike shelter extending from above windows or doors, usually made of canvas or other materials.

Axis: Line around which something rotates or is symmetrically arranged.

B

Backfill: Earth used to fill in areas around foundation walls.

Banister: A handrail for stairs.

Base cabinets: The lower cabinets in kitchens that support the countertops.

Bargeboard: Finish board covering the projecting and sloping portion of a gable roof (Figure 3–23).

Baseboard: Finish board covering the interior wall where the wall and floor meet (Figure 2–39).

Batt: A type of fiberglass insulation designed to be installed between framing members.

Batten: Narrow strip of wood nailed over the vertical joint of boards to form board-and-batten siding (Figure 2–30).

FIGURE 2–30

Board-and-batten siding.

Batter boards: Horizontal boards at exact elevations nailed to posts just outside the corners of a proposed building. Strings are stretched across the boards to locate the outline of the foundation for workers (Figure 3–8).

Bay window: A group of windows extending from an outside wall to form an alcove within.

Beam: Horizontal structural member, usually heavier than a joist.

Beamed ceiling: A ceiling that is supported by exposed beams.

Bearing plate: Metal plate that provides support for a structural member.

Bearing wall: A wall that supports structural weight, such as the roof above.

Bench mark: Mark on some permanent object fixed to the ground from which land measurements and elevations are taken.

Bidet: Low plumbing fixture in luxury bathrooms for bathing one's private parts.

Bird's mouth: A notch cut into a rafter to provide a horizontal bearing on the top plate.

Blind nailing: Method of nailing that will conceal nails, usually used on strip flooring and wood paneling (Figure 2–31).

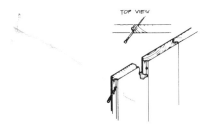

FIGURE 2–31

Blind nailing.

BLOCKING: Small wood pieces in wood framing to anchor or support other major members.
BOARD MEASURE: System of lumber measurement. The unit is 1 bd ft, which is 1 ft square by approximately 1 in. thick.
BOND: Arrangement of masonry units in a wall.
BOND BEAM: Continuous, reinforced concrete block course around the top of masonry walls.
BRACING: Support members in framing that are used to make the major structural members more rigid.
BRICK: Small masonry units made from clay and baked in a kiln.
BRIDGING: Cross bracing or solid blocking between joists to stiffen floor framing (Figure 3–16).
BUILDING LINE: Setback restrictions on property, established by zoning ordinances, beyond which a building must be placed (Figure 3–1).
BUILT-UP ROOF: Roofing for low-slope roofs composed of several layers of felt and hot asphalt or coal tar, usually covered with small aggregate (Figures 3–64 and 3–66).
BULLNOSE: Rounded cabinet trim or ceramic tile edge.
BUTT: Type of hinge allowing edge of door to butt into the jamb; a joint that fastens members end to end (Figure 2–41).
BUTTRESS: Vertical masonry or concrete support, usually larger at the base, which projects from a wall (Figure 2–32).

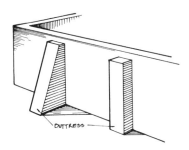

FIGURE 2–32

Buttresses used to strengthen walls.

C

CALLOUT: Note on a drawing with a leader to the relating feature.
CANTILEVER: A projecting beam or structural member anchored at only one end.

CANT STRIP: Angular-shaped member used to eliminate a sharp, right angle, often used on flat roofs (Figure 3–66).
CASING: Trim around window and door openings (Figure 3–58).
CAULKING: Soft, elastic material used to seal small openings around doors, windows, and so on.
CAVITY WALL: Double masonry wall having an air space between the wythes (Figure 2–54).
CHAMFER: Beveled edge formed by removing the sharp corner of a material (Figure 2–33).

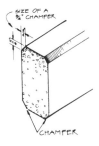

FIGURE 2–33

Chamfer.

CHANCEL: Space, screen, or railing about the altar of a church.
CHASE: Vertical space within a building for ducts, pipes, or wires.
CHORD: Top or bottom member of a truss.
CIRCUIT: Closed wiring or conductor through which an electric current can pass.
CLERESTORY: Windows that are above normal height—usually above a roof level.
COLLAR BEAM: Horizontal member tying opposing rafters below the roof ridge (Figure 2–28).
COLUMN: Vertical supporting member.
CONCRETE: Hardened mixture of cement, sand, gravel, and water; one of our major building materials.
CONDUIT: Round, cross-section electrical raceway of metal or plastic.
CONTROL JOINT: Continuous, vertical joint in masonry walls to control cracking.
COPING: Metal cap or masonry top course of a wall (Figure 2–34).

FIGURE 2–34

Coping is the cap on a wall.

CORBEL: Projection of masonry from the face of a wall; a stepped coursing bracket to support weight above (Figure 3–47).
CORNICE: Molded projection of the roof overhang at the top of a wall.
COVE: Concave molding usually used on horizontal inside corners.
CRAWL SPACE: Shallow space below the floor of a building built above ground, generally surrounded with a foundation wall.
CRICKET: Small gable-like roof structure used to divert water and debris from the intersection of a sloping roof and chimney; also called a saddle (Figure 2–43).
CRIPPLE: Structural member that is cut less than full length, such as a studding piece above a window or door.
CROWN MOLDING: Molding used above eye level; usually the upper trim on interior walls.
CUT STONE: Stone cut to given sizes and shapes.

D

DADO JOINT: Recessed joint on the face of a board to receive the end of a perpendicular board (Figure 2–35).

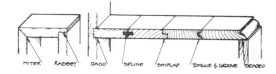

FIGURE 2–35

Wood joints used in cabinet work.

DAMPPROOFING: Material used to prevent passage of moisture.
DEAD LOAD: The weight of the structure and all the permanent installed components.
DECK: Exterior floor, usually extended from the outside wall.
DIMENSION LUMBER: Framing lumber that is 2– nominal thickness.
DISTRIBUTION PANEL: Electrical unit that distributes the incoming current into smaller circuits.
DOMESTIC HOT WATER: Potable (drinkable) hot water that is used for personal needs.
DOOR STOP: Projecting strip around the inside of a door frame against which the door closes.
DORMER: Top-floor projection of a room built out from a sloping roof to allow light and ventilation.
DOUBLE GLAZING: Two panes of glass with air sealed between.
DOWNSPOUT: Pipe for carrying rainwater from the roof to the ground or storm drainage system; also called a leader.

DRESSED SIZE: Dimensions of lumber after planing; also known as finished or actual size.
DRIP: Projecting construction or groove below an exterior member to throw off rainwater (Figure 2–36).

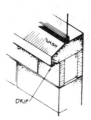

FIGURE 2–36

Drip.

DRY-WALL CONSTRUCTION: Interior wall covering other than plaster, usually referred to as gypsum board or wallboard.
DUPLEX OUTLET: Electrical wall outlet having two plug receptacles.

E

EASEMENT: A legal restriction on a piece of property.
EAVE: Lower portion of the roof that overhangs the walls.
EFFLORESCENCE: Undesirable white stains on masonry walls created by moisture from within.
ELL: Extension or wing of a building at right angles to the main section (Figure 2–37).

FIGURE 2–37

Ell.

EXCAVATION: Cavity or pit produced by digging the earth in preparation for construction.
EXPANSION JOINT: Flexible joint used to prevent cracking or breaking due to thermal expansion and contraction.

F

FACADE: Face or front elevation of a building.
FACE BRICK: Brick of better quality used on the face of a wall.
FASCIA: Outside horizontal face or member on the edge of a roof or cornice (Figure 3–22).
FASTENERS: General term for metal devices, such as nails, bolts, screws, and so on, used to secure structural members within a building.
FENESTRATION: Arrangement and sizing of doors and windows in a building.

FIBERBOARD: Fabricated structural sheets made from wood fiber and adhesive under pressure.

FIRE CUT: Angular cut at the ends of joists framing into a masonry wall (Figure 3–47).

FIRE RATED: A rating given to building materials according to their resistance to fire.

FIRE-STOP: Tight closure material or blocking to prevent the spread of flame or hot gases within framing.

FLAGSTONE: Flat stone used for floors, terraces, steps, and walks.

FLASHING: Sheet-metal work used in roof or wall construction to prevent water from seeping into the building.

FLITCH BEAM: Built-up beam formed by a steel plate sandwiched between two wood members and bolted together for additional strength (Figure 2–38).

FIGURE 2–38

Flitch beam.

FLUE: Vertical opening used to allow smoke and gases to escape, such as within a chimney.

FOOTER: A term sometimes used for the concrete footing below the concrete foundation.

FOOTING: Poured concrete base upon which foundation walls, columns, or chimneys rest; usually has steel reinforcing bars (Figure 3–48).

FRIEZE: Trim member below the cornice that is fastened against the wall.

FROST LINE: Depth of frost penetration in the ground; bottom of footings should always be below this line.

FURRING STRIPS: Thin strips fastened to walls or ceilings for leveling and for attaching finish surface material (Figure 2–39).

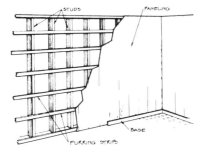

FIGURE 2–39

Furring strips on a wall.

G

GABLE: Triangular-shaped end wall of a gable-roof building.

GLAZING: Installation of glass in windows and doors.

GRADE: (1) Finished surface of ground around a building. (2) Refers to classification of the quality of lumber or plywood.

GRADIENT: Inclination of a road, piping, or the ground, expressed in percent.

GRAVEL STOP: Stirp of metal with a vertical lip used to retain the gravel around a built-up roof (Figure 3–64).

GROUNDS: Wood strips fastened to walls before plastering that serve as edges for the plaster and nailing base for wood trim.

GROUT: Thin cement mortar used for leveling and filling masonry cavities.

GUSSET: Plywood or metal plate used to strengthen joints of a truss.

GUTTER: Metal or wood trough for carrying rainwater to downspouts.

GYP BOARD: Gypsum sheets covered with paper that are fastened to walls and ceilings with nails or screws.

H

HALF-TIMBER: Exterior wall construction having wood frame members exposed and the spaces between filled with stucco or masonry.

HANGER: Metal strap used to support the ends of joists or piping (Figure 2–40).

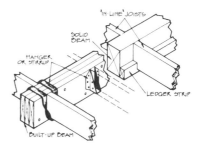

FIGURE 2–40

Framing terms.

HARDWARE: General term used for the metal parts that are used with conventional components of a wood structure, such as hinges and knobs on wood doors.

HEADER: In framing, the continuous joist placed across the ends of floor joists, the double joists at each end of floor or ceiling openings attached to the trimmers, and the structural member above window or door openings. In masonry, exposed ends of masonry units laid horizontally.

HEADROOM: Vertical clearance in a passageway or above a stairway, measured from the edge of the nosing.

HEARTH: The fireproof floor of a fireplace extending in front as well.

Heartwood: Central portion of a tree, which is stronger and more decay-resistant than the surrounding sapwood.
Heat pump: All-electric heating and cooling device that takes heat from outside air or ground water for heating and reverses for cooling.
Hip rafter: Diagonal rafter that extends from the plate to the ridge to form the hip.
Hose bibb: Water faucet made for the threaded attachment of a hose.
House sewer: Watertight soil pipe extending from the exterior of the foundation wall to the sewer main.

I

Incandescent lamp: Lamp in which a filament gives off light.
Interior trim: General term for all the finished molding, casing, baseboard, and so on, applied within the building by finish carpenters.

J

Jack rafter: Rafter shorter than a common rafter; especially used in hip-roof framing.
Jamb: Vertical member of a finished door or window opening.
Jointery: General woodworking term used for better-quality wood-joint construction.
Joist: Structural member that directly supports floors or ceilings and is supported by bearing walls, beams, or girders.

K

Kiln-dried lumber: Lumber that has been properly dried and cured (to 15 percent moisture content) resulting in a higher-grade lumber than air dried.
Knee wall: Low wall in upper story resulting from one-and-a-half story construction.
Knocked down: Unassembled; refers to construction units requiring assembly after being delivered to the job.

L

Lally column: A vertical post used to support floor loads.
Laminated beam: Beam made of superimposed layers of similar material (usually wood) by uniting them with glue under pressure.
Lap joint: Joint produced by lapping and joining two similar members (Figure 2–41).

Lath: Metal or gypsum sheeting used under plaster, stucco, and ceramic tile.
Lattice: Grillwork made by crossing small wooden strips.
Leader: Vertical pipe or downspout that carries rainwater to the ground or storm sewer.
Ledger: Strip of lumber fastened to the lower part of a beam or girder on which notched joists are attached (Figure 2–40).
Lineal foot: One-foot measurement along a straight line.
Lintel: Horizontal support over a window or door opening.
Live load: The combined weight of all movable components the structure will support.
Load-bearing wall: Wall designed to support the weight imposed upon it from above.
Lookout: Short, wooden framing member used to support an overhanging portion of a roof. It extends from the wall to support the soffit (Figure 3–22).
Lot line: Line forming the legal boundary of a piece of property; also called a property line.
Louver: Opening or slatted grillwork that allows ventilation while providing protection from rain, sight, or light.
Luminaire: Interior lighting fixture.

M

Mantel: The decorative shelf above a fireplace opening.
Masonry: General term for brickwork, stonework, concrete blockwork, or similar materials.
Mastic: Flexible adhesive for adhering building materials.
Matte finish: Finish free of gloss or highlights.
Millwork: Finish carpentry work or that woodwork done in a mill and delivered to the site; relates to interior trim.
Miter joint: Joint made with ends or edges of two pieces cut at 45-degree angles and fastened together.
Module: Standardized unit of measure (e.g., 4", 12", or 4' 0") to unify construction.
Monolithic: Term used for concrete work poured and cast in one piece without joints.
Mosaic: Small colored tile, glass, stone, or similar material arranged to produce a decorative surface.
Mudsill: The wood plate resting on the concrete foundation.
Mullion: Structural support member between a series of windows (Figure 2–28).
Muntin: Small bar separating the glass lights in a window sash.

N

Narthex: Enclosed passage between the entrance and nave of a church.
Nominal size: Size of lumber before dressing, rather than its actual or finished size (Figure 2–20).
Nonferrous metal: Metal containing no iron, such as copper, brass, or aluminum.

O

On center: Method of indicating spacing of framing members by stating the distance from the center of one to the center of the next.
Outlet: Any type of electrical box allowing current to be drawn from the electrical system for lighting or appliances.

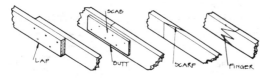

FIGURE 2–41

Wood joints used in framing.

OVERHANG: Projecting area of a roof or upper story beyond the wall of the lower part.

P

PARAPET: Low wall or railing at the edge of a roof; it extends above the roof level.

PARGE COAT: Thin coat of cement mortar applied to a masonry wall for refinement or dampproofing.

PARQUET FLOORING: Flooring, usually wood, laid in an alternating or inlaid pattern to form various designs.

PARTICLE BOARD: Sheets made from compressed wood fiber.

PARTY WALL: Wall common to adjoining buildings that both owners share, such as a wall between row houses or condominiums.

PENNY: Term used to identify nail size.

PERGOLA: Open, structural framework over an outdoor area, usually covered with climbing vines to form an arbor.

PERIPHERY: Entire outside edge of an object or surface.

PIER: Masonry support, usually in the crawl space, to support the floor framing.

PILASTER: Rectangular pier attached to a wall for the purpose of strengthening the wall; also a decorative column attached to a wall.

PITCH: Slope of a roof, usually expressed as a ratio.

PLANK: Lumber 2– thick or over, nominal.

PLATE: Top or bottom horizontal members of a row of studs in a frame wall; also, the sill member over a foundation wall.

PLUMB: Said of a member when it is in true vertical position as determined by a plumb bob or vertical level.

PRIME COAT: First coat of paint applied to wood or metal to prime the surface for succeeding coats.

PURLIN: Horizontal roof members laid over trusses to support roof decking.

Q

QUARRY TILE: Unglazed, machine-made tile used for floors.

QUARTER ROUND: Small molding with a quarter-circle profile.

QUARTER SAWED: Lumber, usually flooring, that has been sawed so that the medullary rays showing on end grain are nearly perpendicular to the face of the lumber.

QUOINS: Large squared stones or brick masonry set in the corners of masonry buildings for achitectural style (Figure 2–42).

FIGURE 2–42

Masonry quoins.

R

RABBET: Groove cut along the edge or end of a board to receive another board (Figure 2–35).

RAFTER: Inclined structural roof member.

RAKE: Inclined edge of a roof that overhangs the gable.

RANDOM RUBBLE: Stonework having irregular-shaped units and no indication of systematic course work.

REBAR: Steel reinforcing bar.

REVEAL: Side of an opening of a window or door (Figure 2–42).

RIBBON: Wood strip let into the studs to provide a bearing for joists (Figure 3–33).

RIDGEBOARD: Horizontal wood framing member to which the tops of rafters are attached (Figure 2–28).

RIPRAP: Stone placed on an incline to prevent erosion.

RISE: Vertical height of a roof or stairs.

ROUGH HARDWARE: All the concealed fasteners in a building, such as nails, bolts, and hangers.

ROUGH OPENING: Any unfinished opening in the framing of a building.

ROWLOCK: Brickwork with exposed ends set vertically.

RUN: Horizontal distance of a flight of stairs, or the horizontal distance from the outside wall to the ridge of a roof.

S

SADDLE: Small gable roof placed in back of a chimney on a sloping roof to shed water and debris; also called a cricket (Figure 2–43).

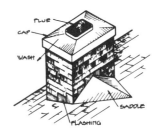

FIGURE 2–43

Saddle.

SASH: Individual frame into which glass is set; the movable part of a double-hung window.

SCARF JOINT: Joint made with diagonal ends (Figure 2–41).

SCHEDULE: Listing of finishes, doors, windows, and so on, on working drawings.

SCUTTLE: Small opening in a ceiling to provide access to an attic or roof.

SEISMIC CODE: Restrictions related to possible earthquake damage.

SETBACK: Distance from the property lines, front, side, and rear, to the face of a building; established by zoning ordinances.

SHAKE: Hand-split wood shingle.

SHEATHING: Rough covering over the framing of a building, either roof or wall, which is not exposed when finish material is applied.

SHOE MOLD: Small rounded molding covering the joint between the flooring and the baseboard (Figure 3–19).

SILL: Horizontal exterior member below a window or door opening (Figure 3–51). In frame construction, the lowest structural member that rests on the foundation (Figure 3–30).

SLEEPERS: Wood strips placed over or in a concrete slab to receive a finished wood floor.

SOFFIT: Underside of an overhang such as the eave, a second floor, or stairs (Figure 3–22).

SOIL STACK: Vertical plumbing pipe that carries sewage.

SOLEPLATE: Horizontal member of a frame wall that is directly under the studs.

SPAN: Horizontal distance between supports for joists, beams, or trusses.

SQUARE: In roofing, 100 sq ft of roofing.

STILE: Vertical framing member of a panel door.

STOOL: Horizontal interior member of the frame below a window (Figure 3–51).

STORY: Space between two floors of a building.

STUCCO: Exterior finish for masonry or wood; made from cement, sand, and hydrated lime mixed with water and applied wet.

STUDS: Vertical framing members in a wall spaced 16″ or 24″ on center.

SUBFLOOR: Material fastened directly to floor joists below the finish floor.

SUSPENDED CEILING: Finish ceiling hung below the underside of the building structure, either floor or roof.

SWALE: A depressed area to divert water away from the building.

T

TAIL JOISTS: Relatively shorter joists that join against a header or trimmer in floor framing (Figure 2–44).

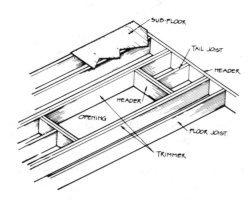

FIGURE 2–44

Floor framing terms.

TENSILE STRENGTH: The greatest longitudinal stress a structural member can resist without adverse effects (breaking or cracking).

TERRAZZO: Wear-resistant flooring made of marble chips or small stones embedded in cement matrix that has been polished smooth.

THERMAL CONDUCTOR: Material capable of transmitting heat.

THERMAL RESISTANCE (R): The ability of a material to resist heat flow.

THRESHOLD: Wood, metal, or stone member placed directly below a door.

TOENAIL: Nailing diagonally through a member (Figure 2–45).

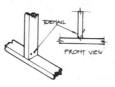

FIGURE 2–45

Studs are toenailed to the soleplate.

T-POST: Post built up of studs and blocking to form the intersection framing for perpendicular walls (Figure 2–46).

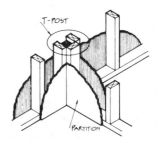

FIGURE 2–46

A T-post is used in framing the intersection of perpendicular walls.

TRAP: U-shaped pipe below plumbing fixtures that provides a water seal to prevent sewer odors and gases from entering habitable areas.

TRIMMER: The longer floor or ceiling-framing member around the rectangular opening into which headers are joined; both headers and trimmers are doubled (Figure 2–44).

TRUSS: Structural unit of members fastened in triangular arrangements to form a rigid framework for support over long spans (Figure 3–65).

TRUSSED RAFTER: Truss spaced close enough (usually 24″ on center) to eliminate the need for purlins.

V

VALLEY RAFTER: Diagonal rafter at the intersection of two intersecting sloping roofs.

VAPOR BARRIER: Watertight material used to prevent the passage of moisture or water vapor into and through walls and under concrete slabs.

VENEER CONSTRUCTION: Type of wall construction in which frame or masonry walls are faced with other exterior surfacing materials.

VENT STACK: Vertical soil pipe connected to the drainage system to allow ventilation and pressure equalization.

VESTIBULE: A small entrance or lobby.

W

WAINSCOT: Surfacing on the lower part of an interior wall when finished differently from the remainder of the wall.

WALL TIE: Small metal strip or steel wire used to bind courses of masonry in cavity-wall construction or to bind masonry to the wood frame in veneer construction (Figure 2–47).

WATER TABLE: Horizontal member extending from the surface of an exterior wall to throw rainwater away from the wall; also, the level of subsurface water.

WEATHER STRIPPING: Strips of fabric or metal fastened around the edges of windows and doors to prevent air infiltration.

WEEP HOLES: Small holes in masonry cavity walls to release moisture accumulation to the exterior (Figure 2–47).

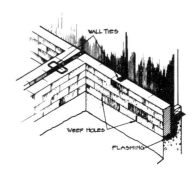

FIGURE 2–47

Wall ties.

WELDED WIRE FABRIC (WWF): Metal fabric made from wire welded together at right angles and used for concrete reinforcement.

WINDER: Stair tread that is wider at one end than the other, allowing the stairs to change direction.

WOOD I BEAM: Preassembled plywood and 2 × 4s, commonly used in contemporary floor framing (Figure 3–67).

WYTHE: Pertaining to a single-width masonry wall.

J. ABBREVIATIONS USED IN ARCHITECTURAL DRAWINGS

1. TERMS

Term	Abbr
Above finished floor	AFF
Above finished grade	AFG
Acoustical ceiling tile	ACT
Acoustical tile ceiling	ATC
Adjacent; adjoining; adjustable	ADJ
Administration	ADMIN
Air condition	A/C
Air conditioning unit	A/C UNIT
Air handling unit	AHU
Air vent; acid vent; audio visual	AV
Alternate; altitude	ALT
Alternating current; armored cable; asbestos cement; asphaltic concrete	AC
Aluminum	ALUM
American wire gauge	AWG
Amount	AMT
Ampere	AMP
Anchor bolt	AB
Angle; liter	L
Anodize	ANOD
Apartment	APT
Architect	ARCH
Architect/engineer	A/E
Architectural woodwork; acid waste; actual weight	AW
Area drain	AD
Asphalt	ASPH
Aggregate base course	ABC
Attention	ATTN
Automatic	AUTO
Avenue	AVE
Average	AVG
Awning window	AWN WDW
Balcony	BALC
Ballast	BLST
Base board radiator	BBR
Base line; building line	BL
Base plate	BPL
Baseboard; bulletin board	BB
Basement	BSMT
Beam; benchmark; bending moment	BM
Beam; wide flange	WF BM

Bearing	BRG
Bedroom	BR
Below ceiling	BLW CLG
Below finish floor	BFF
Benchmark; beam; bending moment	BM
Better	BTR
Between	BTWN
Bevel	BEV
Bituminous	BITUM
Blanket	BLKT
Board; butterfly damper	BD
Board feet (foot)	BD FT
Board measure	B/M
Bookcase; back of curb; between centers; bolt circle; bottom chord; brick color; building code	BC
Borrowed light; built	BLT
Both faces; bottom face	BF
Both sides	BS
Both ways	BW
Bottom	BOT
Bottom of steel	BOS
Boulevard	BLVD
Breaker	BRKR
Bridging	BRDG
British thermal unit	Btu
Bronze	BRZ
Broom closet	B CL
Building	BLDG
Building line	BL
Building paper	BP
Built-up roofing	BUR
Cabinet	CAB
Cabinet unit heater	CUH
Cable television	CTV
Calked joint	CLKJ
Canopy	CAN
Cantilever	CANTIL
Carpet; control power transformer	CPT
Carpet and pad	C&P
Carriage bolt; catch basin; cement base; ceramic base; corner bead	CB
Cased opening; carbon monoxide; Certificate of Occupancy; cleanout; company; cutout	CO
Casement	CSMT
Casement window; chemical waste line; clockwise; cold water piping; cool white	CW
Casework	CSWK
Casing	CSG
Cast concrete	C CONC
Cast-in-place; cast iron pipe	CIP
Cast iron; curb inlet	CI
Cast stone; commercial standard; control switch	CS
Catch basin; carriage bolt; cement base; ceramic base; corner bead	CB
Cavity	CAV
Ceiling	CLG
Ceiling diffuser	CLG DIFF
Ceiling grille	CLG GRL
Ceiling height	CLG HT
Celsius; channel	C
Cement; cemetery	CEM
Cement finish	CEM FIN
Cement plaster	CEM PLAS
Center; contour; cooling tower return	CTR
Center line; class; close	CL
Center to center	C TO C
Centimeter	cm
Ceramic	CER
Ceramic tile; count; current transformer	CT
Ceramic tile base	CTB
Ceramic tile floor	CTF
Certify	CERT
Chalkboard	CH BD
Chilled drinking water	CDW
Circle	CIR
Circuit	CKT
Cladding	CLDG
Class A door	A LABEL
Class B door	B LABEL
Class C door	C LABEL
Classroom	CLRM
Cleanout; carbon monoxide; cased opening; Certificate of Occupancy; company; cutout	CO
Closet	CLO
Closet rod; control relay; control room	CR
Clothes dryer	CL D
Coaxial cable	COAX
Coefficient of performance (heating); coping	COP
Coefficient of utilization; cubic; copper	CU
Column	COL
Combination; combined	COMB
Common	COM
Communication	COMM
Concrete; concentric	CONC
Concrete floor	CONC FLR
Concrete masonry unit	CMU

Condenser, condition	COND	Double hung windows; domestic hot water	DHW
Conference	CONF	Double joist	DJ
Construction	CONSTR	Double strength (glass); disconnect switch; downspout	DS
Construction documents; candela; contract documents	CD	Douglas fir	DOUG FIR
Construction joint; control joint	CJ	Dovetail	DVTL
Construction management; center matched	CM	Downspout; disconnect switch; double strength (glass)	DS
Continue; controller	CONT	Dozen	DOZ
Control	CTRL	Drain tile	DT
Control joint; construction joint	CJ	Drain, waste, and vent	DWV
Control panel; candlepower; concrete pipe	CP	Drawer; domestic water return	DWR
Coordinate	COORD	Drawing	DWG
Counter	CNTR	Dressed four sides	D4S
Counter sunk	CSK	Dressing area	DR AREA
Counterflashing	CFLG	Dressing room; dining room; door; drain; drive	DR
Crossbracing	XBRA		
Cubic feet	CU FT	Drinking fountain; damage free; diesel fuel	DF
Cubic feet per minute	CFM		
Cubic feet per second	CFS	Duplex outlet	DX OUT
Cubic yard	CU YD	Duplicate	DUPL
Curb and gutter	C&G	Dutch door	DT DR
Cut stone	CT STN	Each	EA
Dampproofing	DMPF	Each way	EW
Decibel	dB	Easement	ESMT
Demolition; demonstration	DEMO	East; modulus of elasticity	E
Demountable partition	DPTN	Edge of curb	EC
Department	DEPT	Edge of pavement (paving); electrical panel (panelboard)	EP
Design	DSGN		
Detail	DET	Edge of slab	EOS
Diagonal; diagram	DIAG	Electric	ELEC
Diameter	DIA	Electric heater	EH
Difference; differential; diffuser	DIFF	Electric water cooler	EWC
		Electric water heater	EWH
Dimension	DIM	Elementary; element	ELEM
Dining room; door; drain; dressing room; drive	DR	Elevation; each layer; easement line	EL
Dishwasher; distilled water; domestic water	DW	Elevator	ELEV
		Emergency	EMER
Disposal	DSPL	Enamel	ENAM
Distance; district	DIST	Enclosure	ENCL
Division; divide	DIV	Entrance	ENTR
Domestic	DOM	Equal	EQ
Domestic hot water; double hung windows	DHW	Equipment	EQUIP
		Equivalent	EQUIV
Domestic water heater	DWH	Estimate	EST
Door; dining room; drain; dressing room; drive	DR	Et cetera; and so forth	ETC
		Evacuate	EVAC
Door frame	DR FR	Evaporate	EVAP
Door stop	DRST	Evaporative cooling unit	ECU
Double	DBL	Example	EX
Double acting door	DBL ACT DR	Exhaust; exhibit	EXH
Double glaze	DBL GLZ	Existing	EXIST
Double hung (door, window)	DH	Existing grade	EXST GR

Exit light	EXT LT
Expansion; expand; exposed	EXP
Expansion bolt	EXP BT
Expansion joint	EJ
Exterior; external; extinguisher	EXT
Exterior finish; each face	EF
Exterior finish system	EFS
Exterior grade	EXT GR
Fabric	FAB
Face brick	FC BRK
Face of concrete; face of curb	FOC
Face of masonry	FOM
Face of stud; face of slab; fuel oil supply	FOS
Face of wall	FOW
Face to face	F/F
Fahrenheit; female; fire line	F
Fascia; fire alarm station	FAS
Fascia board	FAS BD
Feet; fire treated; foot; fully tempered (glass)	FT
Feet per minute	FPM
Fiberglass	FGL
Figure	FIG
File cabinet; footcandle	FC
Finish	FIN
Finish floor elevation	FF EL
Finish grade	FIN GR
Finished opening; field order; fuel oil	FO
Fire alarm; face area; final assembly; fresh air	FA
Fire brick	F BRK
Fire extinguisher	FE
Fire extinguisher cabinet	FEC
Fireproof; fire protection; flagpole; freezing point	FP
Flashing	FLASH
Flexible	FLEX
Float glass	FLT GL
Floor; filler	FLR
Floor area ratio	FAR
Floor drain	FD
Floor sink	FLR SK
Flooring; flange	FLG
Fluorescent	FLUOR
Foot; feet; fire treated; fully tempered (glass)	FT
Foot board measure	FBM
Footing	FTG
Foundation	FDTN
Freezer	FRZ
Frosted glass	FRST GL
Full scale; far side; Federal Specification; fire station; full size	FS
Fully tempered (glass); feet; fire treated; foot	FT
Furnace; furnish; furniture	FURN
Furring	FURG
Gage	GA
Gallon	GAL
Gallons per minute	GPM
Galvanized; galvanic	GALV
Galvanized iron	GI
Galvanized steel	GALV STL
Garden	GRDN
Gas fired water heater	GWH
Gauge	—
General; generator	GEN
General contractor	GC
Glass; ground level	GL
Glass block	GL BLK
Glass-fiber-reinforced concrete	GFRC
Glazed concrete masonry unit	GLZ CMU
Glazed structural unit	GSU
Glazed wall tile	GWT
Glazing	GLZ
Glued laminated wood	GLU LAM
Grade beam	GR BM
Granite	GRAN
Ground fault circuit interrupter	GFCI
Ground floor	GR FL
Grout; grease trap; gross ton	GT
Guarantee	GUAR
Gutter	GUT
Gymnasium	GYM
Gypsum	GYP
Gypsum board	GYP BD
Gypsum plaster	GYP PLAS
Gypsum plaster ceiling	GPC
Hand dryer; heavy duty	HD
Handicap; heating coil; heavy commercial; hollow core; hose cabinet	HC
Handicapped	HCP
Handrail	HNDRL
Hardboard	HDBD
Hardware	HDW
Head joint	HD JT
Header	HDR
Headquarters	HQ
Heat-strengthened (glass); hand sink; high strength	HS
Heat absorbing glass	HAGL
Heat treated (glass)	HT TRD

Heating, ventilating, and air conditioning	HVAC	Landing	LDG
Height	HT	Lath and plaster	L&P
Hemlock	HEM	Laundry	LAU
Hexagon; heat exchanger	HEX	Lavatory	LAV
Hollow concrete masonry unit	HCMU	Left hand; latent heat	LH
Hollow core; handicap; heating coil; heavy commercial; hose cabinet	HC	Left hand reverse; latent heat ratio	LHR
		Library	LIB
		Light	LT
Hollow core wood door	HCWD	Light gage	LT GA
Hollow metal	HM	Light switch	LT SW
Hollow metal door; humidity	HMD	Lighting	LTG
Horizontal	HORIZ	Lightweight	LT WT
Horsepower; heat pump; high pressure	HP	Lightweight concrete	LWC
		Lightweight concrete masonry unit	LCMU
Hose bibb	HB	Limestone	LMST
Hospital	HOSP	Limited	LTD
Hot and cold water	H&CW	Linear	LIN
Hot water	HW	Linear ceiling diffuser	LCD
House	HSE	Linear diffuser	LD
Hydrant	HYD	Linear feet (foot)	LF
I beam	IB	Linen closet	L CL
Identification; inside diameter; inside dimension; interior design	ID	Linoleum	LINO
		Living room	LR
		Load-bearing	LD BRG
Illumination	ILLUM	Location	LOC
Incandescent	INCAND	Locker	LKR
Independent; industrial	IND	Loose fill insulation	LF INS
Indoor air quality	IAQ	Louver	LVR
Information	INFO	Louver door	LVDR
Insect screen; island	IS	Lumber	LBR
Inside diameter; identification; inside dimension; interior design	ID	Mahogany	MAHOG
		Mail box; machine bolt; mixing box	MB
Inside face of stud	IFS	Maintenance	MAINT
Insulation	INSUL	Management	MGT
Interior	INT	Manhole	MH
Invert	INV	Manual	MAN
Invert elevation	INV EL	Manufactured	MFD
Iron pipe	IP	Manufacturer; mass flow rate	MFR
Jalousie	JAL	Manufacturing	MFG
Janitor	JAN	Masonry opening; motor operated	MO
Janitor's sink	JS	Master bedroom; member	MBR
Keyway	KWY	Material	MATL
Kickplate	KPL	Maximum	MAX
Kiln dried; knocked down	KD	Mechanical	MECH
Kilowatt	kW	Mechanical contractor; manhole cover; medicine cabinet; metal-clad; moisture content; moment connection	MC
Kitchen	KIT		
Kitchen cabinet	KC		
Knock out panel	KOP		
Laboratory	LAB	Mechanical engineer	ME
Laminate	LAM	Mechanical room	MECH RM
Laminated glass	LAM GL	Medical; medium	MED

Medicine cabinet; manhole cover; mechanical contractor; metal-clad; moisture content; moment connection	MC
Medium; medical	MED
Meeting; mounting	MTG
Membrane	MEMB
Membrane waterproofing	MWP
Metal	MTL
Metal flashing	METF
Metal lath; materials list; monolithic	ML
Metal lath and plaster	ML&P
Metal roof	METR
Meter	m
Microwave; megawatt	MW
Middle	MID
Mill finish; mastic floor	MF
Millimeter	mm
Millwork	MLWK
Minimum; minute	MIN
Miscellaneous	MISC
Modified bitumen	MOD BIT
Modify; model; module; motor operated damper	MOD
Moisture resistant	MR
Molding (moulding)	MLDG
Mop sink; machine screw; motor starter	MS
Mounted; mean temperature difference	MTD
Movable	MVBL
Mullion	MULL
Multiple	MULT
Natural	NAT
Natural gas; girder; ground	G
Negative	NEG
Nickel	NKL
No scale; narrow stile; near side	NS
Nominal	NOM
Normal	NORM
North; newton	N
Not in contract; noise isolation class	NIC
Not to scale	NTS
Number; normally open	NO
Numeral	NUM
Office	OFF
On center	OC
Opaque	OPQ
Opening	OPNG
Operating room; outside radius	OR
Opposite	OPP
Optimum; optional	OPT
Original	ORIG
Ornamental	ORN
Ounce	OZ
Out to out	O/O
Outside air; overall	OA
Outside face of studs	OFS
Overall; outside air	OA
Overhang	OH
Packaged terminal air conditioner	PTAC
Paint; pipe thread; pneumatic tube; post-tensioned; pressure treated	PT
Pair; pipe rail; pumped return	PR
Parallel; parapet	PAR
Parging	PARG
Particleboard	PBD
Partition	PTN
Paved road	PV RD
Paving	PVG
Penny (nail); deep; depth	D
Penthouse; phase	PH
Perforated; perform	PERF
Perpendicular	PERP
Pharmacy	PHAR
Piece; point of curve; polycarbonate; portland cement	PC
Pilaster	PIL
Plaster; plastic	PLAS
Plastic; plaster	PLAS
Plastic laminate	PLAM
Plate	PL
Plate glass	PL GL
Plumbing	PLBG
Plywood	PLYWD
Portland cement, piece; point of curve; polycarbonate	PC
Portland cement plaster	PCP
Position; positive	POS
Post office; purchase order	PO
Post-tensioned concrete	PT CONC
Pounds per cubic foot	PCF
Pounds per square foot	PSF
Pounds per square inch	PSI
Precast concrete; precool coil	PCC
Prefabricate	PREFAB
Preference	PREF
Premolded expansion joint	PEJ
Pressure treated; paint; pipe thread; pneumatic tube; post tensioned	PT
Previous	PREV
Principal	PRIN
Project	PROJ
Property	PROP
Property line	PL

Push/pull; panel point; polypropylene (plastic)	PP
Quadrant; quadrangle	QUAD
Quantity	QTY
Quarry	QRY
Quarry tile	QT
Radiator; radian; return air duct	RAD
Radius; range; riser; thermal resistance	R
Receptacle	RECPT
Recessed	REC
Rectangle	RECT
Redwood	RWD
Reference; refrigerator	REF
Reflected ceiling plan; reinforced concrete pipe	RCP
Refrigerator; reference	REF
Regulation; register	REG
Reinforce	REINF
Reinforced brick masonry	RBM
Reinforced concrete; remote control	RC
Reinforcing steel bars	REBAR
Removable	REM
Required	REQD
Resilient	RESIL
Restroom	REST
Reveal	RVL
Revision; revolutions	REV
Revolutions per minute	RPM
Right of way	ROW
Rigid insulation, solid	RDG INS
Riser; radius; range; thermal resistance	R
Road; refrigerant discharge; roof drain	RD
Roof drain; refrigerant discharge; road	RD
Roofing	RFG
Room	RM
Rough opening	RO
Rough sawn; rapid start	RS
Round	RND
Rubber tile floor	RTF
Saddle	SDL
Sandblast	SDBL
Sanitary	SAN
Sanitary sewer; service sink; standing seam (roof); steam supply; storm sewer	SS
Schedule	SCHED
Schematic	SCHEM
School	SCH
Scored joint; slip joint	SJ
Sealant	SLNT
Section	SECT
Separate	SEP
Sewer	SWR
Sheathing	SHTHG
Shelving	SHV
Shingles; sensible heat; single hung (window)	SH
Shop drawings; smoke detector; soap dispenser; storm drain; supply duct	SD
Shower; sensible heat ratio	SHR
Shutter	SHTR
Sidewalk; switch	SW
Siding	SDG
Similar	SIM
Single	SGL
Single hung (window); sensible heat; shingles	SH
Single acting (door); supply air	SA
Sink	—
Skylight	SKLT
Sliding glass door	SGD
Softwood	SFTWD
Solid core; shading coefficient	SC
Solid core; wood door	SCWD
South	S
Speaker	SPKR
Special	SPCL
Specification	SPEC
Splash block	SB
Spot elevation	SP EL
Sprinkler	SPKLR
Square	SQ
Square foot (feet); safety factor; supply fan	SF
Square yard	SQ YD
Stained glass	ST GL
Stainless	STNLS
Stainless steel	SST
Standard	STD
Standpipe; solid plastic; sump pit	SP
Station	STA
Steel joist	STL JST
Steel plate	STL PL
Storage	STOR
Storeroom	STRM
Street; single throw; stairs	ST
Structural	STRUCT
Structural clay tile	SCT
Subfloor	SUB FL
Substitute	SUB
Sump pump	SMP
Surfaced four sides	S4S
Survey	SURV
Suspended acoustical tile; saturate	SAT

Suspended acoustical tile ceiling	SATC
Tackboard	TK BD
Technical	TECH
Telephone	TEL
Television	TV
Temperature; temporary	TEMP
Tempered glass	TMPD GL
Terra cotta	TC
Terrazzo; telephone equipment room	TER
Thermostat	TSTAT
Thousand; kelvin	K
Thousand board feet	MBF
Threshold	THRES
Tongue and groove	T&G
Top of beam	TOB
Top of concrete; table of content; top of curb	TOC
Top of finish floor	TFF
Top of floor; top of footing; top of frame	TOF
Top of joist	TOJ
Top of masonry	TOM
Top of slab; top of steel	TOS
Topography	TOPO
Transfer grille	TG
Transom; transparent	TRANS
Tread	T
True north	TN
Tub/shower	T/S
Typical	TYP
Ultimate	ULT
Undercut door	UCD
Unexcavated	UNEX
Unfinish	UNFIN
Uniform	UNIF
Universal	UNIV
Unless noted	UN
Unless noted otherwise	UNO
Vacuum; vacuum line	VAC
Vanity	VAN
Vapor retarder; voltage regulator	VR
Variable air volume	VAV
Velocity	VEL
Veneer	VNR
Vent stack; voltmeter switch	VS
Ventilation; ventilator	VENT
Verify in field	VIF
Vertical	VERT
Vestibule	VEST
Vinyl base; vacuum breaker; valve box	VB
Vinyl composition tile; vitrified clay tile	VCT
Vinyl wall covering	VWC
Vinyl wall fabric	VWF
Vitrified clay tile; vinyl composition tile	VCT
Volt	V
Volume	VOL
Wainscot	WSCT
Wall cabinets	W CAB
Wall covering; water closet; water column	WC
Wall to wall	W/W
Water	WTR
Water closet; wall covering; water column	WC
Water heater; wall hung; wall hydrant; weep hole	WH
Waterproof membrane	WPM
Waterproofing, water pump; weatherproof; working point	WP
Weather resistant; water repellent; wire rope	WR
Weep hole; wall hung; wall hydrant; water heater	WH
Weight; water table; watertight	WT
Welded wire fabric	WWF
Welded wire mesh	WWM
West; waste; watt; wide	W
Wide flange; wash fountain	WF
Window	WDW
Window unit	WU
Wired glass	WGL
With	W/
Without	W/O
Wood; wood door	WD
Wood blocking	WBL
Wood door; wood	WD
Wood door and frame	WDF
Wood furring strips	WFS
Wood panelling	WDP
Wrought iron	WI
Yard; yard drain; yard drainage pipe	YD
Year	YR

2. ASSOCIATIONS AND ORGANIZATIONS

American Architectural Association	AAMA
American Concrete Institute	ACI
American Gas Association	AGA
American Institute of Architects	AIA
American Institute of Steel Construction	AISC
American National Standards Institute	ANSI

American Plywood Association	APA	Hollow Metal Manufacturers Association	HMMA
American Society for Testing and Materials	ASTM	Illumination Engineering Society of North America	IESNA
American Society of Heating, Refrigerating, and Air Conditioning Engineers	ASHRAE	International Building Code	IBC
		Marble Institute of America	MIA
Americans with Disabilities Act	ADA	National Association of Home Builders	NAHB
Architectural Woodworking Institute	AWI	National Bureau of Standards	NBS
		National Electrical Code	NEC
Associated Builders and Contractors	ABC	National Fenestration Rating Council	NFRC
Associated General Contractors	AGC	National Fire Protection Association	NFPA
Brick Institute of America	BIA		
Builders Hardware Manufacturer's Association	BHMA	National Institute of Building Sciences	NIBS
Building Officials and Code Administrators Association International	BOCA	Occupational Safety and Health Administration	OSHA
		Portland Cement Association	PCA
Ceramic Tile Institute of America	CTI	Precast/Prestressed Concrete Institute	PCI
Concrete Reinforcing Steel Institute	CRSI	Society of American Registered Architects	SARA
Construction Specifications Institute	CSI	Steel Joint Institute	SJI
Engineered Wood Association	EWA	Tile Council of America	TCA
Engineers Joint Contract Documents Committee	EJCDC	Underwriters Laboratories	UL
		Uniform Building Code	UBC
Environmental Protection Agency	EPA	Western Word Products Association	WWPA
Federal Housing Administration	FHA		
Gypsum Association	GA		

K. READING SCALES AND DIMENSIONS

1. THE ARCHITECT'S SCALE

Most of the views in a set of working drawings must be drawn at a reduced scale by the drafter. We can understand that to represent anything as large as a building on convenient sheets of paper, some consistent method of reduction must be used. Conventional architectural scales are needed for this purpose (Figure 2–48). Some views, such as moldings and interior trim, are often shown at actual size. Other, very small members may even have to be enlarged to make them legible. Most of the scales, however, require the size of the building to be reduced considerably on drawings.

To reduce a drawing from the actual size of a building, the drafter makes every line on the drawing a definite fractional size of the same line on the building. As a simplified example, if a board were 4″ wide on the tentative building and needed to be half-size on the drawing, a line representing the width of the board would be drawn 2″ long. This scale would conventionally be shown on the drawing as 6″ = 1′ 0″. The first figure (6″) represents the length of any line on the drawing that is equal to 1 foot on the actual building. Another example: If we used the conventional scale, one-fourth inch equals one foot ($\frac{1}{4}″ = 1′ 0″$), to draw the plan of a house measuring 28′ 0″ × 36′ 0″, the drawing would then have to be 7″ wide (28 ÷ 4 = 7) by 9″ long (36 ÷ 4 = 9).

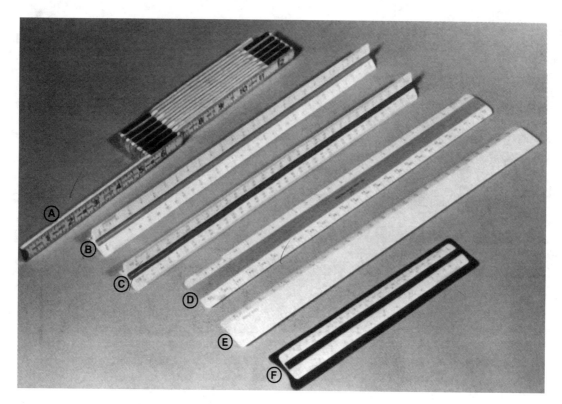

FIGURE 2–48

Measuring instruments. (A) Carpenter's rule. (B) Architect's triangular scale. (C) Civil engineer's triangular scale. (D) Flat four-bevel scale. (E) Flat double-bevel scale. (F) Flat 6" pocket scale.

However, to eliminate the time-consuming calculations, drafters merely use a rule, known as an *architect's scale* (see Figure 2–48B), with the conventional scales on it to do the measuring on the drawing. All that is required is to select the proper scale and then mark off the scaled length of each line. Anyone reading a drawing must, therefore, be able to use the architect's scale in taking off graphic dimensions that may not be given numerically on the drawing. Various scales may be evident throughout a set of drawings; yet each view will be labeled with its correct scale.

The architect's scales for the drafting room are made in several lengths (6", 12", 18") with either flat or triangular profiles. Students usually prefer the triangular type, 12" long, because it combines 11 different scales and can be easily handled. Table 2–3 lists the 11 different scales found on this scale.

Fully divided scales have each main unit throughout the entire scale fully subdivided, such as the full scale labeled 16. *Open divided* scales have the main units undivided, and a fully subdivided extra unit placed at the zero end of the scale. Two scales are combined on each face except the full-size scale, which is fully divided into sixteenths. The combined scales are compatible; one is twice as large as the other, and their zero points and extra subdivided units are on opposite ends of the scale. Architectural drawings, of course, use both feet and inches for units of measurement. Some dimensions are only in inches (16"), some only in feet (24' 0"); others combine both feet and inches (16' 8"). The scale is therefore calibrated into these units with conventional reductions to simplify measurement. The fraction or number near the zero end of each scale indicates the unit length in inches that can be used on the drawing to represent 1 foot on the actual structure. The extra unit near the

TABLE 2-3	**Full Scale ($\frac{1}{16}$″ graduations)**		
Architectural Scales	$\frac{1}{8}″ = 1′\,0″$	$\frac{1}{4}″ = 1′\,0″$	(one-forty-eighth size)
	$\frac{3}{8}″ = 1′\,0″$	$\frac{3}{4}″ = 1′\,0″$	
	$\frac{1}{2}″ = 1′\,0″$ (one-twenty-fourth size)	$1″ = 1′\,0″$	(one-twelfth size)
	$1\frac{1}{2}″ = 1′\,0″$ (one-eighth size)	$3″ = 1′\,0″$	(one-quarter size)
	$\frac{3}{32}″ = 1′\,0″$	$\frac{3}{16}″ = 1′\,0″$	

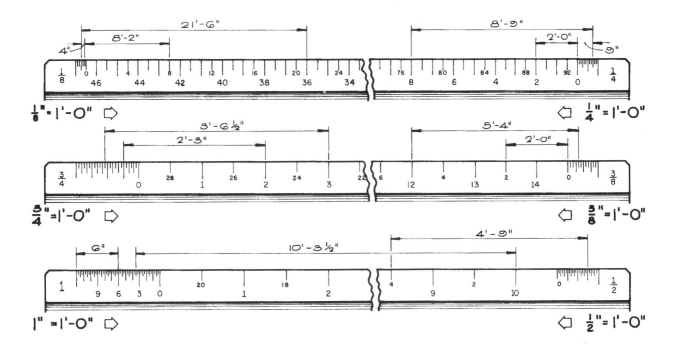

FIGURE 2-49

Measurements with various architectural scales.

zero end of the scale is subdivided into twelfths of a foot or inches, as well as fractional parts of an inch on the larger scales (see Figure 2–49).

In using the scale for taking dimensions from a drawing, select the correct scale and place it on the drawing so that both the total feet and any remaining inches can be read simultaneously (see Figure 2–50). Notice that the full-feet number of the dimension is placed at one extremity of the line and the inches (if the dimension is not in even feet) are read off the extra subdivided unit. If a regular inch-rule or carpenter's rule is used for taking dimensions off a drawing, calculate the feet and inches with arithmetic. For example, if you measure a line $4\frac{3}{4}″$ long, and the scale is $\frac{1}{2}″ = 1′\,0″$, you will notice that there are nine $\frac{1}{2}″$ increments in the measurement, plus half of another. This half-increment must represent $6″$ if the whole increment represents 1 foot. Therefore, the line must represent an actual distance of $9′\,6″$ ($4\frac{3}{4} \div \frac{1}{2} = 9\frac{1}{2}$). Other scaled lines on drawings can be measured and scaled accordingly (see Figure 2–51 for scaling with a carpenter's rule).

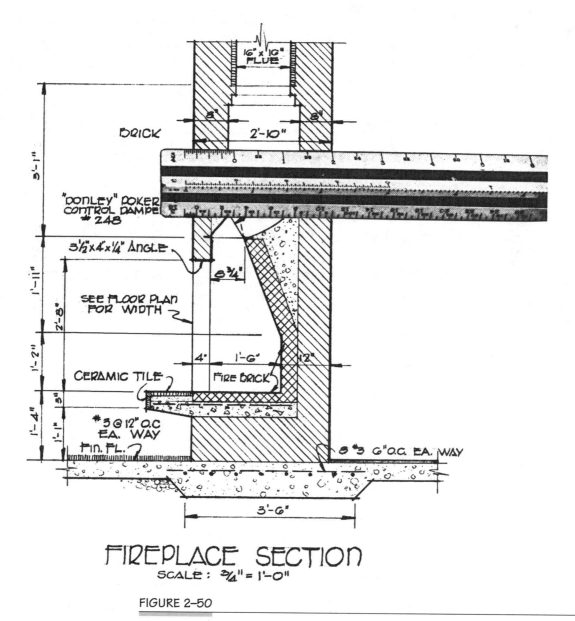

FIGURE 2–50

Measuring a drawing with the architect's scale.

2. THE CIVIL ENGINEER'S SCALE

Land measurement on Site and Plot Plans is slightly different from measurements on buildings. It is conventional to show land measurement in *feet and decimal parts of a foot,* carried out to three places (68.537′). No inches are used. Care must be taken in reading drawings having official land dimensions shown so that the decimal part of a foot is not misinterpreted to be inches. The civil engineer's scales are fully divided and are made in the flat and triangular types (Figure 2–48). Scales of 1″ = 10′, 1″ = 20′, 1″ = 30′, 1″ = 40′, 1″ = 50′, and 1″ = 60′ are found on the triangular type. Notice that these conventional civil engineer's scales differ from the architectural scales (Figures 2–52 and 2–53).

FIGURE 2–51

Measuring scaled drawings with the use of a carpenter's rule.

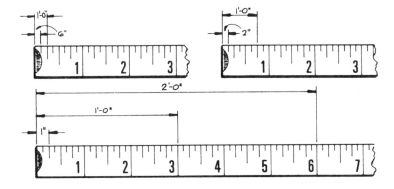

FIGURE 2–52

Using the civil engineer's scale.

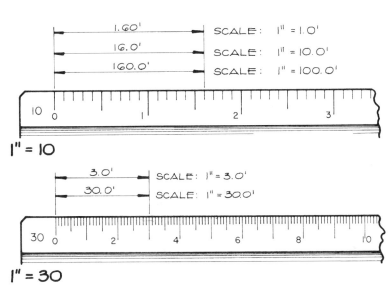

FIGURE 2–53

The 1″ = 30′ civil engineer's scale can be used to measure standard concrete block units on drawings of $\frac{1}{4}″ = 1′-0″$ scale.

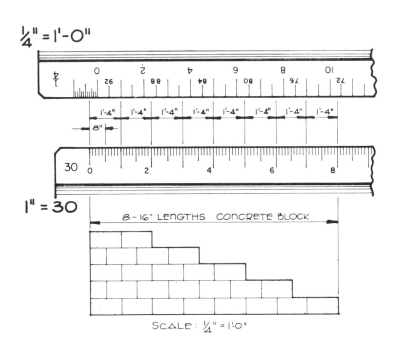

It must be remembered that prints of drawings (especially blueprints) often shrink slightly because of the water baths used; also, they can be affected by humidity changes. This means that direct measurement of dimensions on prints may be slightly off. Moreover, graphic sizes on drawings might occasionally be inaccurate; hence, numerical dimensions take precedence and should be used whenever possible. Sometimes it may be necessary to add several given dimensions or subtract one from another to arrive at the desired dimension, rather than scaling. The important dimensions are usually given in numerical form on working drawings, and the scaling of features should seldom become necessary. Frequently, most required dimensions can be found by close observation and by comparing related views.

The dimension conventions found on working drawings are listed next. Keep in mind that slight variations occur.

1. Both feet and inches are shown (23′–5″). Even if a dimension has no inches, the zero is still used as part of the designation (14′–0″).
2. Dimension lines are continuous, with the number placed slightly above the line and near its center (see Figure 2–50). On engineering drawings, the dimension line is broken and the numerical dimension is placed in the break.
3. Small dimensions, usually those smaller than 1 foot, are shown in inches only.
4. Fractions are shown with a diagonal slash. (Occasionally, fractions are made with small numbers and the slash is omitted.)
5. Arrowheads of various types (see Figure 2–18) are placed at the extremities of the dimension lines to show the limits of the dimension.
6. Dimensions are placed to read from the bottom of the sheet and the right only.
7. Overall dimensions are placed outside the smaller dimensions.
8. Curved or angular leaders are generally used to eliminate confusion with other dimension lines.
9. Location dimensions are given to centerlines of doors and windows on plan views.
10. Dimensions always refer to actual building sizes regardless of the scale used.
11. In light wood-frame construction, dimensions can be given from the outside of the stud walls to the center lines of inside partitions (see Figure 2–54), or they can be continuous dimensions showing wall thicknesses and sizes within rooms from wall surfaces to wall surfaces (see Figure 5–9).
12. Masonry walls are dimensioned from outside surfaces, with the thickness of the masonry given.
13. If grid lines are used on a drawing, such as a modular drawing, only one grid unit is usually dimensioned.
14. Door sizes may be shown on the floor plan symbols, or they may be shown in a Door Schedule.
15. Some features having obvious placement, such as a door in a narrow hall, usually do not have location dimensions.
16. Similar dimensions are not duplicated on the various views.
17. Dimensions for exterior features on the floor plans are generally shown outside the plan, and interior wall dimensions are shown with a continual series in each direction, within the plan.
18. Vertical dimensions on commercial buildings, because they are often established with a transit or level and rod, are sometimes indicated in feet and decimal parts of a foot.
19. It cannot be overemphasized that information is often shown in several places and that all pertinent drawings and views should be consulted before dimensions are taken to be definite. Many costly errors have resulted from failure to check one drawing against the other.

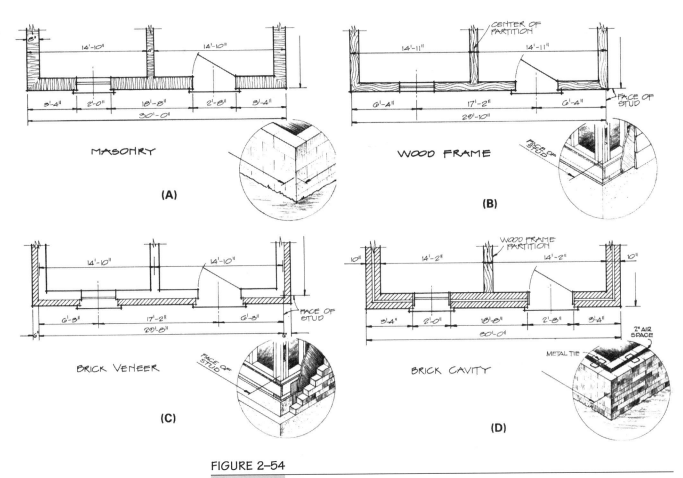

FIGURE 2–54

Conventional dimensioning practice on floor plans. (A) Masonry. (B) Wood frame. (C) Brick veneer. (D) Brick cavity.

FIGURE 2–55

Standard 4" module used in modular dimensioning.

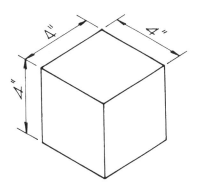

READING MODULAR DIMENSIONS

Some architects use a system called *Modular Coordination* in designing buildings and drawing their working drawings. The system is based on the concept that construction economy will result if the drawings are drawn on gridwork (grid sizes in increments of 4") so that a maximum of modular building materials can be used with minimum amounts of cutting and modification by workers (Figures 2–55 to 2–57). Savings are especially realized in masonry construction where sizes and openings of walls are made compatible with the

FIGURE 2-56

Floor plan layout on 4″ grids. NAHB South Bend research house. Designed by NAHB Research Staff.

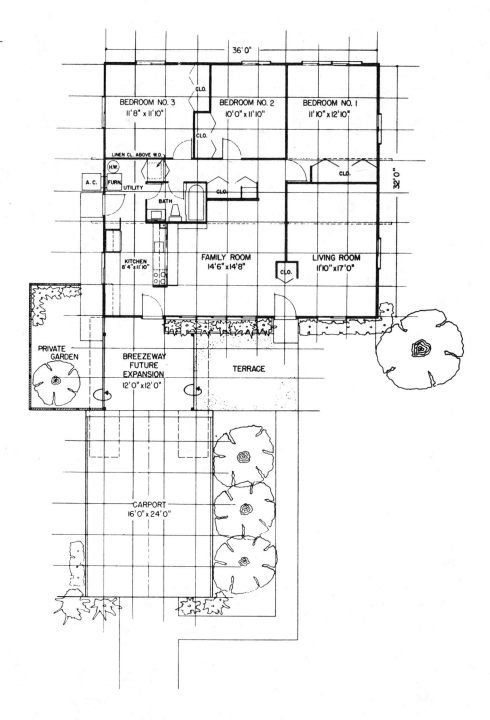

masonry units. The universally accepted standard module is a 4″ cube. Ideally, components and dimensions are multiples of the basic 4″. However, off-grid and nonmodular dimensions are sometimes needed on the modular drawings; Figure 2–61 shows how they are identified. Since fewer are needed, dimensions on modular drawings appear slightly different from those on conventional drawings.

The *actual sizes* of many building materials are slightly smaller than their *nominal sizes* (Figure 2–58). In masonry work, this allows the modular grids to fall in the mortar joints, as shown in Figure 2–57. Nominal sizes of the units fall directly on the grid lines, and few dimensions are needed when we have in mind the standard size for each grid. Notice that the 8″ × 8″ × 16″ size of common concrete block or the 4″ dimensions of 2 × 4

FIGURE 2-57

Modular masonry units.

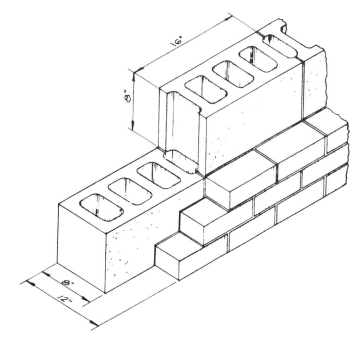

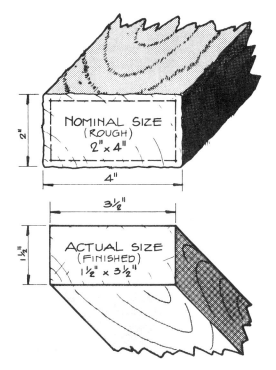

Note: As of September 1970, the various national lumber associations have adopted the American Softwood Lumber Standard, PS 20-70, which now unifies dressed lumber sizes as shown:

NOMINAL SIZE	DRESSED SIZE (S4S) IN INCHES	
	SURFACED DRY	SURFACED GREEN
2 × 4	1½ × 3½	1 9/16 × 3 9/16
2 × 6	1½ × 5½	1 9/16 × 5 5/8
2 × 8	1½ × 7¼	1 9/16 × 7½
2 × 10	1½ × 9¼	1 9/16 × 9½
2 × 12	1½ × 11¼	1 9/16 × 11½

FIGURE 2-58

Nominal and actual size of 2 × 4 lumber.

framing lumber are placed on the grids. Other accepted standards regarding floor and ceiling surfaces as they relate to the grid lines are used (see Figures 2-59 and 2-60).

The system lends itself to prefabricated home construction, where components and wall panels are prebuilt and then merely assembled at the job site. Variations of the designs can be obtained with minor changes in the modular components.

FIGURE 2–59

Modular wood-frame details showing grid lines.

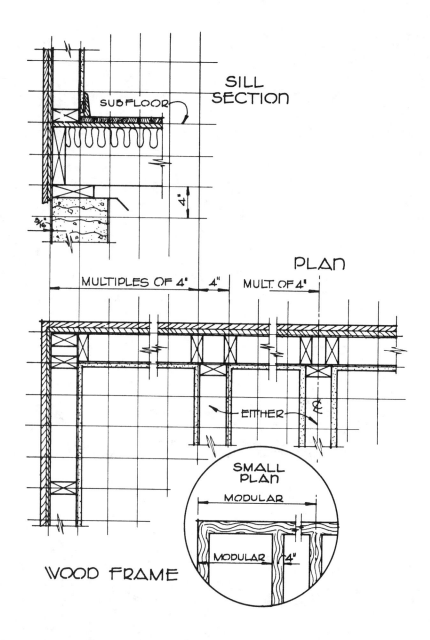

In reading modular drawings, several points must be realized. First, sets of modular-measure drawings are labeled with a note or graphic illustration indicating that they are modular. Also, the dimension lines may have either arrowheads or dots at their ends. The arrowhead indicates that the dimension falls on the grid line; the dot indicates that it falls off the grid line (see Figure 2–61). Dimension lines with arrowheads at each end would then be in 4″ multiples. A dimension with a dot at one end and an arrowhead at the other would be nonmodular. Most modular drawings will have both modular and nonmodular dimensions because certain features on buildings cannot be made entirely modular. But the use of grid lines on modular drawings eliminates the need for many of the dimensions found on conventional drawings. Also, fewer fractional dimensions will be found. However, the presence of grids will allow you to determine many of the sizes quickly by merely counting the grid spaces.

FIGURE 2-60

Modular masonry details showing grid lines.

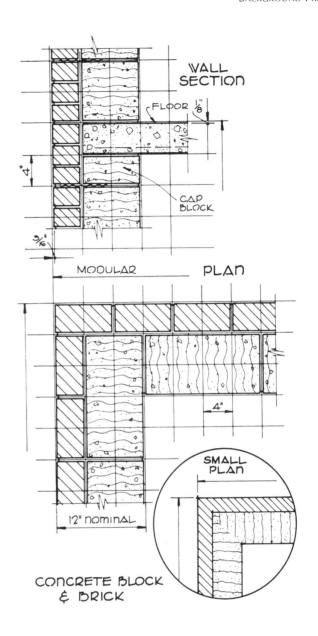

M. METRIC DIMENSIONS

It seems evident that the construction industry will eventually adopt the metric system of measurement. There are a number of advantages to be gained by the conversion: the elimination of fractions on drawings, simpler calculations, international uniformity, and others. But, because the industry has so many interrelated components and various problems to resolve, a complete changeover is not probable in the immediate future (see Figure 2–62).

The American National Metric Council, in its publication *American Metric Construction Handbook,* has made the following recommendations in reference to metric drawings:

1. Architectural working drawings be dimensioned in millimeters (mm) and meters (m).
2. Plot plans and site plans be dimensioned in meters (m) or possibly kilometers (km), depending upon the scale, with accuracy to only three decimal places.

FIGURE 2–61

Modular appliqué that can be attached to a modular working drawing.

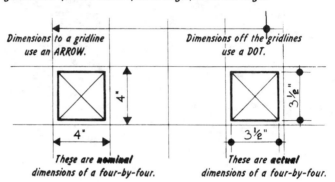

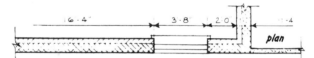

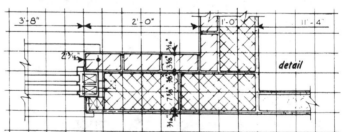

3. No periods after the unit symbols.
4. Scale on drawings be shown by a ratio (1:1, 1:10, 1:50, etc.).

1. CONVERTING ENGLISH DIMENSIONS TO METRIC

Conversion from English to metric dimensions can easily be made with the pocket calculator. First, convert mixed units into single units in decimal form, then multiply by the appropriate factor shown in Table 2–4.

$$2'-4\tfrac{1}{2}'' = 2 \times 12 + 4.5 \times 25.4 = \mathbf{723.9\ mm}$$

Land measurement or site plans scaled with the civil engineer's scale are usually dimensioned in feet, and decimal parts of a foot and can be converted by simply multiplying the appropriate factor, as shown:

$$95.62' = 95.62 \times 0.3048 = \mathbf{29.145\ m}$$

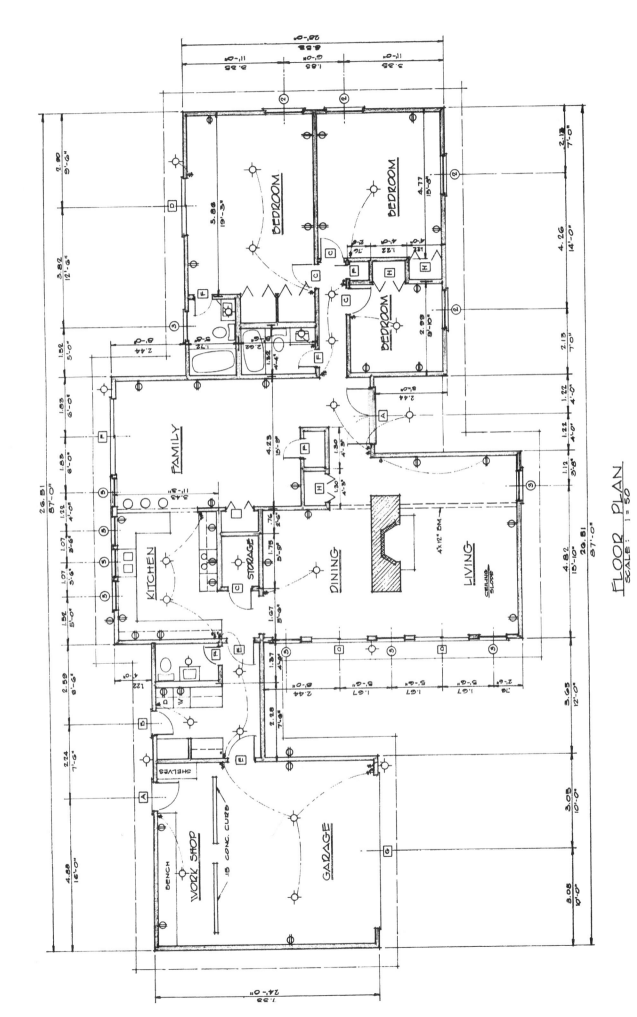

FIGURE 2-62
Floor plan showing the use of both English and metric dimensions.

85

TABLE 2–4

English–Metric Conversions

1 inch = 25.4 millimeters	(0.025 4 meters)
1 foot = 304.8 millimeters	(0.304 8 meters)
1 yard = 914.4 millimeters	(0.914 4 meters)
1 mile = 1 609.34 meters	(1.609 34 kilometers)
1 millimeter = 0.039 370 inch	
1 meter = 3.280 84 feet	
1 meter = 1.093 61 yards	
1 kilometer = 0.621 371 miles	

Notice that a space is used rather than a comma between every three digits on both sides of the decimal.

2. ROUNDING OFF METRIC DIMENSIONS

In construction work, precision cannot possibly be obtained to the same degree as in, say, machine-tool work. The most logical drawings are those devoid of complex fractions and numbers that are difficult for workers to measure. Long dimensions in metric will mainly be found in multiples of 50 or 100 mm, whereas thicknesses and small features may be dimensioned with nondivisible numbers—yet hardly would there be any need for fractional parts of a millimeter. Odd metric numbers will be found with the use of the dual system of dimensioning (both English and metric), simply because conventional scales were used to make the drawing. But when convenient scales are eventually employed in metric drawings, simpler number dimensions will surely be used.

3. SCALES ON METRIC DRAWINGS

Conventional scales are not convenient for metric drawings. For example, $\frac{1}{4}'' = 1'\,0''$ if converted to metric would be 6.35 mm = 304.8 mm. A simple ratio, such as 1:50, is more practical, which incidentally is almost the same as $\frac{1}{4}'' = 1'\,0''$ or 1:48. On dual-dimensioned drawings, if English scales are used, they can be easily converted to simple ratios:

$$\frac{1}{8} = 1'\,0'' \quad 1{:}96 \quad 1'' = 1'\,0'' \quad 1{:}12$$

$$\frac{1}{4} = 1'\,0'' \quad 1{:}48 \quad 3'' = 1'\,0'' \quad 1{:}4$$

$$\frac{1}{2} = 1'\,0'' \quad 1{:}24 \quad \text{Full size} = \quad 1{:}1$$

Eventually, as metric scales are developed and used, more convenient ratio scales, such as 1:200, 1:100, 1:50, and 1:10, will be found on metric-dimensioned drawings.

Reading metric drawings should present no problems to the novice other than understanding the units of measurement being used. Thinking metric will come only after repeated use of the new system.

N. PENCIL SKETCHING

Some basic points on pencil sketching are included here since simple sketches are often needed in reading working drawings. One frequently has to resort to a sketch to clarify a point found on drawings or to make a graphic diagram when discussing features of a detail (Figure 2–74). Words are not entirely adequate in communicating technical information between workers. The use of a simple sketch is usually the difference between understanding

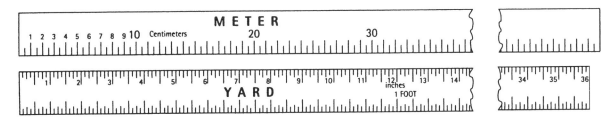

FIGURE 2-63

Comparing a yardstick to a meterstick.

FIGURE 2-64

Hold the pencil in a comfortable position when sketching.

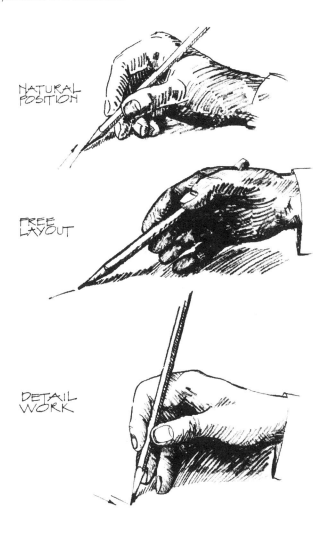

a point clearly and misunderstanding it. If you take the time to learn a few points about sketching, it will become a useful adjunct in reading and understanding drawings. Only a pencil, eraser, and paper are necessary.

Students will become proficient at sketching by being deliberate in early practice. Artistic sketches are not necessary; a revealing sketch that shows information quickly and simply is the objective. Mainly, you should learn to sketch forms and shapes that are commonly encountered on architectural drawings and think of the lines as a means of communicating your ideas on paper. Let the pencil be a useful tool after learning how to control it, using the eye as well as the hand (see Figure 2–64).

Begin with a conical point on the pencil and have an eraser nearby to reinforce your confidence. Avoid a sharp point, yet have a sandpaper pad or some other means of dressing

FIGURE 2–65

Pencil points.

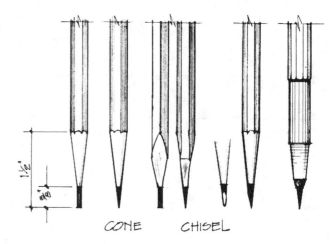

FIGURE 2–66

Beginning practice strokes.

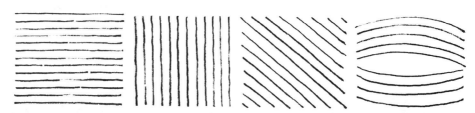

the point at hand when it becomes too blunt. Later, various width lines to give the sketch more expression can be obtained by sharpening the point in a "chisel" fashion, as shown in Figure 2–65. Ordinary typing paper is fine for beginning practice; notice that small sheets are easier to manipulate. Coordinate paper is useful later, when working with larger sketches.

1. PRACTICE STROKES

First practice the horizontal strokes shown in Figure 2–66. The strokes should be only as long as is convenient for the movement of the fingers; the stroke should be a firm, clean-cut line that begins and ends in a uniform manner—usually made with a slight pressure to give it a definite ending. Longer lines will require moving the hand after each short stroke along the path of the required line. Avoid long, uncontrolled strokes that are done without the hand resting firmly on the paper. To ensure the correct layout of beginning linework, use a few points made with a pencil (Figure 2–67) before sketching the finished line. Notice that layout points or targets will ensure a positive beginning and ending. Keep your eye on the target point as you progress along the continuous line.

Vertical lines are slightly more difficult to sketch. If necessary, the paper can be turned so the line will be done with a natural movement of the wrist and elbow, as described for horizontal lines. But practice the vertical strokes so you will become proficient with them

FIGURE 2-67

Sketch long lines with a series of short strokes and a target.

FIGURE 2-68

Sketching angles.

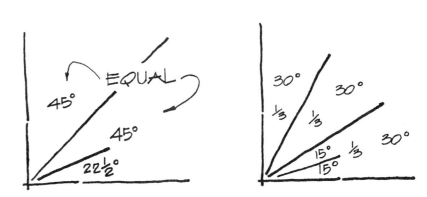

as well. Begin a vertical line at the top and pull the pencil toward the body rather than pushing the pencil away. This allows you to maintain more control. As with horizontal lines, use convenient short, crisp strokes as you progress toward the ending point. Slight breaks between are not objectionable since the strokes will be reinforced with a heavier line after they are found to be correct.

2. ANGULAR LINES

Practice strokes at various angles so the eye and hand develop proper coordination. Usually, sloping lines must be drawn at given angles, and the estimation of commonly used angles is therefore necessary (see Figure 2–68). Estimating angle size is simply done by working from a right angle (90 degrees). Lay out the right angle with light lines; then subdivide as shown in Figure 2–68. Be sure the perpendicular relationship of the right angle is correct before progressing. Forty-five degrees is obtained by dividing the right angle in half visually; 30 degrees is obtained by dividing it into thirds, and so on. Obtuse angles are obtained by combining the right angle with a smaller angle. Practice this technique until nearly perfect angle sizes are obtained. Move the paper around if necessary so that hand and arm movements for angular strokes become comfortable.

3. CIRCLES AND ARCS

Curved lines and circles are more difficult than straight lines for the beginner and therefore require more careful layout work. A true circle relates nicely with a square, so if you sketch the construction square properly, you will be surprised by how good the finished circle will be.

To sketch a *circle,* start with perpendicular center lines (Figure 2–69) to locate the circle in the proper place within the sketch and to help maintain symmetry. Then lightly sketch the square with sides equal to the required diameter of the circle. Make the square as accurate as possible so that the four inner squares that form are identical in size and shape (Figure 2–69). The circle will fit neatly in the larger square. Further help is obtained by lightly constructing triangles that connect the intersections of the center lines and the sides of the square. Within these triangles, place a point at the visual centers. To complete the circle, sketch uniformly curved strokes through these points and the intersections of the center lines and the sides of the larger square. After a few trials an almost perfect circle will be obtained.

Similar construction is needed for the arcs or partial circles that are frequently needed on sketches. The center apex of an arc is located first; then a partial square, similar to the

FIGURE 2-69

To sketch a circle, begin with a square.

FIGURE 2-70

Sketching arcs.

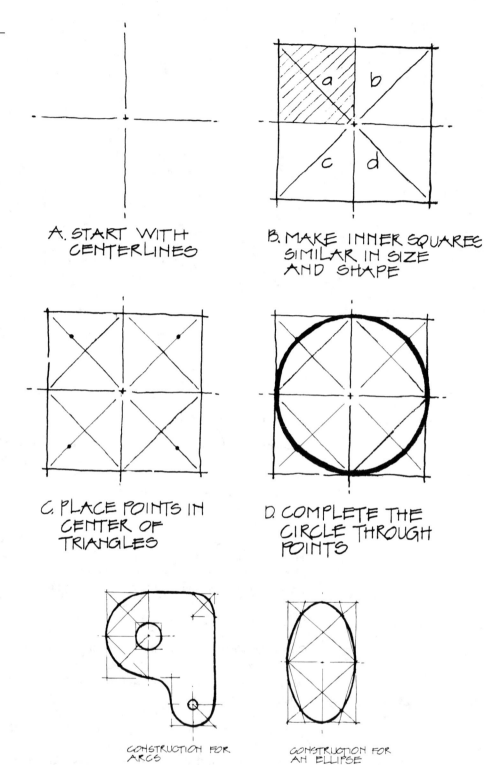

construction for a circle, is sketched in so that the arc will be oriented and have the correct radius, as shown in Figure 2-70.

Larger circles and arcs may require the use of points laid out with the help of a *tick strip*—a piece of scrap paper marked off with the desired radius of the circle or arc (see Figure 2-71). After the center point is located, transfer a number of points from a tick strip, radiating from the center, to locate the uniform curvature conveniently.

FIGURE 2–71

Use of a tick strip for large arcs or circles.

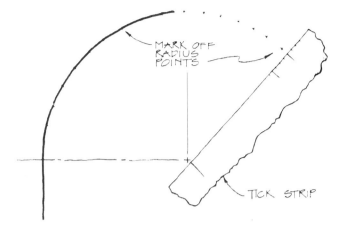

FIGURE 2–72

Locating a center with the use of diagonals.

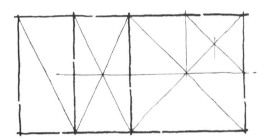

Practice sketching other geometric forms until you become acquainted with their features. Remember to simplify each object by observing the general shapes inherent within it. The secret of successful sketching is being able to see these simple forms and then *blocking in* the shapes accordingly before attempting the final sketch. Notice how useful center lines become in preliminary layout, especially if the sketch is intended to be symmetrical. It helps to maintain balance as well as the similar relationship of one side to the other. Often diagonals are useful in blocking in; they locate center points or other features and help maintain control of the construction (see Figure 2–72).

4. LEARNING GOOD PROPORTION

Developing good proportion in sketches is definitely important, and one must be conscious of this characteristic from the very outset. Proportion is the relationship of height to width to depth. One maintains proportion on a scaled drawing by measuring with the rule, but in sketching it is achieved by training the eye to see correctly and sketching the lines accordingly. After the length of one feature on the sketch is established, the others must be related to maintain conformity.

One method of training the eye to see correct proportion is to look for *squares* within the subject (see Figure 2–73). Overlook details at first and concentrate on general shapes and sizes. If you can simplify the object into squares, or squares relating to the major features, you will have a convenient system for developing proportion in the construction. In pictorial sketches, use the *cube* for maintaining this relationship. Occasionally, only parts of the square or cube need be used when relating the layout to the figure. But continually keep the square in mind, even when refining the finished linework.

Sometimes a *unit* method can be used in developing proportion. After a unit is established on one feature, the relative number of similar units can be laid out for the others. Be sure that centers are located carefully and that one feature is correctly related to the others. Practice sketching different simple objects until a sense of proportion is automatically developed; then progress to more complex objects.

FIGURE 2-73

The use of a square in maintaining proportion.

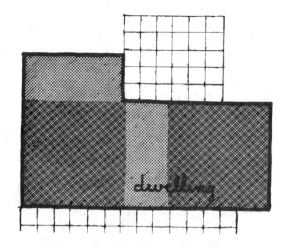

FIGURE 2-74

Sketching over grid paper.

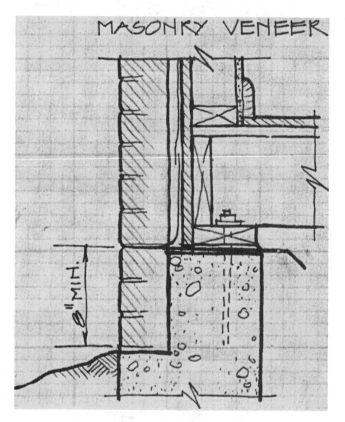

5. SKETCHING OVER COORDINATE PAPER

After acquiring some skill in sketching basic lines and simple shapes freehand, further practice should be devoted to sketching over coordinate paper so more accurate and larger sketches can be done (Figure 2–74). The best procedure is sketching on tracing paper or vellum paper that is taped or fastened to a grid sheet below. Or sketch directly on the grid paper, as is suggested in the beginning sketching exercises following this chapter. With tracing paper, however, the finished sketch will appear more satisfactory with the gridwork removed.

Various types of grid paper are available, but if one is working with fractions, the one-eighth-inch grids having heavier one-inch lines are most convenient. This provides eight squares per inch and lends itself to many architectural scales, and even allows you to sketch a scaled drawing if need be. Grid paper with 10 divisions per inch is convenient if sketching with metric sizes or with the decimal system. Even one-quarter-inch squares are satisfactory for many sketches. Proportion is automatically maintained by merely counting the blocks or grids when developing the sketch, and the horizontal and vertical nature of the sketch is always sustained. Grid paper with one-inch squares subdivided with 12 grids is appropriate for scaled sketches in which feet and inches are involved. An accurate, scaled sketch will result.

6. SKETCHING PICTORIAL DRAWINGS

It is often necessary to see the pictorial nature of an object when reading details of a working drawing. Several types of pictorials are commonly found on drawings: isometric, oblique, and cabinet-oblique. Perspective drawings, although the most realistic, are used mainly for display and presentation renderings and are more difficult for the drafter to execute. Also, a perspective does not lend itself to depicting technical information and, therefore, is not covered in this material. Examples of technical pictorials are found throughout this manual, and the student should be familiar with the nature of each and should be able to sketch any of them when the need arises.

An *isometric* drawing provides the same composite information as three orthographic views. Moreover, the pictorial nature of the drawing is usually easier for the layperson to understand. Section views cut from isometrics are especially revealing. Learn to quickly sketch the various technical pictorials and you will find them useful in understanding and discussing the many involved details found on working drawings.

Begin with *isometric views* (Figure 2–75); suitable beginning exercises are provided at the end of this chapter. Many other figures found throughout this manual are isometric and provide excellent sketching exercises. Use the grid paper provided or ordinary white paper. More complex sketches should be done on isometric grid paper, if available. Further sketches can be done on tracing paper fastened over the grid paper, as mentioned earlier.

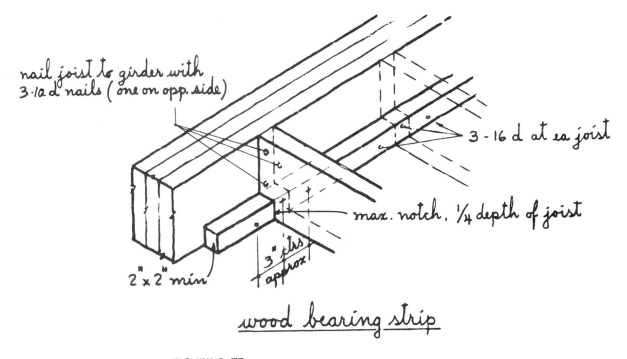

FIGURE 2–75

An isometric sketch (FHA).

Notice that the three isometric planes (Figure 2–16) are vertical, 30 degrees to the right and 30 degrees to the left from horizontal.

To sketch an isometric, block in the overall block of the subject first to form an isometric cage, so to speak, within which further details of the object are added or cut away. On grid paper, count the blocks to correspond with the sizes taken from an orthographic drawing, as shown in the exercises. The isometric cage should be similar to the sizes on the orthographic. However, notice that when a line is not on one of the isometric planes, it will not be identical in length to its orthographic counterpart. Sketch the layout lines lightly and, as the figure develops, gradually increase the darkness of the finished linework (Figure 2–76).

Angular lines and shapes must be formed by locating their points of intersection with the isometric planes (see Figure 2–76), since angular surfaces will not fall on the isometric planes. Locate these intersections on the profile view so they can be transferred to the isometric. This technique will become evident after completing several of the exercises.

7. SKETCHING ISOMETRIC CIRCLES AND ARCS

Circles and arcs lying on isometric planes become ellipses in comparison to their orthographic shapes. Therefore, careful construction must be used to arrive at their correct shape and in locating them on their correct plane. For the beginner it is often confusing to place the circles and arcs on the required plane.

To sketch an *isometric circle,* start with the center lines that locate the circle properly and ensure its correct plane placement. From the center point, lay out an isometric square with light lines similar to the layout of an orthographic circle. Make the square centered on the center lines and its sides equal in length to the diameter of the intended circle (Figure 2–77). Be sure the construction appears to lie on the correct isometric plane before continuing. Then lightly construct triangles that are formed by connecting the intersections of the center lines and the sides of the isometric square. Next place a point on the visual center of these triangles to provide a guide in sketching the extremity of the ellipse that will form the final isometric circle. A neatly curved stroke passing through each point and the center-line intersections at the sides of the square will result in the finished circle. Several trials may be needed to arrive at a satisfactory shape. Erase the construction lines so the circle will dominate and blend in with the remainder of the sketch. Practice various-sized circles on each of the isometric planes to become familiar with their appearance.

In sketching *partial circles,* it is especially important to have the construction carefully in place, since arcs have different curvatures depending on the plane they are to be on. To sketch an arc that is a part of a true circle, first locate the center from which it will radiate. Then complete similar construction as used for a full circle so the arc will have the correct radius, and you can use as much of the construction as needed to attach it to the remainder of the sketch (see Figure 2–78). It is advisable to lightly complete the entire circle construction first, even if only part of it is used. This will ensure proper orientation of the arc. Then sketch in the part of the circle that is needed, darken, and remove excess construction. If a border is needed on the sketch, use the technique shown in Figure 2–79.

Practice sketching cylinders and cones by first sketching in similar construction and center lines of the *entire* figure, as shown in Figure 2–78, even if some of the object may be hidden in the final view. Darken in only the visible features, however, since hidden lines are seldom needed in pictorial views.

Oblique and *cabinet-oblique* drawings are sketched by completing similar blocks, as for isometric sketches, after the correct axes are established (Figure 2–80). Notice that grid paper with only horizontal and vertical lines is convenient if a forty-five-degree axis is used. Merely use the diagonals through the squares. Select a receding plane that will reveal the most important information about the subject. First, block in the surface that will be on the frontal, orthographic plane, which establishes proper size and proportion. Then lay out the receding, parallel lines to determine the depth and the cage for the object. Depth sizes lying on the receding planes are the same as the sizes on the frontal plane. Take the sizes from given dimensions or orthographic views that show depth dimensions.

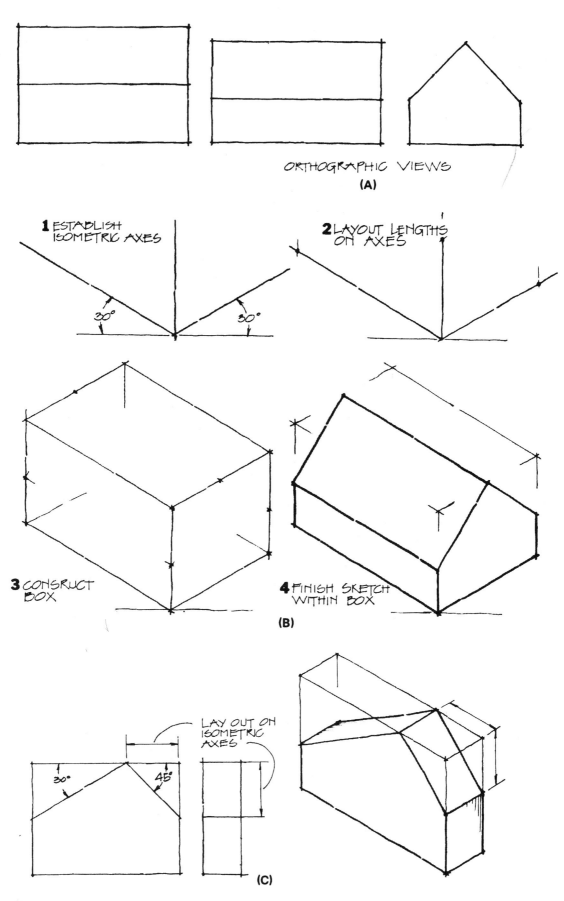

FIGURE 2-76

Blocking in to locate angular shapes.

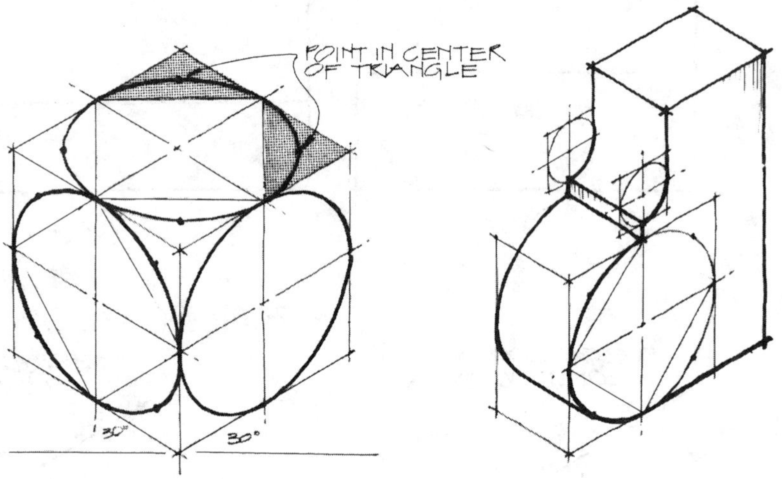

FIGURE 2-77

Sketching isometric circles and arcs.

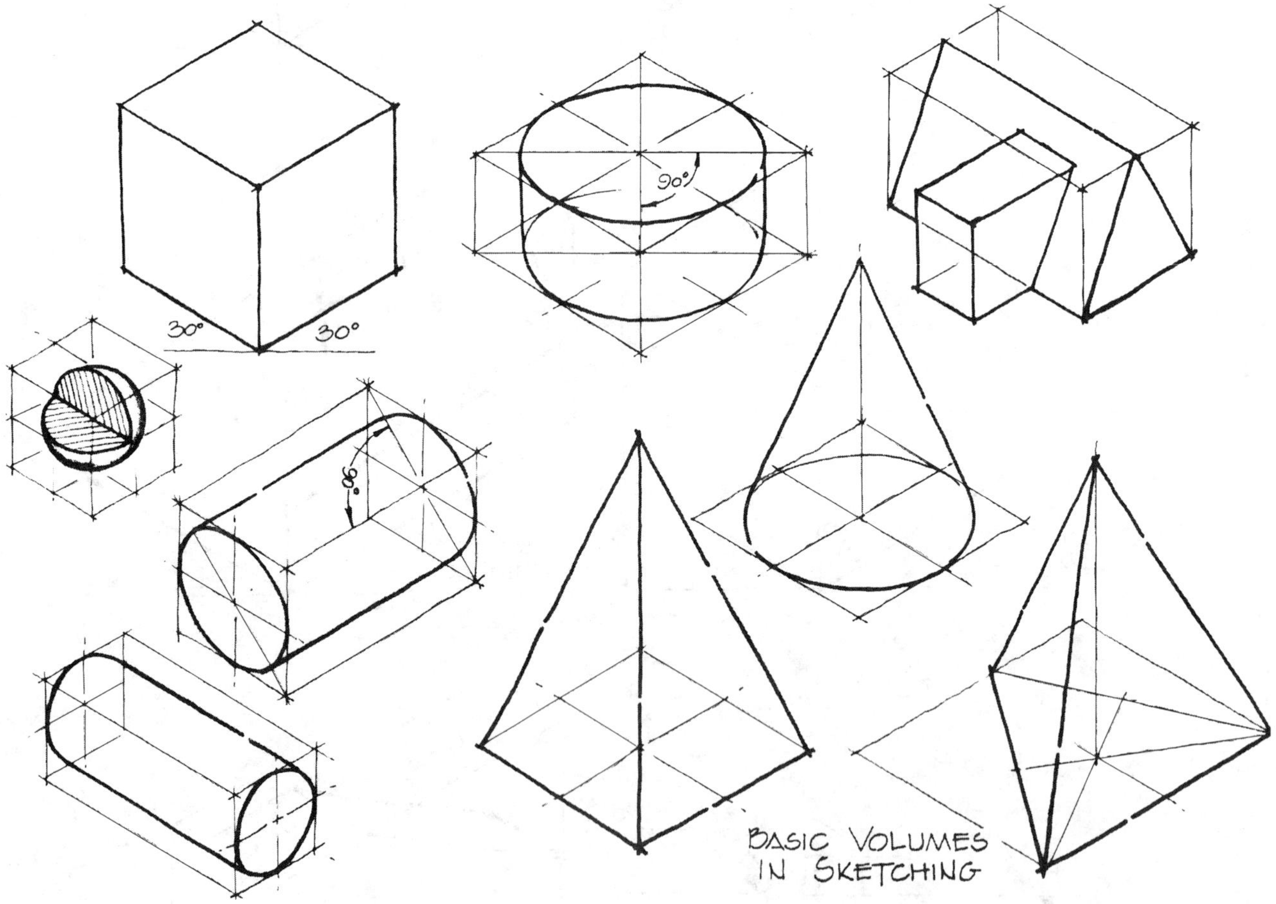

FIGURE 2-78

Sketching isometric volumes.

Study the construction block for size and proportion before continuing. When sketching oblique or cabinet-oblique drawings, turn the object so the important and irregular features fall on the frontal plane, rather than on the receding, oblique planes; this will simplify the sketch and make it more useful, inasmuch as important information placed on receding planes will appear distorted to the viewer (see Figure 2-81).

FIGURE 2-79

Use edge of board for borders.

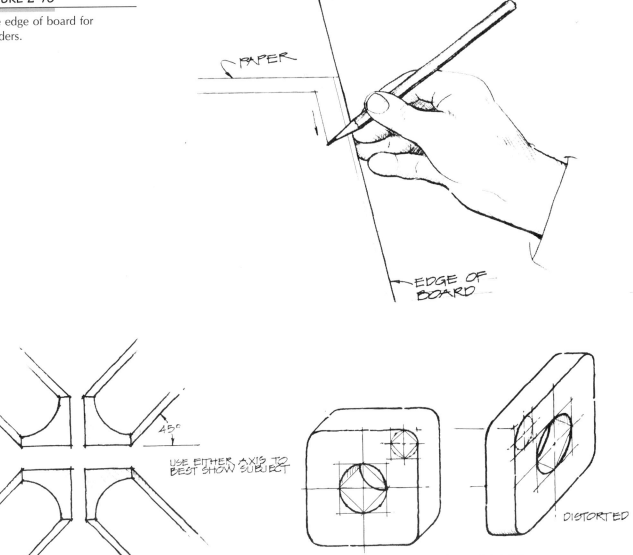

FIGURE 2-80

Sketching oblique views.

    Avoid placing long sides of an object on the oblique planes as well. This minimizes the distortion of foreshortening—the natural appearance of an object at a distance from the viewer, which is drawn smaller. The cabinet-oblique-type drawing is often used by drafters to eliminate this distortion somewhat. Notice that the receding depths on the cabinet-oblique drawing are reduced by one-half, or similar amounts, in comparison to oblique, to result in a more visually pleasing pictorial appearance. Usually, shallow-type figures lend themselves to either oblique- or cabinet-oblique-type drawings.

    Circles and arcs that fall on the receding planes of oblique sketches must be blocked in with construction similar to that used for isometric circles and arcs (see Figure 2–77). Keep the construction on the correct planes and start with center-lines. If the construction appears right, the finished circle or arc will end up right. Circles and arcs that are on the frontal plane will, of course, be true circles, and the construction will be the same as for orthographic circles and arcs.

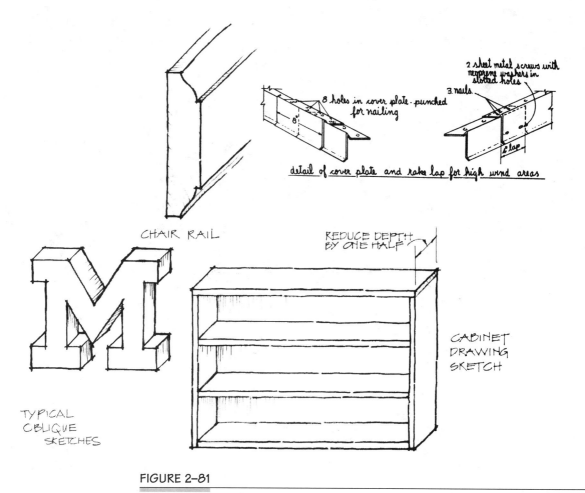

FIGURE 2–81

Typical oblique sketches.

Practice sketching various objects in oblique; try some exercises that convert orthographic or isometric drawings to the oblique type, which will reveal the characteristics and limitations of the oblique and cabinet-oblique drawings.

8. DIMENSIONING PICTORIAL DRAWINGS

Usually, the addition of dimensions on a pictorial sketch provides the required technical information. Rules for dimensioning pictorial sketches are similar to the rules for orthographic dimensioning, except, of course, the dimension numbers and letters are made to appear as though they lie on the same plane as the feature they are dimensioning (see Figure 2–82). Always keep both the extension and dimension lines on these planes. Even the lettering is sloped to make it appear related to the various planes. Avoid placing dimensions directly on the object, which tends to clutter the view. Rather, use extension lines to bring the dimension out to a convenient place so that the pictorial outline is not distracted. Place the shortest dimensions closest to the figure and the overall dimensions farthest away. Use dimensions to locate center points of both circles and arcs. Leaders that dimension circles and arcs are drawn at angles other than the angles that represent the major planes and that point toward or away from their center points. Various sketches shown throughout this manual show how dimensions are related to various features of pictorial drawings.

FIGURE 2-82

Dimensioning pictorial sketches.

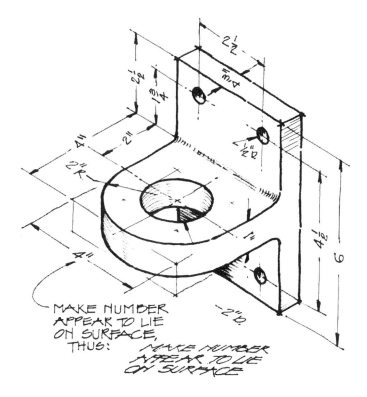

FIGURE 2-83

Technical lettering on sketches.

All notes and dimension numbers must be carefully lettered on a sketch to avoid mistakes. Use the conventional technical alphabet (Figure 2–83). Usually one-eighth-inch guidelines are satisfactory; if you are sketching over grid paper, use the grids on the relating planes as guides to simplify the addition of this important part of the sketch. Feet-and-inch dimensions, as found on architectural drawings, are commonly used.

9. LINE WEIGHTS ON SKETCHES

Keep the pencil quite sharp when developing the beginning construction in sketching; as the sizes and proportion begin to evolve, use heavier and more distinct final strokes. Erase any construction that no longer seems necessary as the sketch develops so that the superfluous linework will not become confusing. When the correct shapes are arrived at, darken the corrected final strokes so that the figure dominates the surrounding construction. Light construction linework may even be left on the sketch to give it a "sketchy" quality, if it does not interfere with the readability of the drawing. Finally, go over the periphery of the sketch with a more blunt point to strengthen its outline and eliminate minor defects. Remember that each darker stroke is meant to perfect the previous strokes, so the sketch continues to improve.

Dress the pencil point again when sketching in the final center lines, extension lines, leaders, and dimensions. Use the worn point to sketch important lines, outlines, and the like.

O. EXERCISES IN BASIC PRINCIPLES

The following pages offer worksheets that are to be completed after the study of basic principles. Some may be used as exercises; others can be taken as examinations to reveal the extent of progress. Often the important information is reinforced when worksheets are completed. Read the directions at the top of each sheet carefully before attempting the work.

1. BEGINNING SKETCHING EXERCISE

With a soft pencil, fill in the empty boxes with strokes shown on the left (after turning the page sideways).

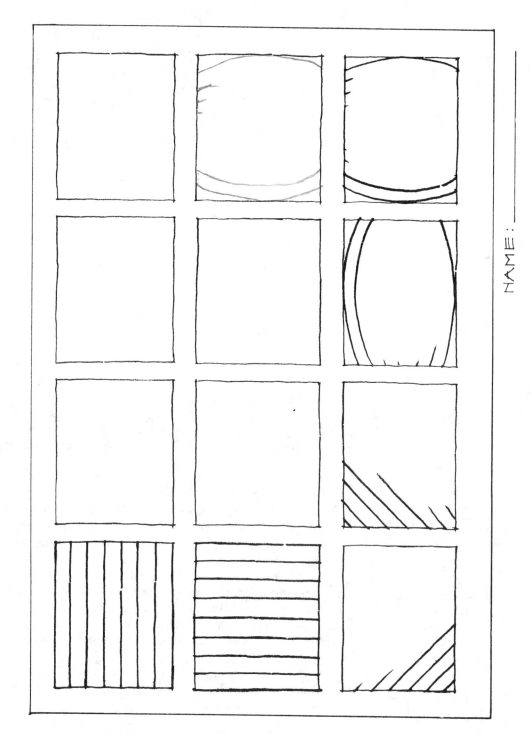

2. SKETCHING ORTHOGRAPHIC VIEWS EXERCISES

To complete the test items on the next pages (Test 1), using a soft pencil, sketch in the required orthographic views of the grid areas, as shown by the following examples. For convenience, the isometric coordinates on the pictorial views are the same size as the orthographic coordinates. Lengths of the features are obtained by merely counting the blocks. Use dashed lines to represent hidden features and allow one block space between relating views, as shown in the examples.

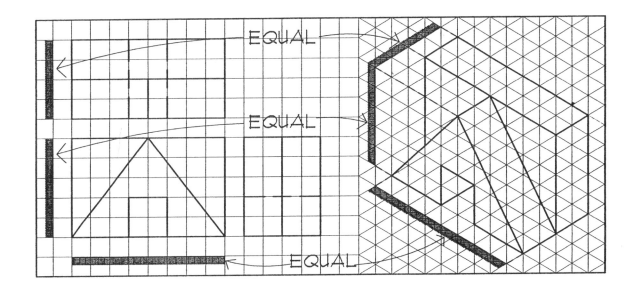

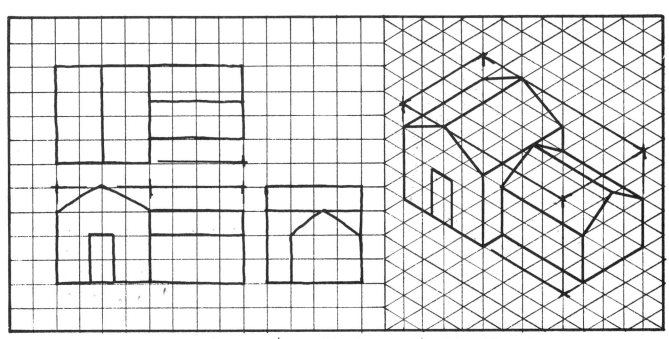

FIRST, LIGHTLY BLOCK IN THE BOX

Architectural Working Drawings

TEST 1

ORTHOGRAPHIC SKETCHING

NAME _____

TOTAL SCORE _____

DIRECTIONS: *With a soft pencil, sketch the three orthographic views of the pictorial drawing on the right. (See instructions on page 101.)*

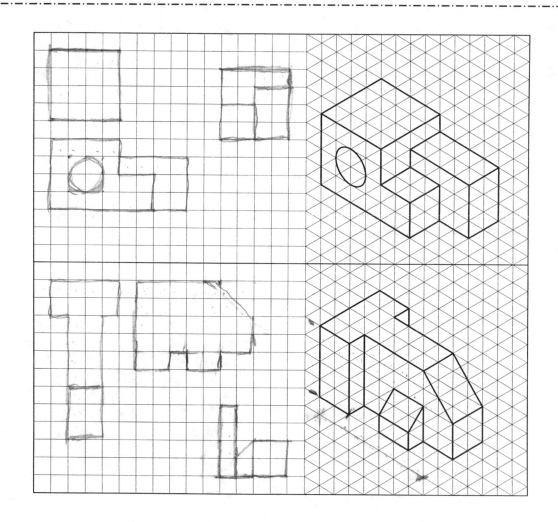

Architectural Working Drawings

TEST 1 (continued)

ORTHOGRAPHIC SKETCHING

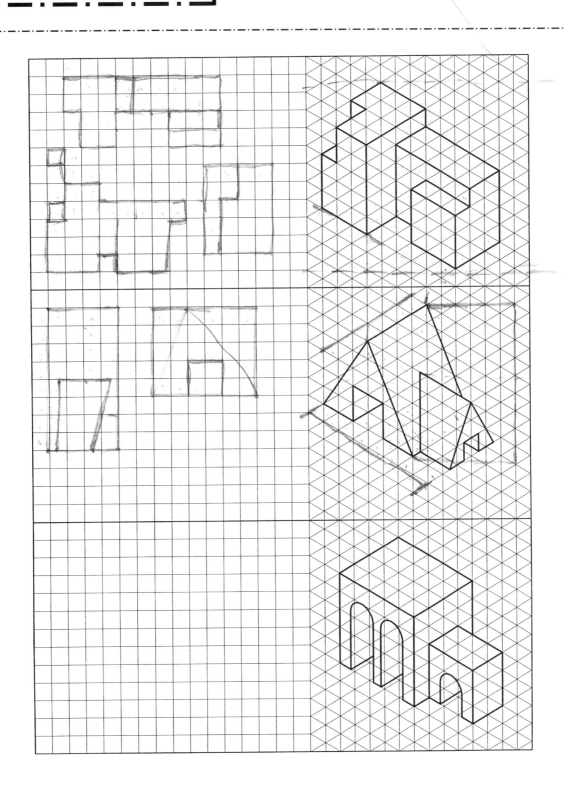

Architectural Working Drawings

TEST 2

ORTHOGRAPHIC SKETCHING

NAME _____

TOTAL SCORE _____

DIRECTIONS: *With a soft pencil, sketch the three orthographic views of the pictorial drawing on the right.*

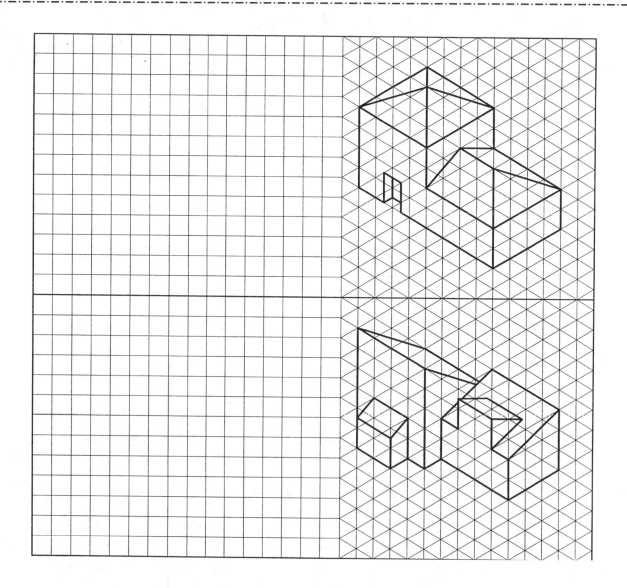

104

Architectural Working Drawings

TEST 2 (continued)

ORTHOGRAPHIC SKETCHING

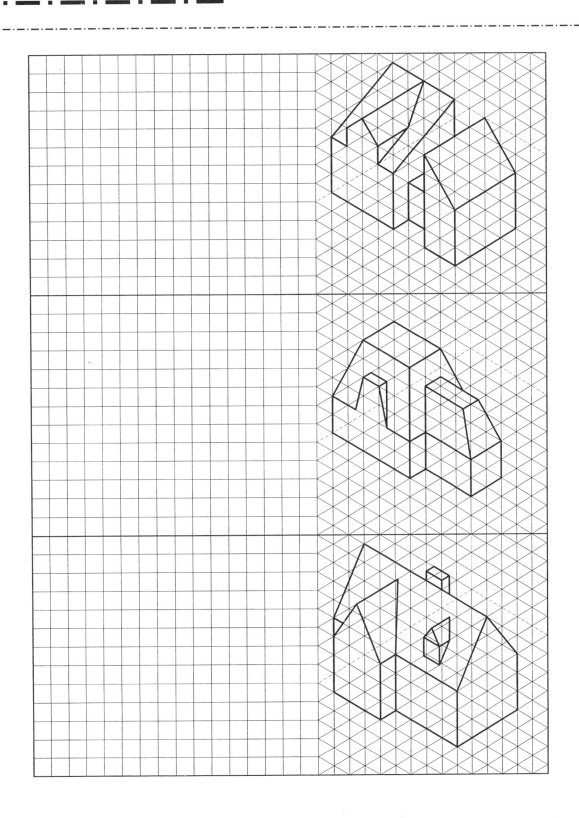

Architectural Working Drawings

TEST 3

MATCHING

NAME _____

TOTAL SCORE _____

Value of each answer __**8.3**__ POINTS

DIRECTIONS: Match the orthographic views with the correct pictorial view. After each letter, place the number of the correct pictorial view.

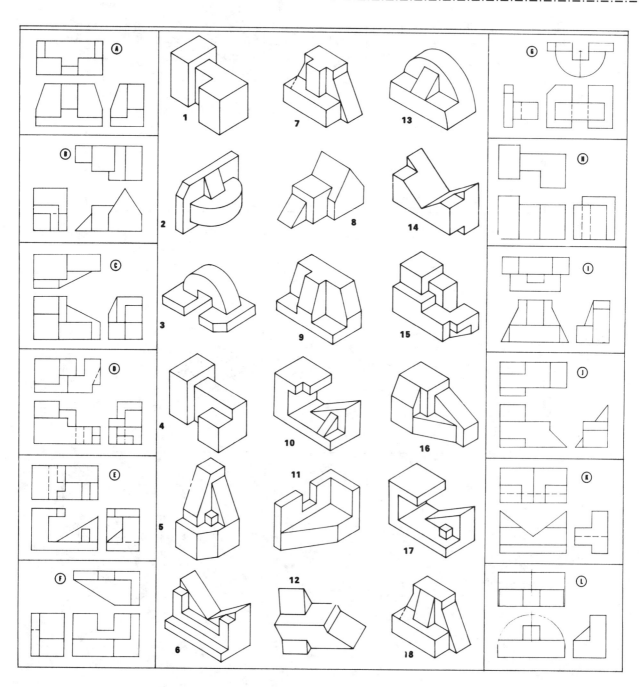

Architectural Working Drawings

TEST 4

SKETCHING ARCHITECTURAL SYMBOLS

NAME _____

TOTAL SCORE _____

Value of each answer __**10** POINTS__

DIRECTIONS: *With a soft pencil, fill in each figure with the correct architectural symbol.*

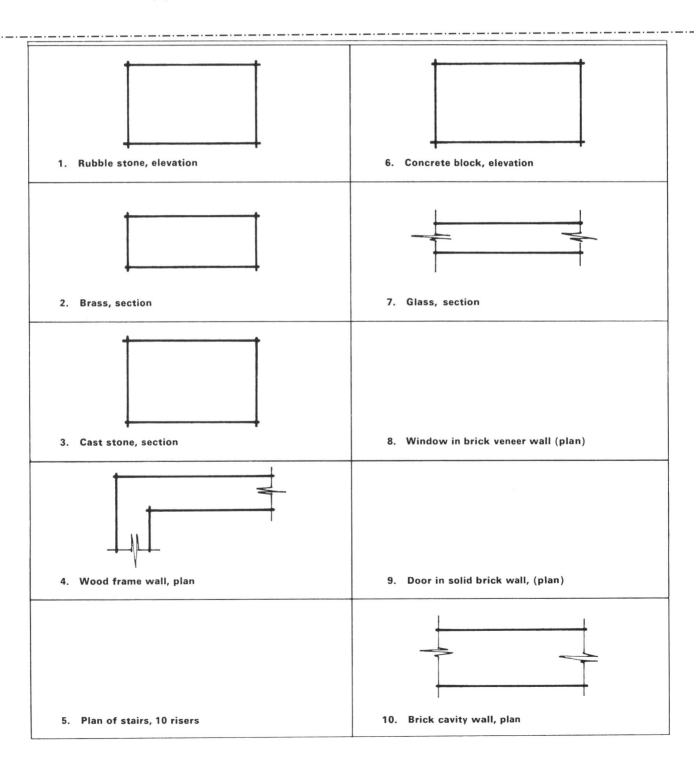

1. Rubble stone, elevation
2. Brass, section
3. Cast stone, section
4. Wood frame wall, plan
5. Plan of stairs, 10 risers
6. Concrete block, elevation
7. Glass, section
8. Window in brick veneer wall (plan)
9. Door in solid brick wall, (plan)
10. Brick cavity wall, plan

Architectural Working Drawings

TEST 5

SKETCHING ARCHITECTURAL SYMBOLS

NAME _____

TOTAL SCORE _____

Value of each answer ____10____ POINTS

DIRECTIONS: *With a soft pencil, fill in each figure with the correct architectural symbol.*

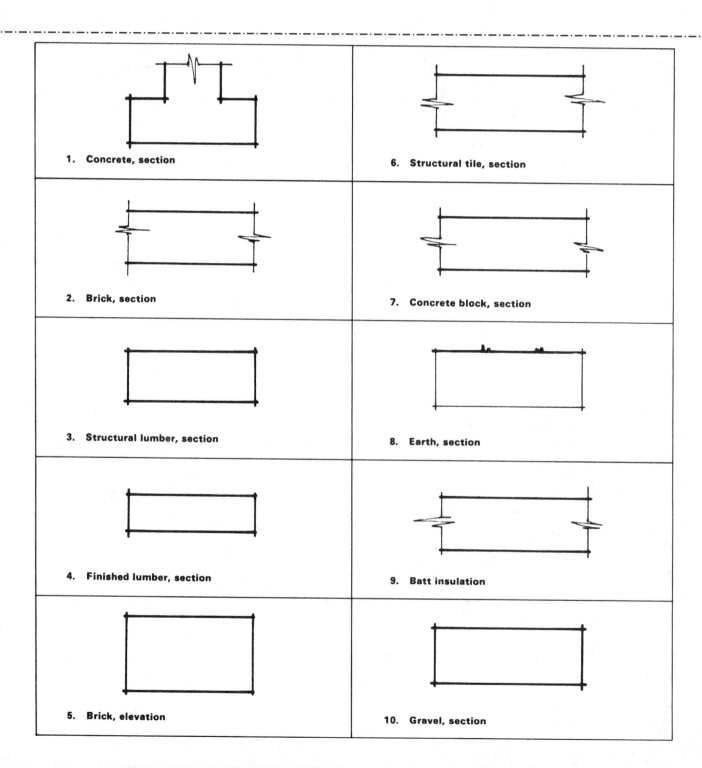

1. Concrete, section
2. Brick, section
3. Structural lumber, section
4. Finished lumber, section
5. Brick, elevation
6. Structural tile, section
7. Concrete block, section
8. Earth, section
9. Batt insulation
10. Gravel, section

Architectural Working Drawings

MATCHING

ELEVATIONS IDENTIFICATION

NAME _____

TOTAL SCORE _____

Value of each answer — **10** POINTS

DIRECTIONS: *From the elevations given below, place the letter of the correct elevation in the proper blank to match the building shown.*

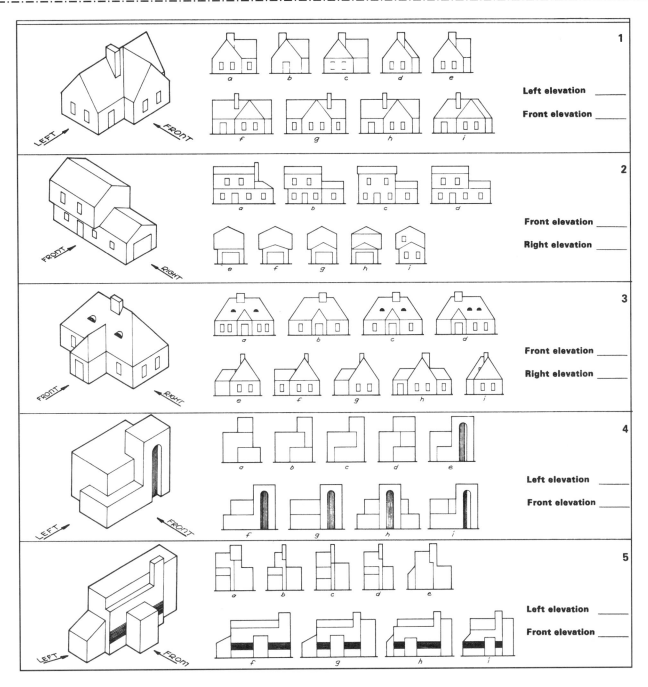

1. Left elevation _____
   Front elevation _____

2. Front elevation _____
   Right elevation _____

3. Front elevation _____
   Right elevation _____

4. Left elevation _____
   Front elevation _____

5. Left elevation _____
   Front elevation _____

109

Architectural Working Drawings

TEST 7
MATCHING

SYMBOL IDENTIFICATION

NAME _____

TOTAL SCORE _____

Value of each answer __**5** POINTS__

DIRECTIONS: *Place the identifying letter of the term opposite the correct symbol.*

A. Outlet bracket with pull switch 1. _____

B. Telephone outlet 2. _____

C. Special purpose outlet 3. _____

D. Three-way switch 4. _____

E. Duplex outlet 5. _____

F. Bell 6. _____

G. Wall outlet 7. _____

H. Gate valve 8. _____

I. Buzzer 9. _____

J. Supply duct 10. _____

K. Shower head 11. _____

L. Range outlet 12. _____

M. Fan outlet 13. _____

N. Ceiling outlet 14. _____

O. Motor outlet 15. _____

P. Pipe elbow 16. _____

Q. Water heater 17. _____

R. Roof-slope diagram 18. _____

S. Floor outlet 19. _____

T. Four-way switch 20. _____

Architectural Working Drawings

TEST 8

READING THE SCALE

THE ARCHITECT'S SCALE

NAME _____

TOTAL SCORE _____

Value of each answer ____**10** POINTS____

DIRECTIONS: *Shown below are various lines drawn at different scales. With the use of an architect's scale, measure each line and indicate the dimension it represents at the given scale.*

1. $\frac{1}{8}'' = 1'-0''$

    1. _____

2. $\frac{1}{2}'' = 1'-0''$

    2. _____

3. $1'' = 1'-0''$

    3. _____

4. $\frac{1}{4}'' = 1'-0''$

    4. _____

5. $3'' = 1'-0''$

    5. _____

6. $\frac{3}{4}'' = 1'-0''$

    6. _____

7. $12'' = 1'-0''$ (full size)

    7. _____

8. $\frac{3}{8}'' = 1'-0''$

    8. _____

9. $\frac{1}{4}'' = 1'-0''$

    9. _____

10. $1\frac{1}{2}'' = 1'-0''$

    10. _____

Architectural Working Drawings

TEST 9

MULTIPLE CHOICE

CONSTRUCTION TERMS

NAME _____

TOTAL SCORE _____

Value of each answer ____**7.14** POINTS____

DIRECTIONS: *Various statements are given after each construction term. Select the statement to make the sentence correct and place its letter in the parentheses.*

Question   Answer

1. An ELL is   ( )

   a. a metal structural member.
   b. a wing of a building at right angle to main portion.
   c. a room at right angles to another.
   d. the abbreviation for elevation.

2. A GUSSET is a   ( )

   a. gasket in a pipe joint.
   b. short strut in roof framing.
   c. plate used to strengthen the joints of a truss.
   d. diagonal brace in a wall.

3. An ASTRAGAL is   ( )

   a. a trim mold on a cornice.
   b. a vent in a crawl space.
   c. a molding between meeting doors.
   d. a recessed area below grade.

4. A CRIPPLE is   ( )

   a. a short structural wall member.
   b. a broken rafter.
   c. a strut in a trussed rafter.
   d. the extension of a rafter over a wall.

5. A DOORSTOP is   ( )

   a. a molded strip on a door jamb.
   b. a rubber bumper.
   c. a metal latch on the edge of the door.
   d. a half-round molding in back of the door.

6. The HEARTH is   ( )

   a. the fireplace opening.
   b. a wooden beam above the fireplace.
   c. the firebrick lining in a fireplace.
   d. the masonry floor in front of a fireplace.

7. A MONOLITHIC SLAB   ( )

   a. a concrete floor with steel reinforcement.
   b. a concrete floor cast in one pour.
   c. a concrete floor finished with a steel trowel.
   d. a concrete floor with one-size aggregate.

Architectural Working Drawings

TEST 9
(continued)
MULTIPLE CHOICE

CONSTRUCTION TERMS

8. GLAZING ( )
    a. placing glass in windows.
    b. polishing terrazzo floors to a smooth finish.
    c. varnishing a natural wood finish.
    d. waxing wood floors.

9. A PERGOLA is ( )
    a. the colonial trim on a cornice.
    b. an open framework over an outdoor area.
    c. a rectangular pier attached to a masonry wall.
    d. a low railing around a roof.

10. The SOFFIT is ( )
    a. the edge of a stair flight.
    b. the flooring in an attic.
    c. the underside of a roof overhang.
    d. the vent pipe of a basement water closet.

11. A DUPLEX OUTLET ( )
    a. a double ceiling outlet.
    b. two outlets side by side.
    c. a faucet for hot and cold water.
    d. a wall receptacle for two electrical plugs.

12. A TRANSVERSE SECTION ( )
    a. vertical section across the width of a building.
    b. horizontal section.
    c. section through the long dimension of a building.
    d. drawing showing the thickness of a wall.

13. GROUNDS are ( )
    a. cement blocks below grade.
    b. wood strips at the edges of plaster.
    c. anchor rods deep in the ground.
    d. wood fasteners on concrete slabs.

14. A HOSE BIBB is ( )
    a. a hanger for the garden hose.
    b. flashing for water protection.
    c. an exterior water outlet for a threaded hose connection.
    d. a splash block at the bottom of a leader.

Architectural Working Drawings

TEST 10

IDENTIFICATION

CONSTRUCTION TERMS

NAME _____

TOTAL SCORE _____

Value of each answer __**4** POINTS__

DIRECTIONS: Listed below are definitions of common construction terms. In the blank provided, write the correct term.

A horizontal member tying opposing rafters below the ridge in roof framing. 1. _____

Recessed area below grade around the foundation to allow light and ventilation into a basement window. 2. _____

In framing, the joists placed at the ends of a floor opening and attached to the trimmers. 3. _____

A stair tread that is wider at one end than the other, allowing the stairs to change direction. 4. _____

An opening or slatted grill allowing ventilation while providing protection from rain. 5. _____

The structural member between a series of windows. 6. _____

A strip of metal with a vertical lip used to retain the gravel around the edge of a built-up roof. 7. _____

The face or front elevation of a building. 8. _____

A low wall resulting from one-and-a-half-story construction. 9. _____

Horizontal interior trim member below the stool of a window. 10. _____

The vertical height of a roof or stairs. 11. _____

A rectangular pier attached to a wall for the purpose of strengthening the wall. 12. _____

A plywood or metal plate used to strengthen the joints of a truss. 13. _____

Architectural Working Drawings

TEST 10 (continued)
IDENTIFICATION

CONSTRUCTION TERMS

An exterior water faucet made for the threaded attachment of a hose. 14. _____

The outside horizontal member on the edge of a roof or overhang. 15. _____

A rafter shorter than a common rafter; especially used in hip-roof framing. 16. _____

The forming of white stains on masonry walls from moisture within the walls. 17. _____

A low wall or railing; usually around the edge of a roof. 18. _____

A horizontal member extending from the surface of an exterior wall so as to throw off rainwater from the wall. 19. _____

Gable molding attached on the incline of the gable. The molding must be a different profile to match similar molding along the remaining horizontal portions of the roof. 20. _____

Wood strips placed over or in a concrete slab to receive a finish wood floor. 21. _____

The narrow strip of wood nailed vertically over the joints of boards to form a type of wood siding. 22. _____

A vertical space within a building for ducts, pipes, or wires. 23. _____

The size of lumber before dressing, rather than its actual size. 24. _____

An open court within a building. 25. _____

Architectural Working Drawings

TEST 11

IDENTIFICATION

CONSTRUCTION TERMS

NAME _____

TOTAL SCORE _____

Value of each answer _____ **4** POINTS

DIRECTIONS: Listed below are definitions of common construction terms. In the blank provided, write the correct term.

A projection of masonry from the face of a wall, or a bracket used for support for weight above. 1. _____

The finish board covering the projecting portion of a gable roof. 2. _____

A mark on some permanent object fixed to the ground from which land measurements and elevations are taken. 3. _____

A pipe for carrying rainwater from the roof to the ground or sewer connection. 4. _____

A small bar separating the glass lights in a window. 5. _____

Horizontal roof members laid over trusses to support rafters. 6. _____

A U-shaped pipe below plumbing fixtures to create a water seal and prevent sewer gases from being released into the habitable areas. 7. _____

Small holes in masonry cavity walls to release water accumulation to the exterior. 8. _____

Flooring, usually of wood, laid in an alternating or inlaid pattern to form various designs. 9. _____

A relatively shorter joist that joins against a header or trimmer in floor framing. 10. _____

The small molding covering the joint between the flooring and the baseboard on the inside of a room. 11. _____

A masonry pillar usually below a building to support the floor framing. 12. _____

A small opening in a ceiling to provide access to an attic or roof. 13. _____

116

Architectural Working Drawings

TEST 11 (continued)
IDENTIFICATION

CONSTRUCTION TERMS

The triangular end of a gable-roofed house. 14. _____

A built-up beam formed by a metal plate sandwiched between two wood members and bolted together for additional strength. 15. _____

A short, wooden framing member used to support an overhanging portion of a roof. It extends from the wall to the underside surfacing of the overhang. 16. _____

Devices, usually metal, used in building construction to secure one material to another. 17. _____

Interior trim used around window and door openings. 18. _____

The shallow space below the floor of a house built above the ground. Generally, it is surrounded with the foundation wall. 19. _____

A type of insulation designed to be installed between framing members. 20. _____

An angular board used to eliminate a sharp, right angle, usually on roof decks. 21. _____

Any unfinished opening in the framing of a building. 22. _____

A small gable roof placed in back of a chimney on a sloping roof to shed water and debris. 23. _____

The longer floor-framing member around a rectangular opening into which a header is joined. 24. _____

A horizontal support member across the head of a door or window opening. 25. _____

Architectural Working Drawings

TEST 12

ABBREVIATIONS

ABBREVIATIONS RECALL

NAME _____

TOTAL SCORE _____

Value of each answer _____ **5** POINTS

DIRECTIONS: In the space provided, write the abbreviation for the following terms, based on the standard abbreviations on pages 65-73.

1. Plate
2. Room
3. On center
4. Glass
5. Standard
6. Molding
7. Finish
8. Thousand
9. Wood
10. Column
11. Dimension
12. Apartment
13. Floor
14. Ceiling
15. Concrete
16. Exterior
17. Drawings
18. Kiln-dried
19. Celsius
20. Footing

1. _____
2. _____
3. _____
4. _____
5. _____
6. _____
7. _____
8. _____
9. _____
10. _____
11. _____
12. _____
13. _____
14. _____
15. _____
16. _____
17. _____
18. _____
19. _____
20. _____

Architectural Working Drawings

TEST 13

IDENTIFICATION

CONSTRUCTION TERMS

NAME _____

TOTAL SCORE _____

Value of each answer __1.66__ POINTS

DIRECTIONS: *From the figure below, identify the numbered parts and place the correct terms in the corresponding blanks.*

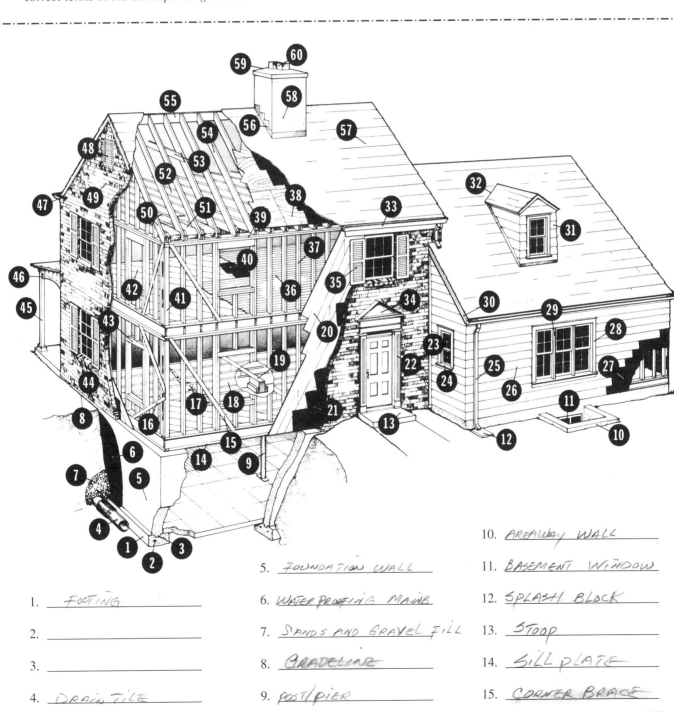

1. FOOTING
2. _____
3. _____
4. DRAIN TILE
5. FOUNDATION WALL
6. WATERPROOFING MAMB
7. SAND AND GRAVEL FILL
8. GRADELINE
9. POST/PIER
10. AREAWAY WALL
11. BASEMENT WINDOW
12. SPLASH BLOCK
13. STOOP
14. SILL PLATE
15. CORNER BRACE

119

Architectural Working Drawings

TEST 13 (continued) IDENTIFICATION

CONSTRUCTION TERMS

16. KNEE BRACE
17. BRIDGING
18. FLOOR JOIST
19. BEAM / GERDER
20. SHEATHING
21. BUILDING PAPER
22. TRIM PILASTER
23. ___
24. ___
25. ___
26. ___
27. ___
28. ___
29. ___
30. ___
31. ___
32. ___
33. ___
34. ___
35. ___
36. ___
37. ___
38. ___
39. ___
40. ___
41. ___
42. ___
43. ___
44. ___
45. ___
46. ___
47. ___
48. ___
49. ___
50. ___
51. ___
52. ___
53. ___
54. ___
55. ___
56. ___
57. ___
58. ___
59. ___
60. ___

Architectural Working Drawings

TEST 14

IDENTIFICATION

CONSTRUCTION SYMBOLS

NAME _____

TOTAL SCORE _____

Value of each answer ____8 POINTS____

DIRECTIONS: From the figure below, identify the lettered features and place the correct terms in the corresponding blanks.

A. _____

B. _____

C. _____

D. _____

E. _____

F. _____

G. _____

H. _____

I. _____

J. _____

K. _____

L. _____

121

Architectural Working Drawings

TEST 15

FEET AND INCHES

ARITHMETIC REVIEW

NAME _____

TOTAL SCORE _____

Value of each answer __**4** POINTS__

DIRECTIONS: *Place the correct answer in the blank. Simplify answers to feet and inches.*

A. Add the following mixed dimensions found on working drawings.

Example: 5′ 4″ + 8′ 2″ = 13′ 6″

(a) 4′ 6″ + 5′ 2″ = _____ (b) 3′ 7″ + 1′ 4″ = _____

(c) 12′ 5″ + 8′ 1″ = _____ (d) 29′ 6″ + 1′ 7″ = _____

(e) 3′ 8″ + 2′ 6″ = _____ (f) 6′ 11″ + 7′ 10″ = _____

(g) 2′ 6″ + 42″ = _____ (h) 4′ 3″ + 39″ = _____

(i) $5' 6\frac{1}{2}'' + 4' 3\frac{1}{2}''$ = _____ (j) $3' 8\frac{1}{4}'' + 2' 7\frac{1}{8}''$ = _____

(k) $2' 9\frac{5}{8}'' + 7' 4\frac{3}{4}''$ = _____ (l) $4' 11\frac{5}{8}'' + 1' 3\frac{1}{2}''$ = _____

(m) $6' 3\frac{3}{16}'' + 8' 10\frac{7}{8}''$ = _____

B. Subtract the following dimensions found on working drawings.

Example: 12′ 7″ − 8′ 5″ = 4′ 2″

(n) 6′ 8″ − 4′ 4″ = _____ (o) 5′ 0″ − 3′ 7″ = _____

(p) 4′ 2″ − 1′ 6″ = _____ (q) 12′ 4″ − 6′ 9″ = _____

(r) 47″ − 2′ 6″ = _____ (s) 6′ 3″ − 45″ = _____

(t) $9' 2\frac{1}{2}'' - 3' 10\frac{1}{4}''$ = _____ (u) $5' 8\frac{1}{8}'' - 2' 5\frac{3}{4}''$ = _____

(v) $10' 2\frac{1}{4}'' - 4' 7\frac{5}{8}''$ = _____ (w) $163'' - 9' 5\frac{1}{8}''$ = _____

(x) $7' 3\frac{7}{8}'' - 1' 1\frac{3}{16}''$ = _____ (y) $12' 0\frac{3}{8}'' - 4' 10\frac{9}{16}''$ = _____

Architectural Working Drawings

TEST 16
BOARD FEET

ARITHMETIC REVIEW

NAME _____

TOTAL SCORE _____

Value of each answer ____**10** POINTS____

DIRECTIONS: *Place the correct answer to the following problems in the blank provided.*

A. Calculate the board feet in the following quantities of lumber:

(a) 10 pcs. 2″ × 4″ × 8′ −0″ long .... _____

(b) 36 pcs. 1″ × 6″ × 12′ −0″ long .... _____

(c) 25 pcs. 4″ × 4″ × 10′ −0″ long .... _____

(d) 6 pcs. 2″ × 10″ × 14′ −0″ long .... _____

(e) 400 pcs. 2″ × 8″ × 16′ −0″ long .... _____

B. Calculate the cost of the following quantities of lumber using the cost per thousand board feet given:

(f) 28 pcs. 1″ × 8″ × 12′ −0″ long

@ $125.00/M .... _____

(g) 16 pcs. 2″ × 12″ × 14′ −0″ long

@ $95.00/M .... _____

(h) 145 pcs. 2″ × 4″ × 8′ −0″ long

@ $68.50/M .... _____

(i) 38 pcs. 4″ × 6″ × 12′ −0″ long

@ $115.00/M .... _____

(j) 18 pcs. 2″ × 8″ × 10′ −0″ long

@ $78.00/M, and

(k) 24 pcs. 2″ × 10″ × 12′ −0″ long

@ $82.50/M .... _____

Architectural Working Drawings

TEST 17

RATIO AND PROPORTION

ARITHMETIC REVIEW

NAME _____

TOTAL SCORE _____

Value of each answer __**20** POINTS__

DIRECTIONS: *Place the correct answer to the following word problems in the blank provided.*

A. Two gable roofs have the same slopes. If roof number 1 has a span of 32′ and a rise of 8′, what is the rise of roof number 2 if its span is 24′?

Ans. _____

B. Two metal drums have the same diameter. If the first is 5′ long and holds 100 gal of liquid, how many gallons can the second drum hold if it is 12′ long?

Ans. _____

C. A flagpole casts a shadow 92′ long on the ground. How high is the flagpole if a nearby post 5′ high casts a shadow 14′ long? (Give answer in feet and inches.)

Ans. _____

D. If the floor tile for a room with 96 sq ft costs $132.00, how much will similar tile cost for a room with 160 sq ft of floor area?

Ans. _____

E. How many bundles of shingles will be needed to cover a roof with 1900 sq ft if it takes 9 1/2 bundles of the same shingles to cover a similar roof with 1250 sq ft?

Ans. _____

Architectural Working Drawings

TEST 18
PLANE FIGURES

ARITHMETIC REVIEW

NAME _____

TOTAL SCORE _____

Value of each answer __**10** POINTS__

DIRECTIONS: *Place the correct answer for the following problems in the blank provided. Show answers in feet or square feet. If necessary, refer to PLANE FIGURES in the Appendix.*

A. Find the (a) AREA, (b) PERIMETER, and (c) length of the DIAGONAL of the following two plane figures.

Square

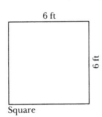

(a) Area _____

(b) Perimeter _____

(c) Diagonal _____

Rectangle

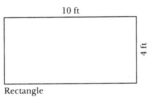

(d) Area _____

(e) Perimeter _____

(f) Diagonal _____

B. Find the AREA of the following plane figures.

(g)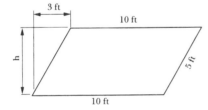

(g) Area _____

(h)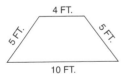

(h) Area _____

(i)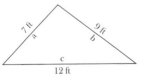

(i) Area _____

(j)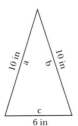

(j) Area _____

125

Architectural Working Drawings

TEST 19
PLANE FIGURES

ARITHMETIC REVIEW

NAME _____

TOTAL SCORE _____

Value of each answer _____ **10** POINTS

DIRECTIONS: *Place the correct answer for the following problems in the blank provided. If necessary, refer to PLANE FIGURES in the Appendix.*

A. Find the AREA of the right triangle *ABC* as shown.

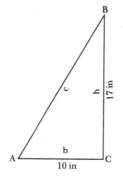

(a) Area _____

B. Find the LENGTH of side *c* of the right triangle *ABC* shown in Problem A.

(b) Length _____

C. Find the LENGTH of side *h* of the right triangle *ABC* as shown.

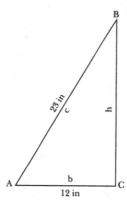

(c) Length _____

D. Find the AREA of the right triangle *ABC* shown in Problem C.

(d) Area _____

Architectural Working Drawings

TEST 19
(continued)
PLANE FIGURES

ARITHMETIC REVIEW

E. Find the AREA and CIRCUMFERENCE of a circle with a 3″ radius. (e) Area _____

(e) Circumference _____

F. Find the AREA and CIRCUMFERENCE of a circle with a 5′–6″ diameter. (f) Area _____

(f) Circumference _____

G. Find the AREA of the ring between the two circles in the following figure. (g) Area _____

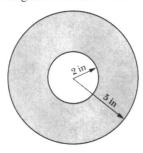

H. Find the AREA of the circular sector shown in the following figure. (h) Area _____

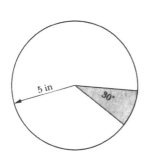

Architectural Working Drawings

TEST 20
SOLID FIGURES

ARITHMETIC REVIEW

NAME _____

TOTAL SCORE _____

Value of each answer ____**20** POINTS____

DIRECTIONS: *Place the correct answer to the following problems in the blank provided. If necessary, refer to SOLID FIGURES in the Appendix.*

A. Find the VOLUME of a 4′ cube (consider as inside dimension).   (a) Volume _____

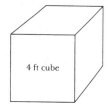

B. Find the VOLUME contained by a rectangular box with inside dimensions of 10′–6″ × 3′ × 5′ (state in cubic feet).   (b) Volume _____

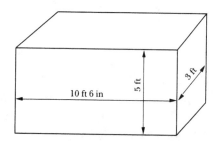

C. Find the exterior surface AREA of a plywood box (with cover) as shown.   (c) Area _____

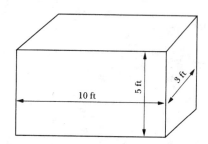

Architectural Working Drawings

TEST 20
(continued)
SOLID FIGURES

ARITHMETIC REVIEW

D. Find the VOLUME of the hollow prism as shown (in cubic feet).

(d) Volume _____

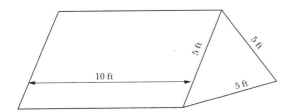

E. Find the VOLUME in cubic feet of the cylindrical storage tank as shown.

(e) Volume _____

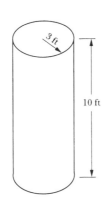

Architectural Working Drawings

TEST 21
SOLID FIGURES

ARITHMETIC REVIEW

NAME _____

TOTAL SCORE _____

Value of each answer _____ **20** POINTS

DIRECTIONS: *Place the correct answer to the following problems in the blank provided. If necessary, refer to SOLID FIGURES in the Appendix.*

A. Find the AREA in square feet of sheet metal required to construct the tank as shown (omit area for seams).

(a) Area

B. Find the surface AREA of the square-base pyramid as shown.

(b) Area

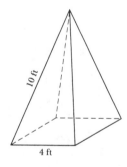

C. Find the VOLUME in cubic feet displaced by the given cone.

(c) Volume _____

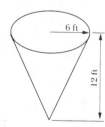

Architectural Working Drawings

TEST 21
(continued)
SOLID FIGURES

ARITHMETIC REVIEW

D. Find the VOLUME in the given building in cubic feet. (d) Volume _____

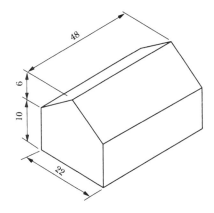

E. Find the VOLUME in the given building in cubic feet. (e) Volume _____

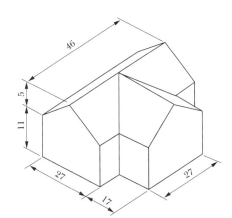

Architectural Working Drawings

TEST 22

METRIC DIMENSIONS

NAME _____

TOTAL SCORE _____

Value of each answer __**6.66**__ POINTS

DIRECTIONS: *Place the correct answer in the blanks provided.*

1. Six cubic yards of concrete would be similar to _____ m³.

2. A 2400 sq ft building would be similar to a building _____ m² in area.

3. A 160 sq ft room would be equivalent to a _____ m² room.

4. A piece of lumber 2″ × 6″ × 8′–6″ would be equal to a similar piece _____ mm × _____ mm × _____ mm. (Use nominal size.)

5. If a room is 12′–0″ wide and has 340 sq ft of floor area, what is the length of the room in mm? _____

6. A residential lot is 100′–0″ × 460′–0″. What is its size in meters? _____

7. A building 56 ft high casts a shadow 67 ft long; an adjoining building casts a shadow 18 meters long. How high is the adjoining building in meters? _____

8. A box is 80 mm × 100 mm × 48 mm. What is its volume in mm³? _____

9. A flat-roof building is 25 ft × 68 ft × 97 ft. What is its volume in m³? _____

132

Architectural Working Drawings

TEST 22 (continued)

METRIC DIMENSIONS

10. The finish size of a 2 × 4 is $1\frac{1}{2}'' \times 3\frac{1}{2}''$. What is its finish size in mm? _____

11. Change the length of a 16'–8" floor joist to cm. _____

12. The length of a foundation wall is $32' - 6\frac{1}{2}''$. What is the length in mm? _____

13. If 8 pieces of lumber are labeled as 47 board feet, what would one piece be labeled in mm²? _____

14. How many mm² would be equivalent to 14 m²? _____

15. A driveway is 12'–6" wide. What is its width in mm? _____

Chapter 3

RESIDENTIAL WORKING DRAWINGS

Two complete sets of working drawings for residences are included in this material to give you a variety of representations to study. One is a two-story colonial home popular in many areas of the country. The second one is a larger home involving more architectural features commonly found in luxury homes.

As you read this chapter, take the time to refer occasionally to the working drawings for clarification of points that are not entirely clear. The drawings are discussed in the same order as they generally appear in a set. However, the easiest way in which to orient yourself to an unfamiliar set of drawings is to look over the elevations first, then the plan views, and, finally, the sections and details.

A. SURVEY PLATS AND PLOT PLANS (SITE PLANS)

The character of a building depends to a great extent on the land on which it is built. A house that has been found enjoyable to live in is the result of a perfect blending of the architecture and the natural characteristics of the property. Consequently, an important part of the information given in a set of drawings includes how the house is to be placed on the property, how it is to be approached from the road, how the surface of the land may have to be modified to utilize it best, and, finally, a complete legal description of the lot. This, along with other information, is shown on the *Plot Plan* (Figure 3–1). If the legal description of the property is incomplete, a land surveyor is employed to survey the property and prepare a *Survey Plat,* or topographic drawing. From the Survey Plat information, the architect develops beginning design studies and later, the finished Plot Plan.

Several systems of land measurement for legal purposes are used throughout the United States. Most of the eastern states use a system retained from colonial days called the *Metes and Bounds* system, which identifies a parcel of land by locating a boundary point, usually a corner of the property, and then describing the direction and distances of the property lines until the perimeter has been completely traced around to the starting point (Figure 3–2). Each leg of the perimeter is described first by the direction, then by its length.

Curved lot lines are described by giving their radius, length of the arc, and the direction of the curve. The system establishes correct lot lines by identifying a point-of-beginning (P.O.B.) corner of the property from a nearby prominent feature, such as a marker, curb, or section lines. Such features are given in the legal description, which is obtained from the local Deed Registry Office. In some states it would be the office of the Clerk of Superior Court. Angular bearings, reading from north or south, establish each

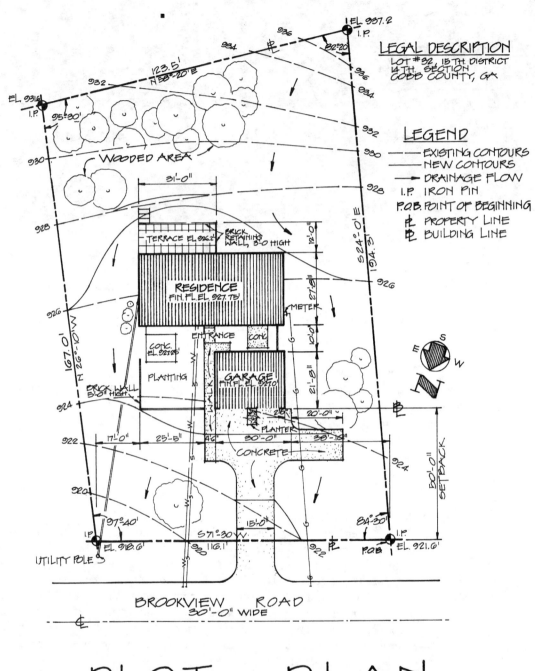

FIGURE 3–1

Typical residential Plot Plan (Site Plan).

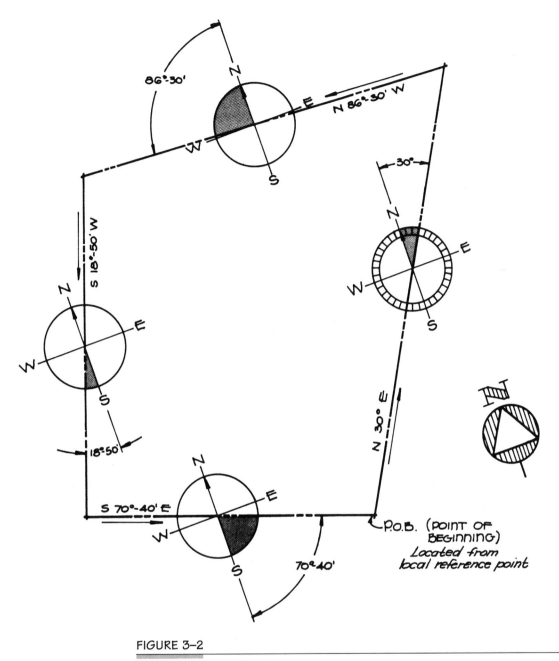

FIGURE 3–2

Line bearings on a plat map.

lot-line direction starting from the P.O.B. Notice in Figure 3–2 that the first lot line, labeled N30°E, indicates that it is 30 degrees from true north toward the east. This angle is called the *bearing of the line,* and it can be taken from either the north or south compass direction, whichever keeps the angle less than 90 degrees. The length of each line is measured in feet and decimal parts of a foot (civil engineer's scale). At the next iron pin or stake, the next bearing shows N86°30′W, which is 86°30′ from true north toward the west. Each line around the property is identified in this manner. Notice that the last two lines in the figure indicate bearings from the south rather than north to avoid obtuse angles.

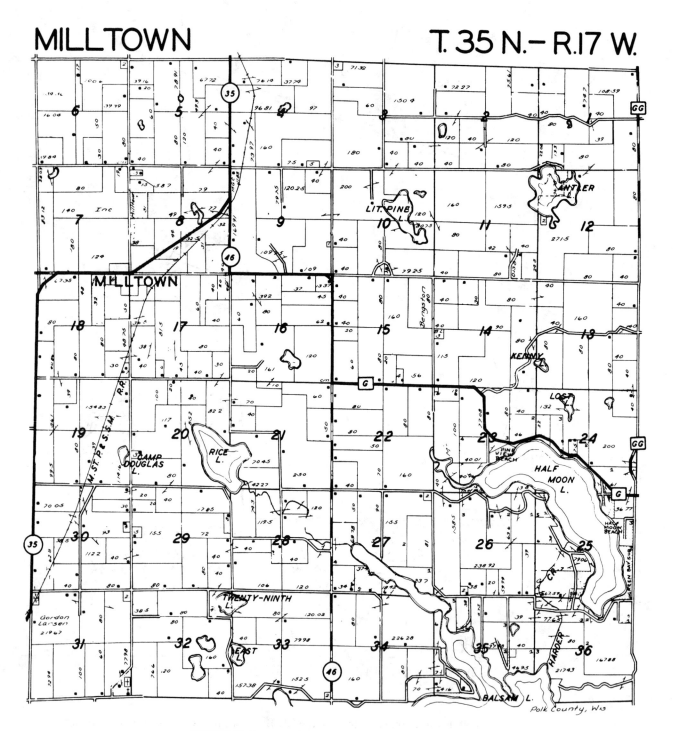

FIGURE 3-3

Plat map of a township.

In the Midwest and Far West states, the federal government devised a rectangular system for describing land, which utilizes lines of latitude and longitude to form rectangular *Townships* and smaller *Sections* (Figures 3–3 and 3–4). A section is 1 square mile, and of course, even smaller divisions are made to accommodate smaller pieces of property; yet, the legal description is identified from the major township and section divisions located either north or south of established base lines and either east or west of a principal meridian

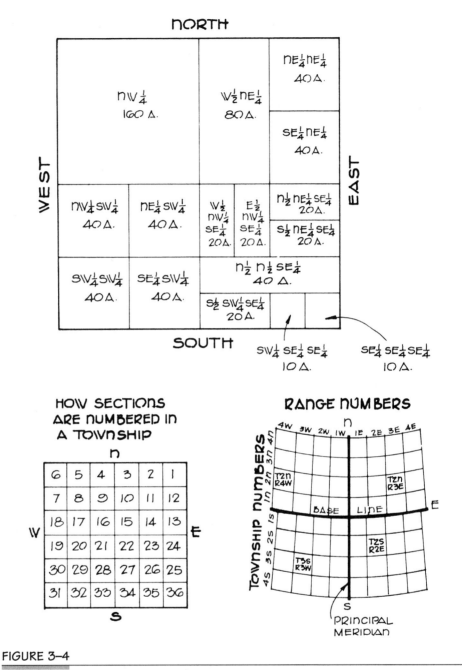

FIGURE 3-4

Legal land description of the township-section system. The sample section also shows acreage.

(Range Numbers). Public-owned water areas are referred to as *meandered water* on maps. Old acreage parcels of land resulting from converging northerly divisions are sometimes called *Government Lots* and given a lot number as description. Property within incorporated cities and towns is usually identified by section numbers and local references.

In addition, the survey establishes the elevations (heights) at the corners of the property and contours of the land if complete topographic information is necessary. Drawings with this information are called *Contour Maps* or topographic drawings. Elevations are determined from a local *datum point*. The U.S. Geological Surveys use sea level as a datum; but, for convenience, local surveys use a permanent point in the neighborhood—a fire hydrant, a concrete pillar, or a monument—and it is given a zero elevation. All elevations, plus or minus, are relative to the datum. Points used by surveyors for establishing elevations of property are sometimes known as *bench marks*.

The variations in elevation are shown by the use of *contour lines* (see Figure 3–1). Points of similar elevation are connected to form the lines, shaped according to the character of the surface. The water's edge around a lake forms a contour line at the water-level elevation. The vertical distance between contour levels is called the *contour interval*. Intervals on maps may vary from 1′ to 10′, depending on the scale of the map and the accuracy intended. To facilitate the reading of contours, every fifth line is usually made heavier, and its numerical elevation is labeled on it. After examining the contours in Figure 3–1, observe that the right rear corner of the lot is the highest, the front left corner the lowest, and the contour interval is 2′. Lines close together indicate a steep rise, whereas lines farther apart indicate a more gradual slope. Other surface characteristics soon become evident once you are familiar with the drawings. Lines over streams always point upstream, and each line will be continuous and close on itself, even though this fact may not be evident on a small map. Contour lines may be shown either dashed or solid. When surface changes are shown on the same drawing, original contours are made with broken lines and finish contours with solid lines. As mentioned earlier, survey drawings also show the complete legal description, which includes subdivision name and lot number, section number, county, state, north-point arrow, scale, name and width of street, setback lines, easements, zoning regulations, and certification by a registered land surveyor.

The Plot Plan (Figure 3–1) is made by the architect or designer from the survey drawings; it orients the proposed building to the lot. The exact location of the foundation of the building is prominently outlined and dimensioned with respect to the lot lines. Otherwise, much of the information from the survey is included. Thus typical information found on the Plot Plan includes:

1. Legal description of the property.
2. Location of any accessory buildings.
3. Location and dimensions of drives, walks, patios, retaining walls, steps, fences, and so on.
4. Elevations at floor levels, grade at corners, and slab levels.
5. Setback lines (restrictions of the location of the building within lot lines).
6. Easements (rights to the property by someone other than the owner).
7. Utilities connections.
8. Trees to be saved.
9. Location of septic tank and seepage beds, if applicable.
10. Invest elevations at manholes (level of drain at manhole).
11. Scale of the drawing and north-point arrow.

LANDSCAPE PLANS: Landscape plans are similar to plot plans except that they are prepared by a landscape architect for the purpose of showing proposed planting and natural changes required to enhance the grounds. Symbols are used to represent the different trees and shrubs suggested. Botanical names are included in a schedule (Figure 3–5). Some plans are *phased* to show a step-by-step planting schedule over a period of years; some indicated plants are scheduled for planting the first year, others the following year, and so on. Often surface drainage, garden areas, walls, and walks that complement both the grounds and the building are the important aspects of the landscape plan.

B. FOUNDATION (BASEMENT) PLANS

The Foundation Plan is used mainly by masons and concrete workers, who build the foundation. Like other plan views, it is a horizontal section view taken through the masonry walls of the foundation and showing all relative information. Most residential foundations

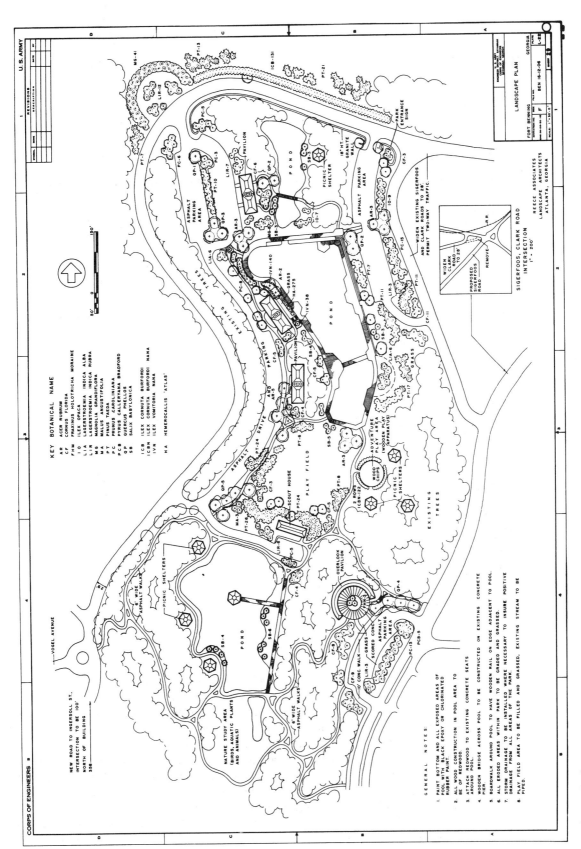

FIGURE 3-5
A typical landscape plan.

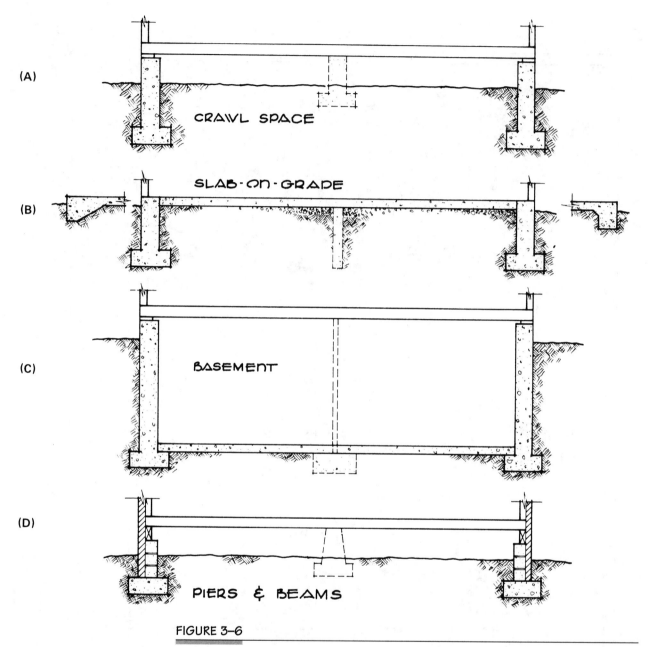

FIGURE 3-6

Foundation types found in residential construction.

are composed of concrete block, poured concrete, solid brick, or perhaps some other structural, masonry material (Figure 3–6). The concrete block is presently the most popular material (Figure 3–7). The reasons for its widespread use are that it is readily available, economical, lays up fast, and allows the carpentry work to commence almost as soon as it is in place. The 8″ × 8″ × 16″ and 8″ × 12″ × 16″ block sizes are mainly used. Basement blockwork is usually waterproofed on the exterior with a $\frac{1}{2}$″ coat of cement plaster covered with a coat of asphalt pitch; the asphalt is applied only below grade because of its appearance. Figure 3–8 shows the conventional method of laying out a foundation; the top edges of the batter boards are leveled with a civil engineer's level before they are secured.

If the house is to have a basement, the Basement Plan would serve the same purpose, of course, as the Foundation Plan and would include all information about the foundation walls as well as any other information about the layout of the basement. Generally, you will en-

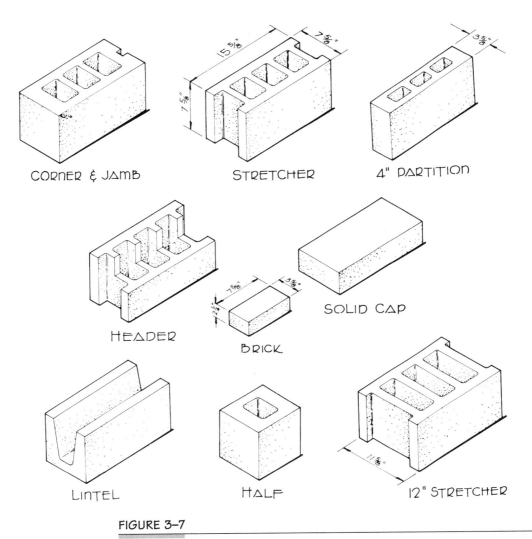

FIGURE 3–7

Typical concrete block units.

counter symbols representing pipe columns, stairs, lighting, areaways, and floor structure above. Columns in basements carry concentrated loads from the floor framing and hence must have enlarged footings—normally 24″ × 24″ × 12″ thick under the basement floor.

To provide natural light and ventilation, basement windows, which may often have to be placed in the walls below the level of the grade line, are included. In this case, areaways are indicated. These are recessed areas below grade surrounded with concrete or steel (see Figure 3–43F).

Footings, upon which the walls rest, are shown with broken lines parallel to the wall lines on the plan. These poured concrete "feet," so to speak, carry the entire weight of the building and are made wider than the foundation walls in order to spread the weight over a larger earth area for stability. Cross-sectional descriptions of footings are discussed later under Sections and Details.

Houses without basements may have *crawl spaces* below the first floor (Figure 3–6A) or they may be built over *concrete slabs* resting directly on the ground (Figure 3–6B). Both are satisfactory methods of residential construction. However, be sure that you understand this important aspect of the construction when reading a set of drawings.

Note: Figures 3–9 through 3–24 illustrate the structural and design features of the Colonial Home discussed in this section. This is conventional construction; many contemporary builders may use slightly different details in their construction.

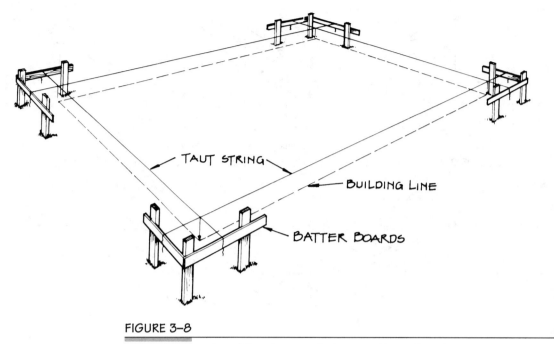

FIGURE 3-8

Laying out a residential foundation with the use of batter boards.

FIGURE 3-9

Foundation construction in the Colonial Home.

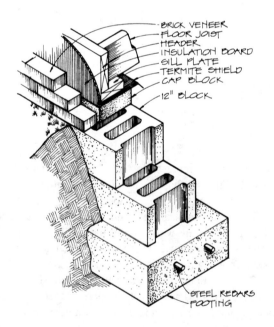

FIGURE 3-10

Pier construction in the Colonial Home.

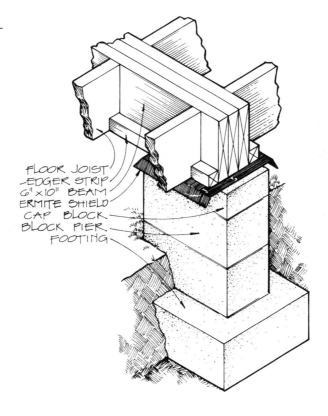

FIGURE 3-11

Brick-screen wall construction in the Colonial Home carport.

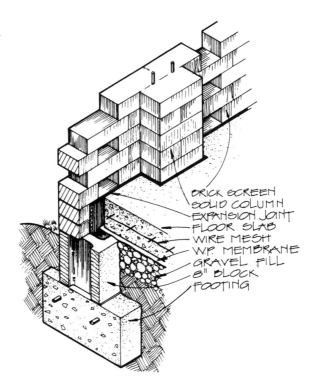

FIGURE 3-12

Perspective view of the Colonial Home.

FIGURE 3-13

Front elevation of the Colonial Home, showing where floor-plan cutting planes are theoretically taken.

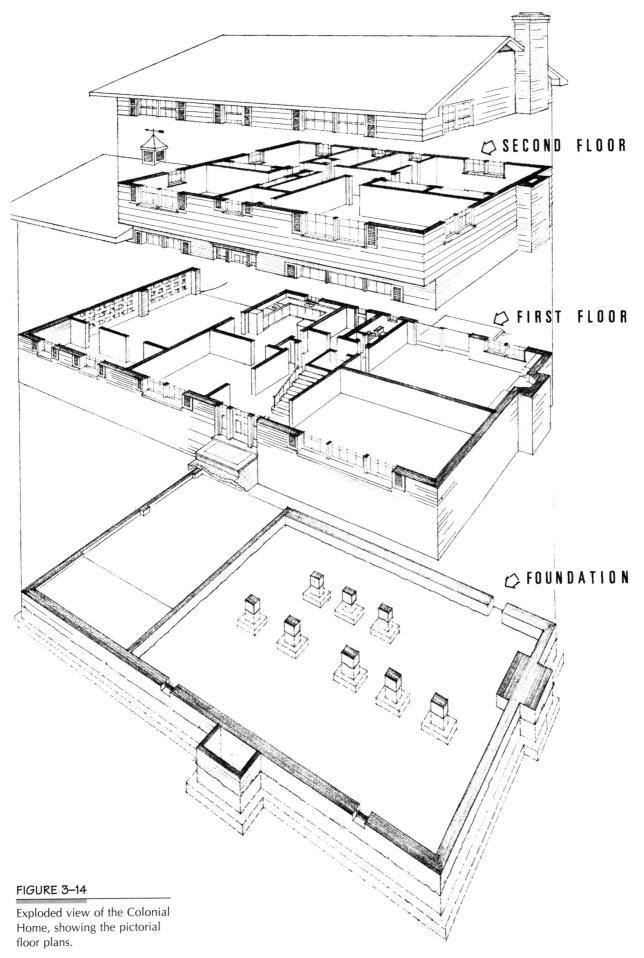

FIGURE 3-14

Exploded view of the Colonial Home, showing the pictorial floor plans.

FIGURE 3–15

Double-joist construction between low partition walls.

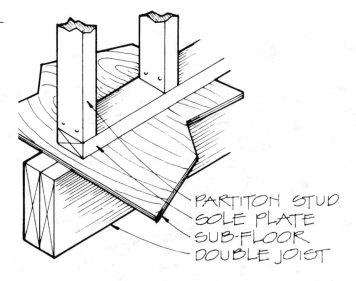

FIGURE 3–16

Continuous X-bridging used in conventional floor framing.

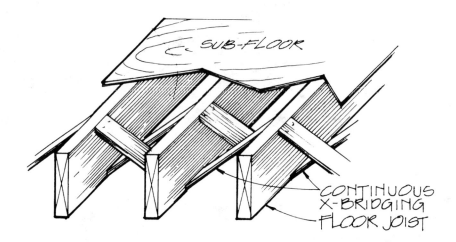

In examining a Foundation Plan in which crawl-space construction has been used, you will notice that usually *piers* are shown throughout the crawl-space area on the plan. These piers, either masonry or wood, are used to support heavy beams or concentrated dead loads under wood floors. Beams (sometimes called girders) support the floor joists, which otherwise would have to be too large and long for practical construction. The columns in basements previously mentioned serve the same purpose as the piers. All beam sizes are indicated with a note, and the piers and columns are located with dimensions to their center lines. If necessary, double joists for the support of walls above are noted.

If the house has a fireplace or chimney, the masonry must continue up from footings in the gound; therefore, it must appear in its correct position on the Foundation Plan and must be dimensioned.

Foundation Plans of buildings with slab-on-ground construction appear slightly different from those with basements or crawl-space construction. If a slab is supported or surrounded with masonry foundation walls, the walls are shown in sections, similar to other construction, and footings are indicated with broken lines parallel to the walls. Areas intended to be covered with slabs are indicated with notes giving the thickness and reinforcement required. Often areas intended for slabs are shown with diagonals through the

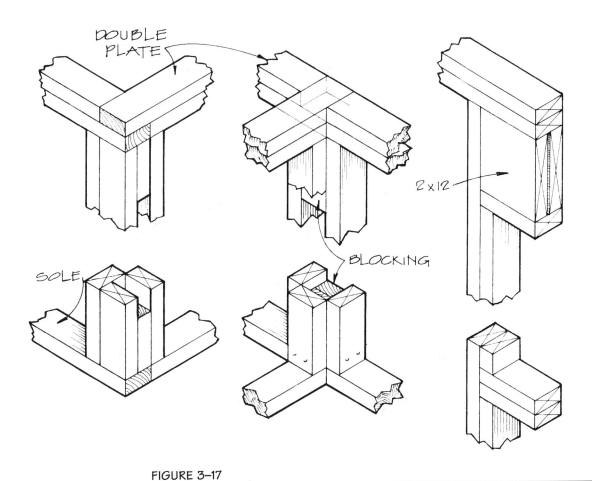

FIGURE 3-17

Construction at corners, intersections of partitions, and window openings in conventional wood frame.

area. If an area of a plan is labeled "Unexcavated," it means that the area is not to be dropped below grade and that a slab is usually intended over it.

Thickened areas of a slab (generally below load-bearing walls) are also indicated with broken lines. Occasionally, concrete piling or sunken piers below the slab are necessary and must also be shown with broken lines throughout the slab area. Complete information about these lines, however, may require further reference to other details. In short, look on the Foundation Plan for information about the foundation walls and the structure of the first floor regardless of the type of construction.

Study the various foundation plans herewith to see how dimensions are placed on the different features. Notice that the Foundation Plan of the Colonial Home (Figure 3–25) indicates crawl-space construction. On it, the concrete-block walls are veneered with brick for a better exterior appearance. To support the beams and first-floor joists, block piers are shown with individual footings, and the joist size, which will properly span the distances between supports, is indicated with a conventional note. Notice that double joists are required directly below major partitions above. Vents in the walls allow ventilation throughout the entire crawl-space area. In the carport and storage area, a concrete floor is specified with woven-wire reinforcement, and the floor in the storage area is 4″ higher than the carport floor. The brick-screen wall is to have a thickened pilaster near its center and a thickened end at the corner for stabilization. Section A–A indicates that the floor slab is to be thicker along the edge. Solid masonry is necessary to build up a foundation for the fireplace and hearth shown on the First-Floor Plan (Figure 3–26).

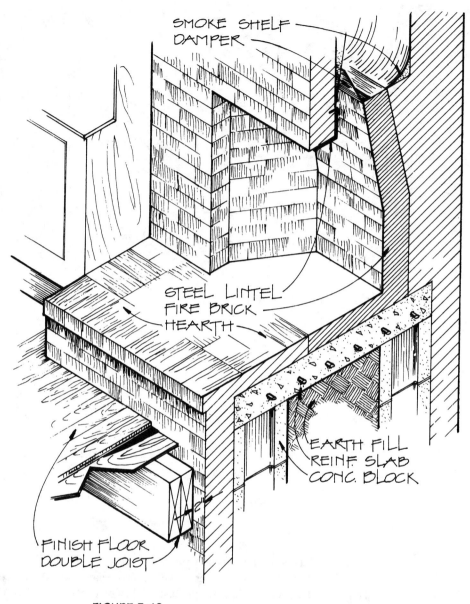

FIGURE 3-18

Section through Colonial Home fireplace.

FIGURE 3-19

Trim and construction at exterior door openings.

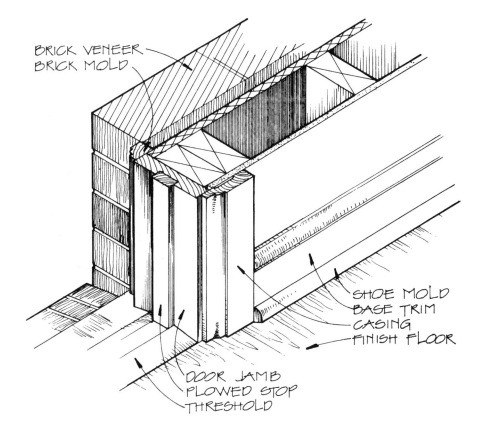

FIGURE 3-20

Construction of cantilevered second-floor joists in front wall of the Colonial Home.

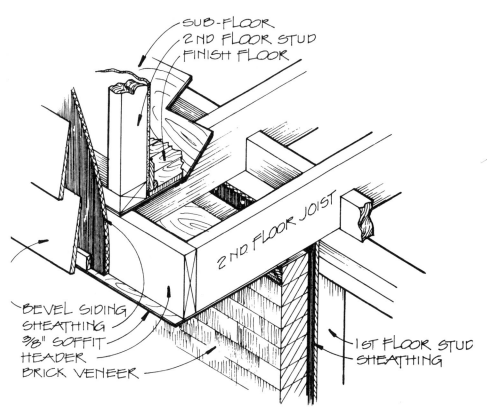

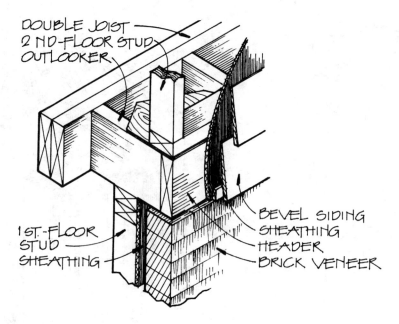

FIGURE 3–21

Second-floor framing necessary to bring the wood siding flush with the brick veneer in the Colonial Home.

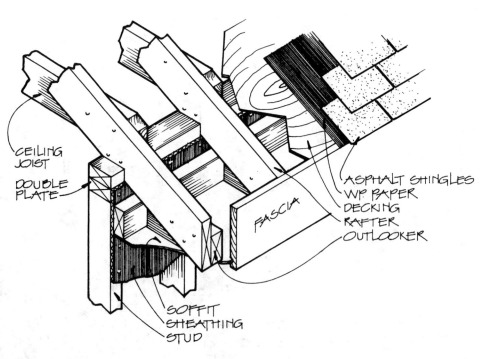

FIGURE 3–22

Construction at the roof overhang.

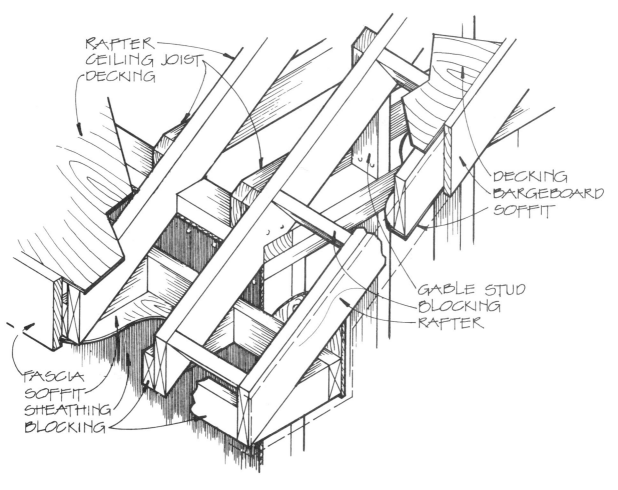

FIGURE 3–23

Construction at corner of the roof overhang.

FIGURE 3–24

Photo of Colonial Home model, showing construction at rear.

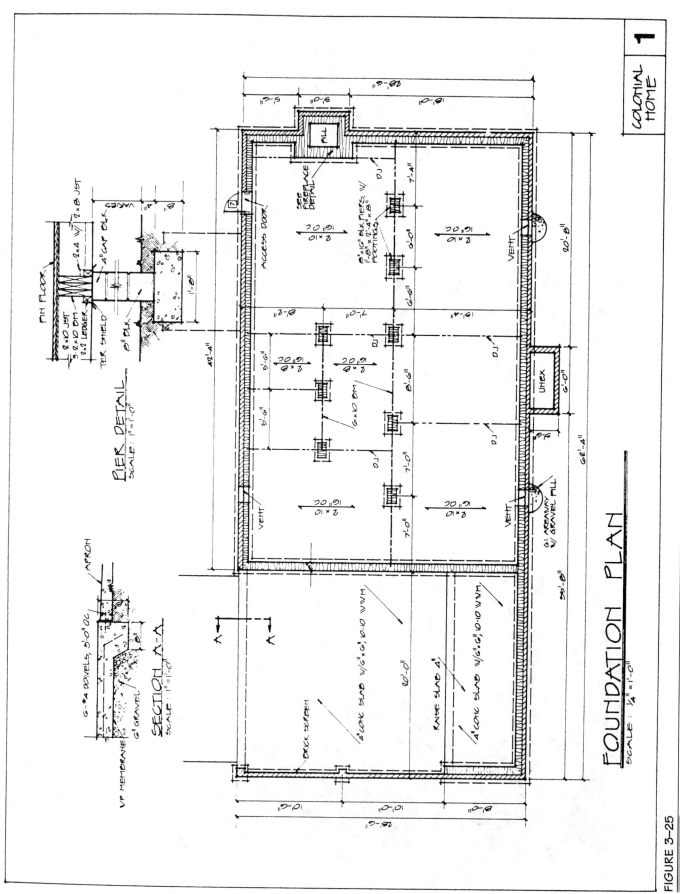

FIGURE 3-25

Foundation plan of the Colonial Home.

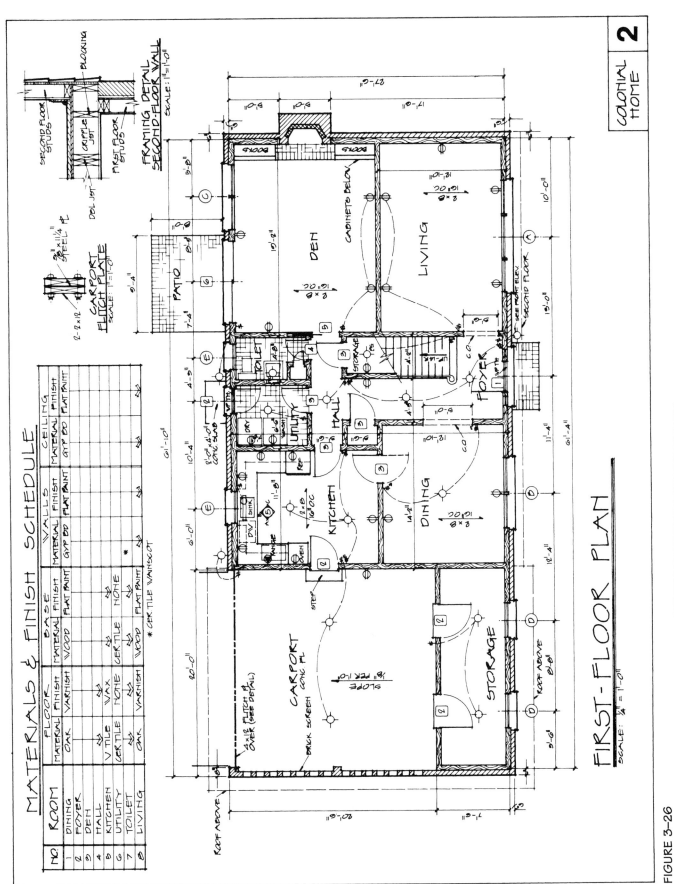

FIGURE 3-26
First-floor plan.

C. FLOOR PLANS

The Floor Plan contains more information than all the other working drawings and is the one most often consulted by workers. It represents considerable time spent by the designer in arranging rooms and traffic patterns, devising door and window placements, and locating the many components within the optimum workability of the entire structure. It could be said that the other drawings merely evolve around this plan and amplify it.

As mentioned earlier, a floor plan is actually a horizontal section view of the building viewed from above. The theoretical plane of the section is about 4′ above the floor so that it conveniently passes through the window and door openings. However, the theoretical plane may be offset in places to include other important features, which may be above or below the plane, to make the plan more inclusive. For example, small, high windows in a wall would be shown even though they fall above the plane; or fireplaces that have openings often only several feet above the floor are shown as though the plane passes through their openings. Regardless of their heights within the story level, important features are represented with symbols, walls are shown in their correct positions and thicknesses, and materials are explained with symbol hatching to make the plan a very informative drawing. Occasionally, broken lines are used to represent features that are obviously higher or lower than the plane, thereby simplifying the impression.

The Floor Plan shows the layout of the rooms, the placement of doors and windows, stairs, fireplaces, chimneys, cabinets, and plumbing fixtures, the approximate locations of lighting outlets and switches, and other structural information. Notice that notes relative to joist framing on a plan relate to the framing *above* the level of the plan. That is, a note on the Foundation Plan pertains to the first-floor framing, whereas a note on the First-Floor Plan refers to the ceiling framing overhead, and so on.

Wood-frame walls are scaled and drawn as close as is convenient in width to include the wood framing and the inside and outside wall coverings. Brick-veneer walls (Figure 3–26) include the brick wythe, wood frame and coverings, and the air space between; yet the symboling gives evidence of only brick and wood frame in the representation. Frequently, no symboling is used on wood-frame walls. These conventions must be understood.

In observing the dimensions of wood-frame walls, you will notice that the extension lines are placed slightly within the wood-frame symbol. This is meant to indicate the *outside* edge of the wall studs so that the carpenters have definite dimensions for laying out the walls. Inasmuch as the brick in a veneer wall is merely an exterior wall covering for the wood frame, the veneering is not usually dimensioned on the floor plan. Other solid masonry, however, is dimensioned from outside edges, which are the surfaces masons work from. Masonry thicknesses are also shown. Thus, when we look at a plan, we can see what material a wall is to be made of and what its length and thickness dimensions are to be.

Similarly, door and window symbols in the walls are scaled the same approximate width as the actual unit, and the symbol gives us an idea as to the type of window or door intended (see Figures 3–43 and 3–46). Although the drafter places symbols as close to their correct positions as possible, exact positions of doors and windows are given by dimensions to their center lines. Also, for brevity, drafters often omit minor dimensions. For example, the placement of doors in narrow halls or closets obviously would not always need dimensions, for exact locations are not critical, and carpenters will center them in standardized positions regardless. Carpenters must calculate the rough openings for windows and doors unless they are given in schedules. Notice that the door symbol shows the direction of the swing. This is important for clearance as well as for the general traffic patterns in the building. Various types of doors, such as double-acting, folding, and sliding, can be recognized by their symbols. Doors seldom swing into hallways; the exception would be closet doors.

Basically, two methods are used for dimensioning wood-frame walls on the plan. The more conventional method shows continuous dimensions from the outside edge of the exterior wall studs to the center line of inside partitions, completely through the building to the outer edge of the opposite exterior wall. The other method shows full thicknesses of walls and dimensions from surface to surface between the walls. This method gives exact inside room dimensions and exact thicknesses of walls in a series of continuous dimensions.

Chimney symbols show the horizontal cut through the masonry, indicating the vertical flue and the intended masonry materials. The fireplace shows the cut through the opening with its firebrick lining; if a flue is required for a fireplace in a lower floor, the flue **symbol is shown off to the side.** Each fireplace requires a separate flue. A masonry hearth, about 18″ wide, is necessary in front of a fireplace opening.

In rooms such as kitchens and baths, which require permanently installed cabinets and plumbing fixtures, the symbols must be carefully scaled in size and placed in their correct positions. Tradespeople installing this equipment depend on the plan for the accurate positions. Bathroom fixtures, in particular, need to be accurately shown so that the layout of the bath will be workable. Plumbing walls (usually near the water closet fixture) are normally scaled 8″ wide to accommodate 4″ vent stacks within. Cabinetwork in kitchens is standardized to fit major appliances and body measurements. Notice that the width of kitchen base cabinetwork is always 2′ wide on a plan and that the wall cabinets above (shown with broken lines) are scaled 12″ or 13″ wide. Refer to detail drawings for additional information about both kitchen and bath layouts.

Stair symbols, which have been first worked out on a detail drawing, represent the actual number of risers necessary to ascend from one floor level to another. On the plan, the total run (horizontal length) of the stairs becomes important in relating upper hallways to those of the lower floors. Most stairs open into hallways. Others, such as basement stairs, often open directly into kitchens and are therefore surrounded by walls and a door at the landing. For obvious reasons, the door swing should not be shown over the stair symbol. Part of the stair symbol is the note indicating the number of risers and the arrow indicating the direction of either ascent or descent.

In many buildings, the stairs are placed one above the other in *stairwells* for efficient headroom layout. Each stair symbol is then shown with a diagonal break near its center to represent partial layouts of the two stairs adjoining each floor level (see Figure 3–56). Refer to the partial stair symbol on the First-Floor Plan of the Colonial Home. If a basement stair had been required, it might have been placed directly below the first-floor stairs, and its symbol would have replaced the storage room space. If railings, newel posts, or other typical stair features are needed, indications of these features would be shown on the symbol. More complete information about stairs is found under Stair Details.

On drawings of small homes, the floor plan also shows the necessary electrical information. Drafters use a small circle as a basic electrical outlet symbol with slight variations of the circle for various types of outlets. An S near a wall indicates a switch. The placement of symbols is approximate, for the electrical contractor must have some latitude in the final placement of outlets and switches. To show which outlet a switch operates, the drafter merely connects both symbols with a broken line. If several outlets are connected with a broken line, it indicates that both outlets are to be operative together. Outlets connected by two switch symbols indicate the use of three-way switches allowing the outlets to be operated from two different points—for instance, at the top and bottom of stairs or at different entrance doors to a room. If an outlet is to be operated from three different points, notice that it is connected with two three-way switch symbols and one four-way symbol. This arrangement provides a very convenient control of the lighting in a room that is accessible from several entrances.

Some houses are now equipped with a *low-voltage switching system* that allows an outlet to be controlled from a number of switches. Master switches, located perhaps in the kitchen and master bedroom, are made to control all the outlets in the house. Other lighting controls are also possible for convenience. The system utilizes low-voltage switching circuits, which make it comparable in cost with conventional wiring methods. On a plan requiring the installation of this low-voltage system, symbols appear similar to those of a conventional system except that notes would be used to indicate master control switches and other switching arrangements.

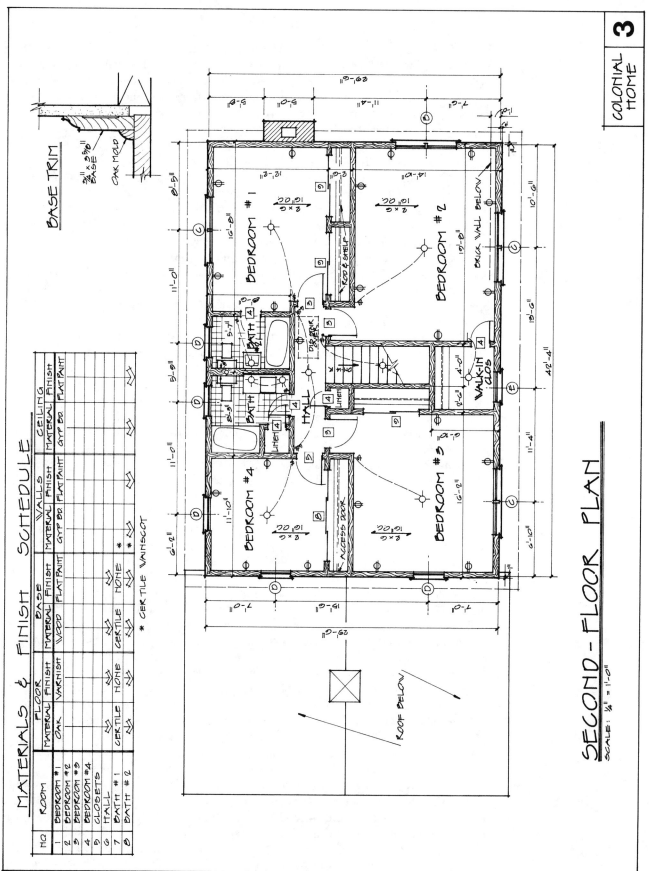

FIGURE 3-27

Second-floor plan.

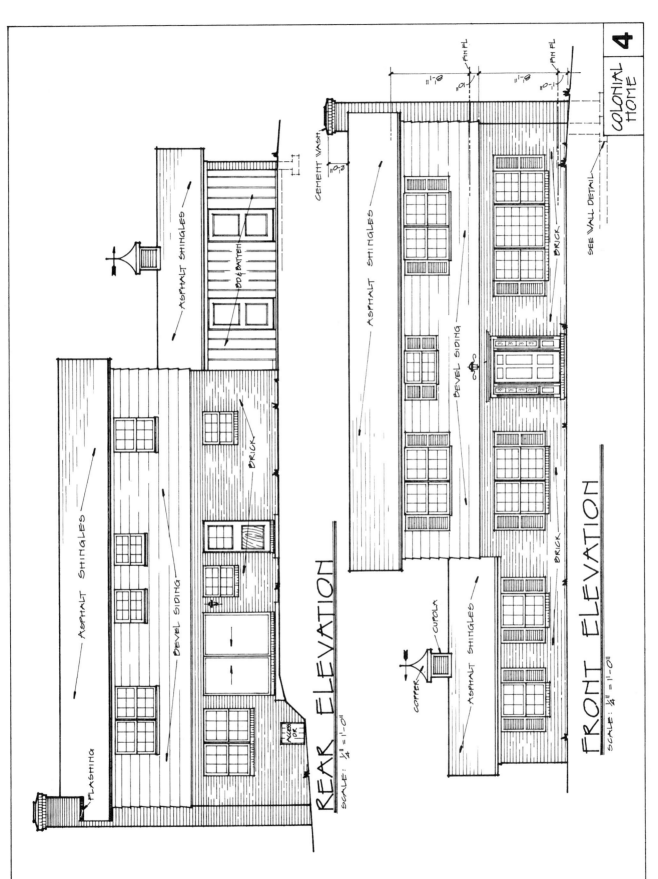

FIGURE 3-28
Elevation views.

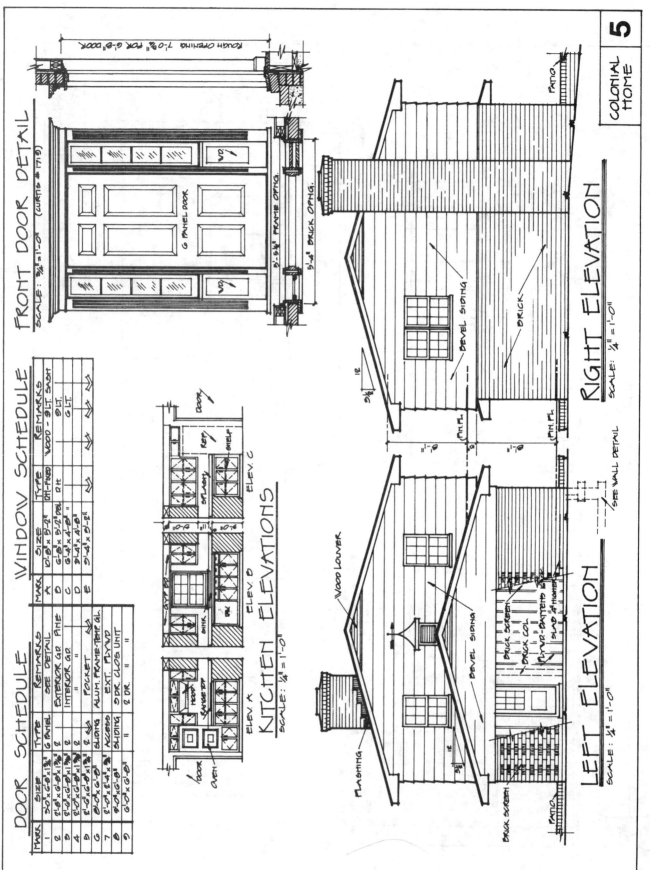

FIGURE 3-29

Elevation views, schedules, and details.

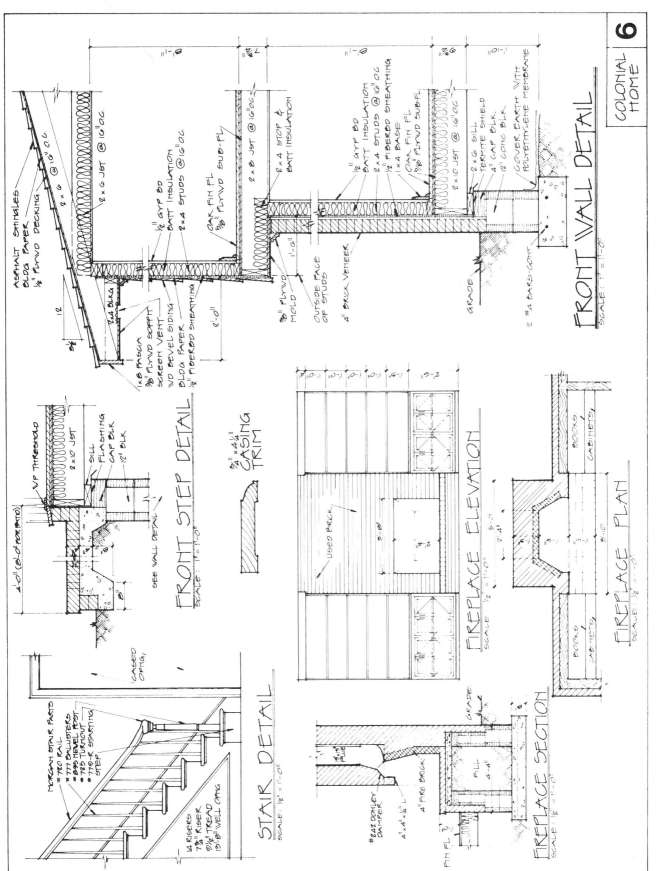

FIGURE 3-30
Wall sections and details.

Even though electrical circuits must be planned by the contractor in house wiring, very little other than the location of outlets and switches is evident on the plan. Occasionally, the location of the *entrance panel box* is established with a symbol and labeled with a note; the box is typically located near the point where the utility lead wires enter the building. Refer to Electrical Drawings later in this chapter for further information.

1. SECOND-FLOOR PLANS

A Second-Floor Plan of a two-story house is similar to the First-Floor Plan except that it shows, of course, the layout of the second-story level. Similar representations are encountered; in reading the upper plan, however, carefully orient the second floor to the first. As a rule, the stairs, outside walls, location of vertical plumbing stacks, chimneys, and similar features will aid in this orientation. Frequently, major partitions are superimposed over partitions of the first floor and windows are superimposed over first-floor windows. Occasionally, one finds a second-floor wall cantilevered beyond a first-floor exterior wall (see Figure 3–20), thus requiring further reference to elevation views or wall sections for positive orientation. If an attic above the second floor is to be accessible, a *scuttle,* or disappearing stair symbol made with broken lines, is shown in the hall. All enclosed structural cavities of this type in wood-frame buildings should have some type of access doors.

Second-Floor Plans of *one-and-a-half-story houses* appear smaller than their first-floor counterparts because roof slopes prevent habitable space throughout. Sloping ceilings and *knee walls* are generally necessary to provide satisfactory living space in these upper levels (see Figure 2–9). The Second-Floor Plan then shows the outline of the knee walls as well as the outline of the full-height walls, thereby forming the enclosed living space. The lower roof beyond the knee wall is also outlined on the plan to show the relative position of the lower walls. *Dormers* must often be used to provide natural light and ventilation to the entire upper floor areas; the walls then outline the dormer areas and show symbols for the dormer windows. Windows in second floors are usually smaller than those in the lower floors, which results in a more pleasing exterior appearance.

2. SPLIT-LEVEL PLANS

A Split-Level House has three or more floor levels connected by short flights of stairs rather than by full-height stairs (see Figure 3–31). As the name implies, two floor levels are *split* by another level between the two. The arrangement allows desirable isolation between rooms of different levels, and this type of house lends itself to a sloping lot with the floor levels somewhat following a sloping grade. Many variations are possible; one variation is called a *Split Entrance* when the front-entrance floor level is between the several living levels. Others may have a basement under part of the plan; still others may have a continuing roof over several levels, producing higher ceilings in some areas—usually the living area.

In reading a floor plan of a Split-Level House, keep in mind the *volumes* of the different rooms, for several levels must be included on each plan. This feature is more difficult to visualize than in conventional plans. The upper-level plan in Figure 3–31 shows the living–dining, kitchen, and entrance areas as well as the bedroom area, which is seven steps above the living–entrance area. The lower-level plan also shows the living, kitchen, and entrance areas even though it is higher than the recreation–garage area. The short flights of stairs between each level provide a clue for easy identification of floor levels when reading this type of plan. When rooms communicate and there is no stair symbol between, we know that they are on the same level. However, when a stair symbol is shown between several areas (ordinarily stairs communicate between halls), we know that there is a change in floor level, and we can tell the approximate change by counting the number of risers in the symbol. Residential stair risers are commonly between 7″ and $7\frac{1}{2}$″ high. Some codes restrict riser heights to 7″ however. The note on the stair symbol, as mentioned earlier, also tells us their direction of either ascent or descent. If a full-section drawing showing the different

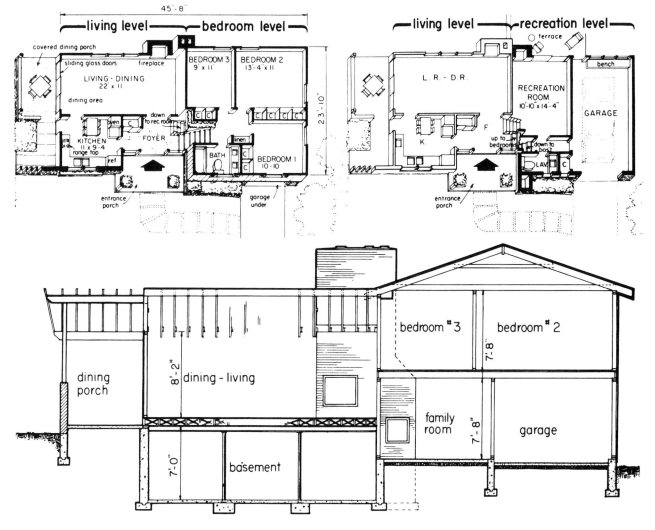

FIGURE 3-31

A Split-Level House.

floor levels is included in the set, take the time to study it carefully for positive floor-level orientation.

3. SCHEDULES

Because schedules are closely associated with information found on the floor plan, they are ordinarily found on the same sheet. These lined enclosures (Figure 3–29) provide a convenient method for organizing extensive information for easy reference. Information on windows and doors is usually given in schedules. Also, information about the desired finish required in each room, such as walls, ceilings, floors, and trim, is shown in a *Finish Schedule*. Numbers or letters are used to identify each item vertically in the first column, and all pertinent information about each item is shown horizontally in the other columns. Lines running through spaces in the columns indicate a repeat of the "ditto" abbreviation. If no Finish Schedule is shown, such information as finish-floor materials is given in the form of a note directly on the plan drawing.

4. LIGHT-FRAME CONSTRUCTION

The largest percentage of American homes by far are built mainly of wood. Because wood is abundant and easily worked, it has become the most economical material for small buildings through the years. Much of the construction has become standardized (even though houses vary in design and size), with several types of wood construction being the most universal. The most popular type is known as *Platform Frame Construction,* also known as "Western Frame Construction" (see Figure 3–32). Notice that the wall framing is placed upon the subfloor and that each story has a separate set of studs. Another standardized type of construction, known as *Balloon Frame,* often called "Eastern Frame," is used mainly for multilevel wood construction (see Figure 3–33). Its identifying features are the studs resting directly on the sill plate and continuing uninterrupted throughout the height of the structure. Although the continuous studs minimize the amount of vertical shrinkage in walls, long stud members are more costly and more difficult to handle; hence platform frame has become more popular in current construction.

Another method, called *Post-and-Beam Construction* (Figures 3–34 through 3–41), utilizes heavier but fewer structural beams in the floors and roofs with thick planks over them to span the greater distances. The concentrated loads on the beams are transferred directly to the foundation by the use of posts or columns in the walls below them, a concept similar to the one used in large steel-frame buildings. The system lends itself to contemporary construction where generous glass walls are used and exposed-beam ceilings are visible. Floor plans must show the location of all beams in floors and ceilings (see Figure 3–36). Together these types of construction, as used in home building or in small commercial buildings, are commonly referred to as *Light Wood-Frame Construction.*

Basically, as is commonly known, wood-frame members are chiefly fastened together with nails to form the structure of the floors, walls, and roof. Structural members, other than heavy beams, are mainly 2″ planks of various widths conventionally spaced 16″ apart for modular convenience. *Joists* are the members found in the floors and ceilings, *studs* are the members in the walls, and *rafters* are the members in the roofs. Wood boards, plywood, or other sheet materials are nailed directly to these structural members to give them rigidity and enclosure.

Framing members, when erected in a parallel fashion, such as the studs in a wood wall, are subject to lateral forces that the parallel arrangement in itself has little resistance to. This tendency for the parallel members to collapse when a force such as wind pressure is applied against the upper line of the members is called *racking action,* and special bracing known as corner bracing must be used to maintain a rigid structure. Subwalls and other materials nailed to the studs partially reduce this racking tendency, yet approved corner bracing is required by codes. Commonly, 1″ × 4″, let-in (cut into the studs) braces, which are nailed at 45-degree angles the full height of the wall, are used (see Figure 3–32). Other techniques that satisfy local codes are used to ensure a rigid structure. Some codes allow the use of $\frac{1}{2}''$

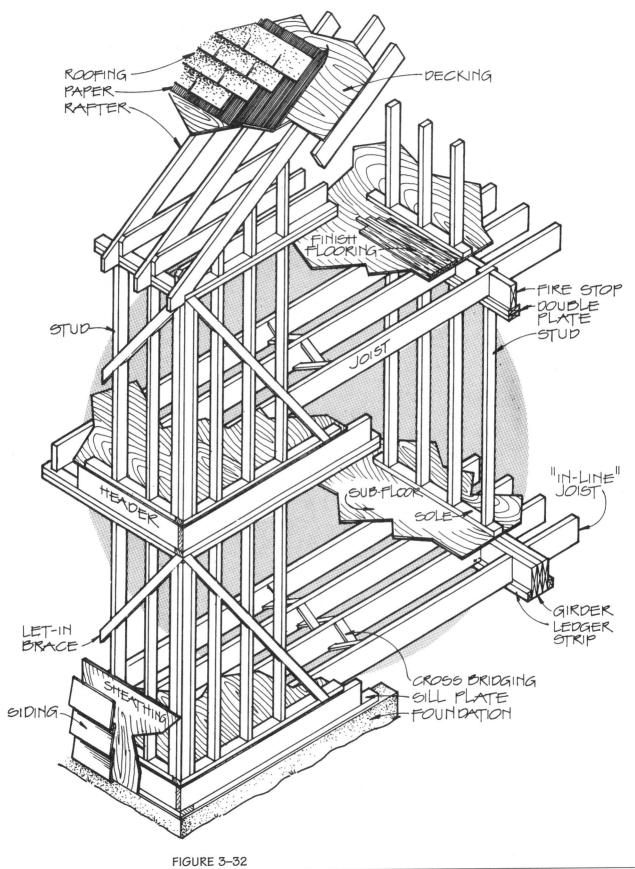

FIGURE 3–32

Platform Frame (Western Frame) construction. Notice that each floor has a separate set of studs.

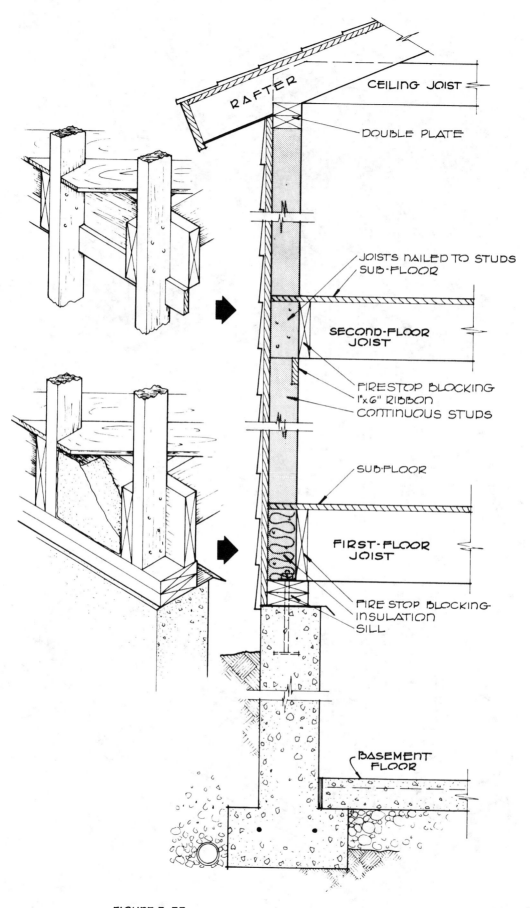

FIGURE 3-33

Balloon Frame (Eastern Frame) construction, showing continuous studs from sill to top plate.

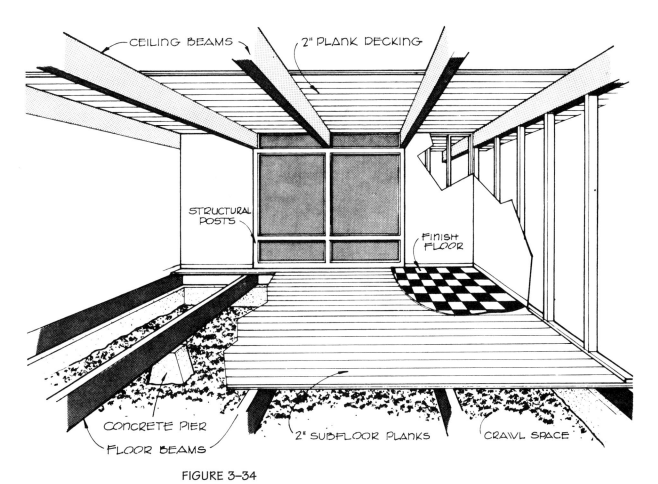

FIGURE 3-34

Post-and-beam floor and ceiling framing.

plywood panels, securely fastened at all corners and extending from the top plates to the lower plate. Metal plates fastened at the corners may also be used for this purpose.

Then various materials are applied over this skin to provide the exterior characteristics needed for weather protection, endurance, wear, or beauty. Inside wall surfaces may be plaster, gypsum board, paneling, or prefinished plywood; any one of these materials is applied to the interior of the stud walls.

Window and door units are fastened into wall openings, and electrical wiring and plumbing piping are run through the wall and floor cavities to conceal them from view. Heating and cooling ductwork is also placed within the cavities insofar as possible. The structure, of course, must rest on a solid foundation, usually constructed, as mentioned previously, of concrete block, poured concrete, stone, or brick. In simple terms, this is how most small wood buildings are constructed. Variations in materials and minor details are reflected in many of the drawings found throughout this manual. The more familiar you become with standard construction methods, the easier drawings will be to read.

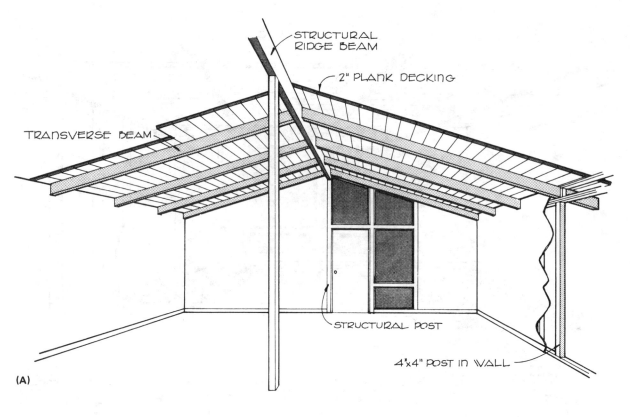

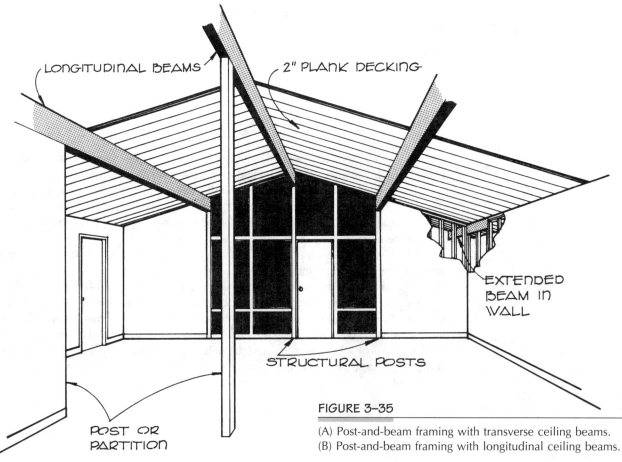

FIGURE 3-35

(A) Post-and-beam framing with transverse ceiling beams.
(B) Post-and-beam framing with longitudinal ceiling beams.

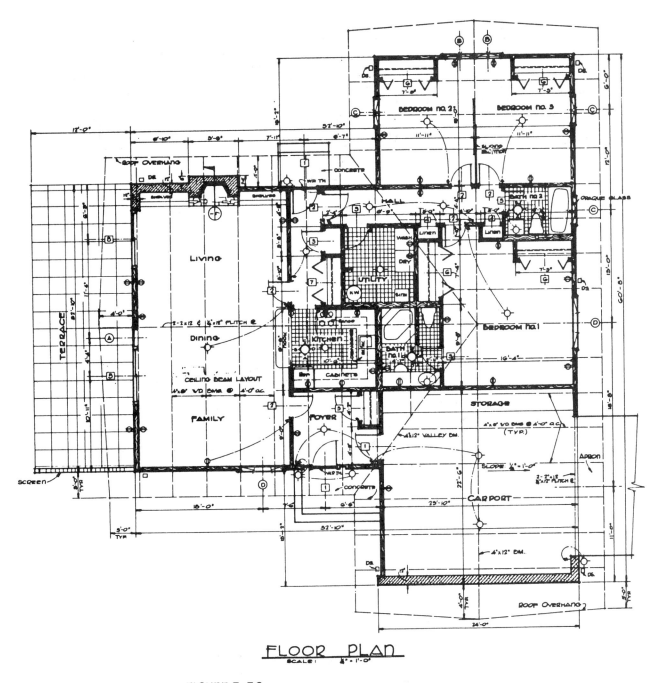

FIGURE 3-36

Floor plan of a post-and-beam dwelling.

FIGURE 3–37

Front elevation of the home shown in Figure 3–36.

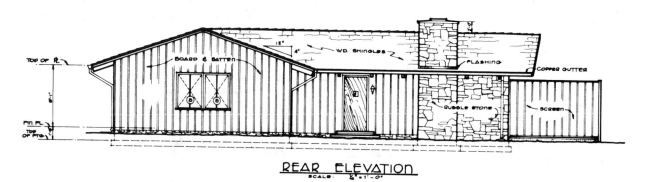

FIGURE 3–38

Rear elevation.

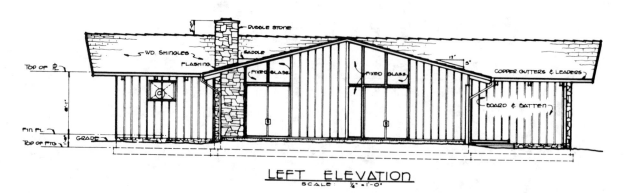

FIGURE 3–39

Left elevation.

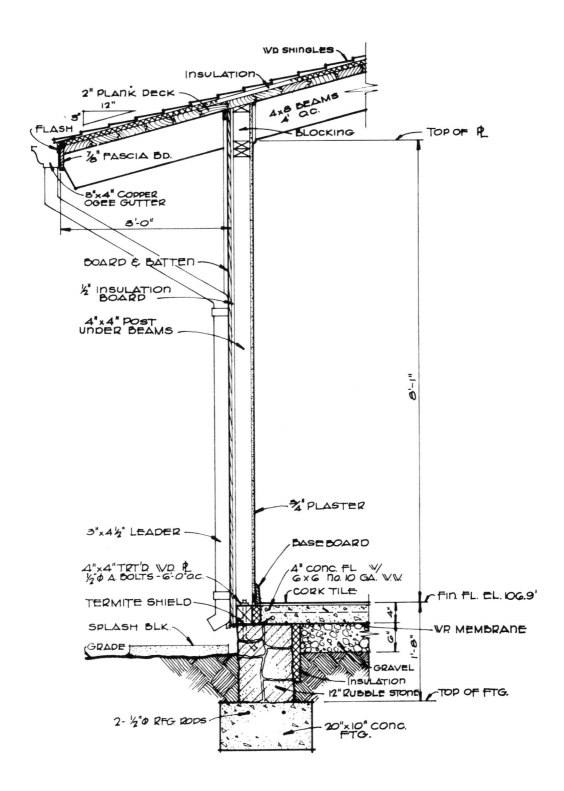

FIGURE 3-40

Wall section of the home shown in Figure 3-36.

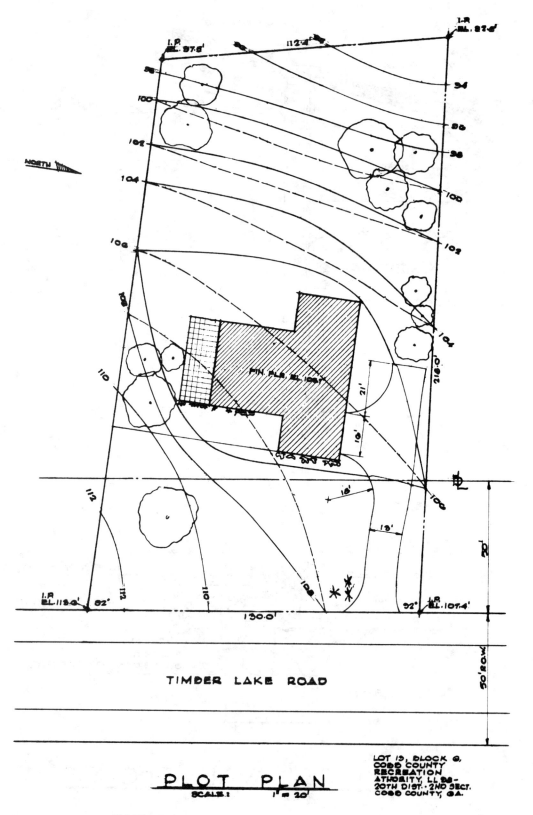

FIGURE 3-41

Plot plan of the home shown in Figure 3-36.

5. READING THE COLONIAL HOME FLOOR PLANS (FIGURES 3–25 TO 3–30)

Looking at the First-Floor Plan (Figure 3–26), we see that the major outside walls are brick veneer and that the interior partitions are wood frame. If we enter through the front door, the foyer provides a separate area and allows us to enter the living room on the right or the dining room on the left, through cased openings. The note and broken line indicate that these openings are to be cased out with trim. We may also go directly upstairs to the second floor from the foyer, taking the colonial-type stairs as shown by the symbol, or we may proceed straight ahead and enter the back hall to the kitchen or den. The brick fireplace in the den shows cabinets and bookshelves on each side; an outside patio is reached through sliding glass doors. The pocket-door symbol between the hall and the den indicates that the door slides into the partition rather than swings into the room. We find a linen closet in the small toilet, and the utility room with washer and dryer has a rear entrance door with a concrete stoop. The U-shaped kitchen has base and wall cabinets with major appliances built in. In going through the kitchen to the carport, a step is shown, which tells us that the carport concrete floor is lower than the kitchen floor. Lines through the storage room doors indicate that there is a change in floor level here also. Concrete garage floors conventionally slope slightly toward the entrances, thus allowing any water dripping from parked cars to drain to the exterior.

We see by notes throughout the plan that the ceiling framing is to be 2″ × 8″ joists spaced 16″ on center; directions of the arrows on the notes tell us in which direction the joists are to be installed—usually the direction of the shorter span. The broken line parallel to the front wall of the house tells us that the second floor is to cantilever beyond the first floor, so we must remember this fact when reading the Second-Floor Plan.

Looking at the Finish Schedule, we can quickly tell what floor and inside coverings are required in each room. Dimensions are shown from the outside edge of the stud framing on exterior walls to the centerline of interior partitions, which is conventional. The 6″ dimension at corners of the veneer gives a relationship between the dimensions of the stud framing and the outside masonry on the foundation. We can locate placements of windows by the dimensions to their centerlines. All electrical outlets and switches are shown with symbols. Notice, for example, that the kitchen lights can be operated from two different switches.

Symbols showing the locations of mechanical equipment, such as furnaces, cooling units, and register outlets for both heating and cooling, are normally placed directly on the floor plans of small buildings. If this information is extensive, however, a separate floor plan, called a Heating and Cooling Plan, is used. (It will be discussed later.) No heating or cooling information is given on the Colonial Home Plan.

When you begin reading a plan, first orient yourself (possibly begin at the front door as we have) by making mental notes of the major features as you progress from one room to another. Place yourself within the reality of the building and visualize the completed construction as you observe the symbols on the plan. The ability to visualize is part of reading drawings. Even the north-point arrow shown on the Plot Plan will prove helpful in keeping the compass directions oriented with the plan. Remember that a plan view is conventionally shown with the *front wall down* near the bottom of the sheet (see Figure 3–26).

In relating plan views and elevation views, we might say that the plan shows *interior* information and that the elevations show *exterior* information. Horizontal measurements on elevations are direct projections from their plan views. If necessary, find a door on the plan, note its identification mark, and then find the same door on the elevation, or take other prominent features from one drawing and orient them to the other. It is helpful to visualize ridges and roof shapes lying over the plan during orientation. But the mental image of the exterior appearance of the building is necessary when reading a plan, even if it becomes necessary to clarify some points by placing the corresponding elevation adjacent to the wall of a plan and by referring to other detail drawings shown elsewhere in the set.

D. ELEVATIONS

As we have already seen, elevations are the drawings that show the true exterior appearance of the building. They are usually drawn to the same scale as the plan and thus are orthographic views of each side, giving us exterior "pictures" of the house. Doors, windows, roof shapes, cornice treatment, trim, chimneys, and exterior materials are all shown as realistically as convenient on the elevations so as to provide the builders with a finished appearance. Few dimensions are given on these views; only those vertical dimensions that, of course, cannot be shown on the plan are given.

In general, four elevations are needed in each set for a complete exterior description. More complex buildings, on the other hand, may require more (see the Contemporary Home drawings).

Outlines of footings and foundation walls are mainly below grade on elevation views and hence are shown with broken lines. The heavy *grade line* indicates the approximate level of the ground adjacent to the house. Heavy dashed lines at the floor levels are used to relate the interior finish-floor levels with exterior features. The lines should coincide with the bottom of exterior doors. If the floor line is near grade and no basement is shown, possibly the house is to be built on a concrete slab. If it is up several feet above grade, then crawl-space-type construction is probably required. Wood-floor framing must be placed well above ground to prevent its eventual destruction in most areas of the country.

Floor lines above grade require exterior steps leading up to the level of the doors. Exterior step risers are usually 61/2″ to 7″ high, slightly lower than interior steps. Notice the relationship between the finish floor and the grade line in the Front Wall Detail of the Colonial Home (Figure 3–30). Elevations of wood-frame construction also show *plate height* (ceiling) of the walls. Standard room heights for small homes are slightly over 8′, which allows economical use of 4′ 8′ sheeting materials and gives carpenters a basis for cutting stud lengths. Notice that the window and door heads are conventionally placed at the same level to unify the exterior appearance and simplify the wall framing.

Occasionally, the elevations are shown at a smaller scale than the plans. The reason may be because of the space restriction on the sheet, or possibly the drafter felt that the elevations were not complex enough to require larger views. Although the scale of the elevations may vary from that of the plan, their relationship is nevertheless the same. Be careful to observe the scale of drawings before orienting various views.

Elevations also show important roof information. Various types of roofs are shown in Figure 3–42. Profiles of gable roofs are accurately drawn on the elevations and projected to the other views. The ridgeline forms the highest part of the roof. Roof slopes have direct relationships to the type of roofing specified, to the roof framing, and to the general appearance of the building. Some roofing materials cannot be used on low-slope roofs. Slopes are usually indicated with a slope-ratio diagram (see Figures 2–23 and 2–24 on roof slopes) adjacent to each different roof profile. Roofing material is indicated with a note.

Generally, louvers are necessary to ventilate the roofs of wood-frame houses. This is an important consideration because the area directly below roof decks must have continual ventilation. The typical wall section may also show vent openings under the eaves to supplement the ventilation of the louvers. Some roofs may even require vents placed directly on the roof. Other information about the roof layout may be found on a separate Roof Plan, if one is included in the set.

Window symbols shown on elevation views closely represent the character of the windows intended (see Figure 3–43). Sizes are given in window schedules. Many different types of residential windows are available in today's market. The major ones are listed next so that you can identify them when you see the symbol.

> DOUBLE-HUNG or vertical slide window is the most widely used in this country; it has two sashes that operate vertically, allowing a maximum of 50 percent ventilation. Windows placed adjacent to each other are divided by a structural member known as a *mullion* (see Figure 2–28). Divided lights (small panes) in windows are usually associated with colonial exteriors.

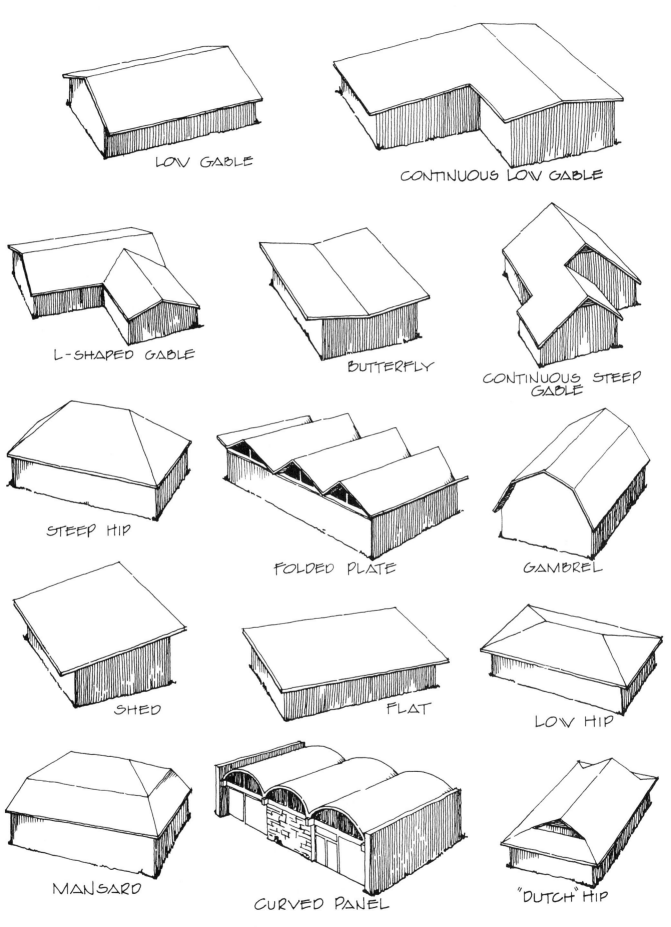

FIGURE 3–42

Residential roof types.

FIGURE 3–43

Typical residential windows showing their symbols on elevation and plan views.
(A) double-hung,
(B) casement, (C) sliding,
(D) awning, (E) hopper and fixed, (F) basement,
G) bow or bay.

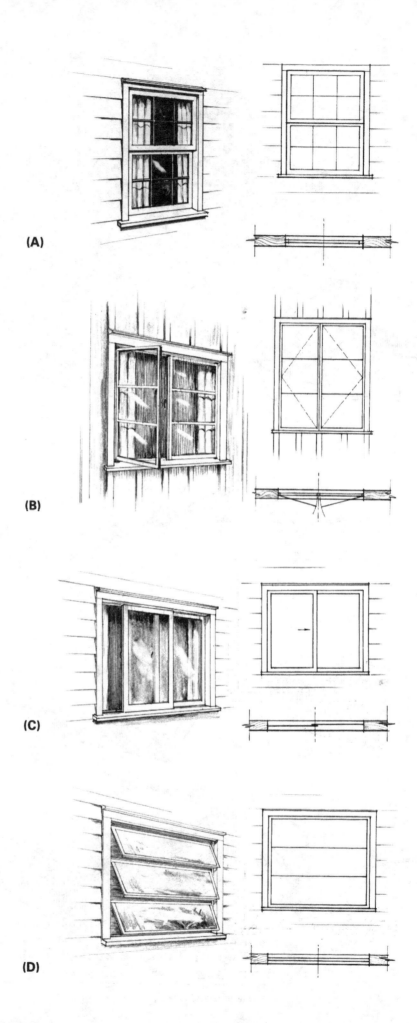

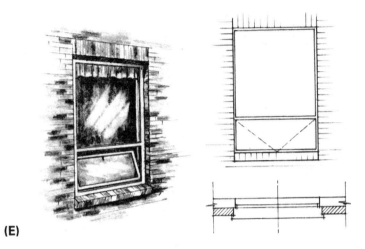

(E)

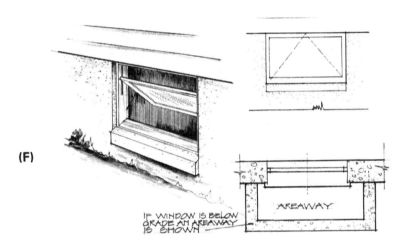

(F)

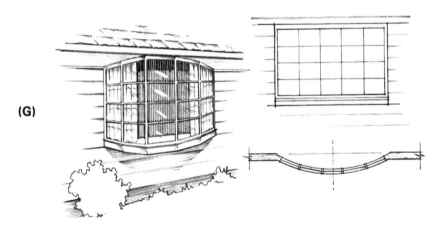

(G)

FIGURE 3-43

Continued.

CASEMENT has its sash hinged on the side and can be opened either in or out; both types have their operating hardware on the inside. They allow 100 percent maximum ventilation.

HORIZONTAL SLIDE has sashes sliding horizontally like sliding doors. Similar to double-hung windows, they allow only 50 percent maximum ventilation. Units are often combined with fixed, picture-window sashes.

AWNING has its sashes hinged at the top so that they open out horizontally, thereby allowing 100 percent ventilation as well as protection from rain even when open. Interior cranks and hardware operate all the sashes in each unit simultaneously.

HOPPER has a small sash hinged on the bottom that usually opens in. Hoppers are generally combined with the larger, fixed-glass units found in many contemporary homes.

JALOUSIE has small, frameless glass panels that operate simultaneously with a crank similar to awning windows. They are seldom weathertight and therefore are mainly used on enclosed porches or in homes in warmer climates.

SKYLIGHTS are actually roof windows that are fixed into the roofing of roofs of little or no slope. These plastic "bubbles" are commonly placed over interior rooms that have no natural lighting. As a rule, their symbol is made with broken lines shown on the upper floor plan and a note indicating their description.

FIXED GLASS are windows affixed directly to the framing, which allow large window areas to be installed with minimum expense. No opening, hardware, or screens are required, yet opening sash units can be combined with the fixed glass to provide ventilation if necessary.

All types of residential windows made from wood, steel, or aluminum are available. The term *fixed glass* indicates glass panels that have been set directly into the rough framing and trimmed out to resemble other window units. Although fixed glass is more economical than window units purchased and set into the rough framing, it does not allow ventilation into the interior unless movable sashes are combined with it.

Window symbols on elevations are not always to exact scale, for only their general appearance is necessary; dimensions are more important on the plan. On some symbols, diagonal broken lines are used to indicate the swing of the opening sash; the converging point side of the symbol is the hinged side (see Figure 3–44). *Rough-opening* sizes given in a window schedule indicate the sizes of the rough framing into which the window units are to be placed.

Door symbols in elevation, like window symbols, are drawn to show their general appearance (Figure 3–45). Standard, exterior, residential doors are 6'-8" or 7'-0" high and vary from 2'-8" to 3'-0" wide. Double-entrance door units often measure 5' or 6' wide, however. Generally, *flush* or *panel* doors are used on exteriors, and these and other types in interiors (see

FIGURE 3–44

Symbol for swing of window sash.

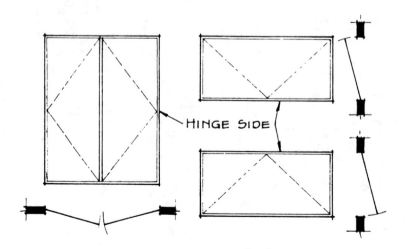

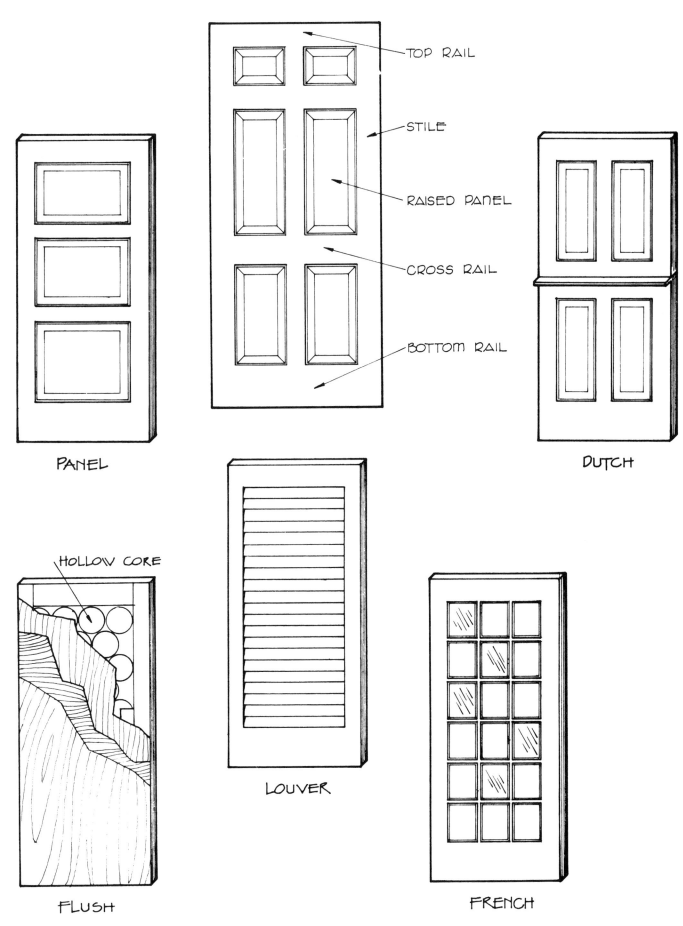

FIGURE 3–45

The popular six-panel door and other typical residential doors.

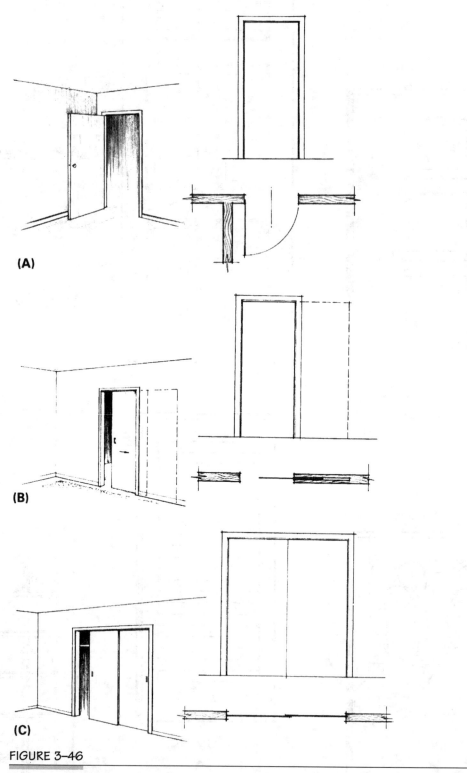

FIGURE 3-46

Typical residential doors showing their symbols on elevation and plan views. (A) Flush, interior-swing, (B) pocket, (C) sliding closet unit, (D) accordian, (E) bifold.

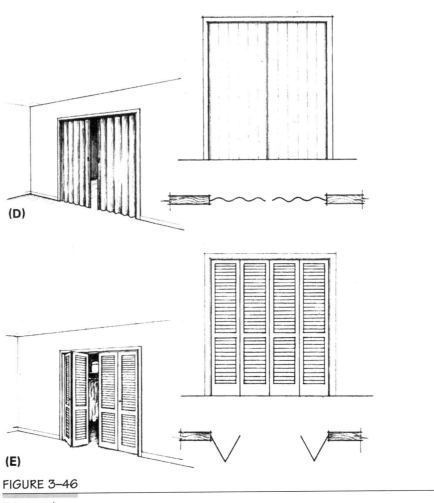

(D)

(E)

FIGURE 3-46

Continued.

Figure 3–46). Exterior doors are ordinarily specified $1\frac{3}{4}''$ or even $2\frac{1}{4}''$ thick to provide insulation and stability; interior stock doors are generally $1\frac{3}{8}''$ thick. No hardware symbols are shown on doors in elevation.

Dormers of one-and-a-half-story houses must be carefully shown on elevations. Placement of the dormers on the roof must agree with interior room heights so that their windows will be satisfactory and the framing can be worked out. Roof treatment and trim on dormers are accurately represented on elevation views and are usually similar to the treatment found on the other main roofs.

1. READING THE COLONIAL HOME ELEVATIONS

In interpreting the elevations of the Colonial Home (Figures 3–28 and 3–29), notice that horizontal linework represents brick and that brick sills are shown below lower-level windows and doors to shed water from the masonry surface. The wide bevel siding is also shown with horizontal lines; yet the lines are farther apart than the brick to give the symbol realism. Notes also clarify the major material symbols. On the Rear Elevation, vertical lines allow us to distinguish the board-and-batten siding required on the interior of the carport from the other symbols. Notice that the same symbol shows the material on the Left Elevation where the brick screen has been omitted. Broken-out portions on views are frequently necessary to show important features that are otherwise hidden. By referring back to the floor plan we see that a solid brick column (pilaster) is required in the screen wall, shown by the vertical broken lines, to give the screen more stability. A dashed line also shows the exact height of the storage room floor. The brick screen provides privacy, yet allows ventilation through the carport. Vertical dimensions give the exact heights of finished floors to the plate levels of both the first and second floor. The chimney height is shown to be 2′ above the upper ridge on the Front Elevation, which is standard procedure if a complete chimney detail is not included. The elevation views also show the masons how the chimney cap is to be finished. Other minor details about louvers, cornice returns, corner treatment, and approximate grade levels are easily seen on these views.

E. SECTIONS AND DETAILS

Sections and various other smaller drawings found in a complete set are commonly known in broad terms as *details*. They are usually drawn at larger scales than the plans and elevations, thereby showing the size, shape, and arrangement of individual structural members within the construction more easily. Plans and elevations show the composite of the building, but the details show, as well as dimension, critical points throughout the construction. Some details of conventional construction that are typical to many similar buildings are known as *standard details;* others, depending on the complexity of the dwelling, are special to each job and are conceived for that job only. The architect utilizes his or her knowledge and background to determine which details are necessary and which should be included in each set of drawings. Almost every set for a custom-designed building would undoubtedly contain details unlike those of other sets. Details then provide isolated information about the construction, whereas the plans and elevations show how all the details are related.

Notice that many sections and details contain *break* symbols for the purpose of deleting areas of little importance, especially in long members. This convention on structural details allows the important part of each detail to be drawn larger, thereby possibly cutting down on drafting time. As a rule, dimensions shown running through the deleted areas give the true size, if such is necessary. However, try to visualize the true length of each member when reading details with the breaks in them.

The first important requirement in reading sections is knowing *where the section is taken from*. Usually, a cutting plane relating it appears somewhere on a plan or other larger drawing. Section identification symbols (Figure 2–18) provide this relationship. If a section is titled "typical," no cutting plane is necessary, for the detail obviously relates to uniform construction throughout. Most sets of drawings contain a *typical wall detail* (see Figures 3–47 and 3–48), which are vertical sections through a typical wall showing all the information from the footing to the roof. Actual dimensions of Norwegian brick are shown in Figure 3–49. Notice that the important points of the section are (1) the lower foundation and footing area, (2) the intersection of the foundation and floor, called the *Sill Detail,* and (3) the intersection of the roof and wall, called the *Cornice Detail* in wood-frame construction. These are critical areas during construction, and occasionally the three parts of the wall section are shown as individual drawings and isolated on the sheet.

1. READING A TYPICAL WALL DETAIL

In reading a wall detail, we see at the bottom that the footing is generally of poured concrete made wider than the foundation wall to disperse the weight of the building over a wider soil area for stability. If uneven settling of the foundation is anticipated, reinforcing rods are introduced into the footings near the top. The depth at which footings are placed below ground depends not only on the stability of the soil but also on the *frost line* in the locality. The bottom of the footing must be below the deepest penetration of frost. Frost depths vary from a few inches in southern regions to 5′ or 6′ in the extreme north. Freezing and thawing of the soil below a footing would soon rupture the foundation and eventually the entire building. Check the local codes in your area for positive frost-line depths. A note, "Place Footings on Undisturbed Soil," specifies that the footings must rest on naturally stable soil and that no fill dirt may be used below them.

Foundation walls in residences may be either poured concrete like the footings, or they may be constructed of any of the structural masonry materials. As mentioned previously, concrete block is the most universal. Many block foundations have brick or stone veneer-

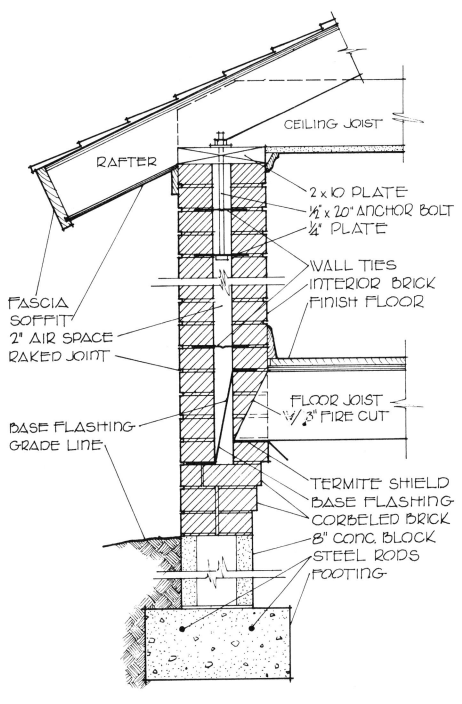

FIGURE 3-47

Brick-cavity wall section.

ing above grade to improve their appearance. If moisture is an anticipated problem, block foundations are waterproofed, as shown by a note on the detail. Round drain tile placed near basement footings on some details is used to relieve hydraulic pressure from forcing moisture through foundation joints onto the basement floor. The tile is surrounded with loose gravel and slopes to an outfall away from the building.

Conventional basement floor slabs are 4″ thick, showing a waterproof membrane directly below and a 6″ gravel or cinder fill below the membrane. The slabs usually bear directly on the footings. Between the slab edge and the wall, expansion-joint material provides a flexible joint. Expansion joints are evident on many masonry details. Long

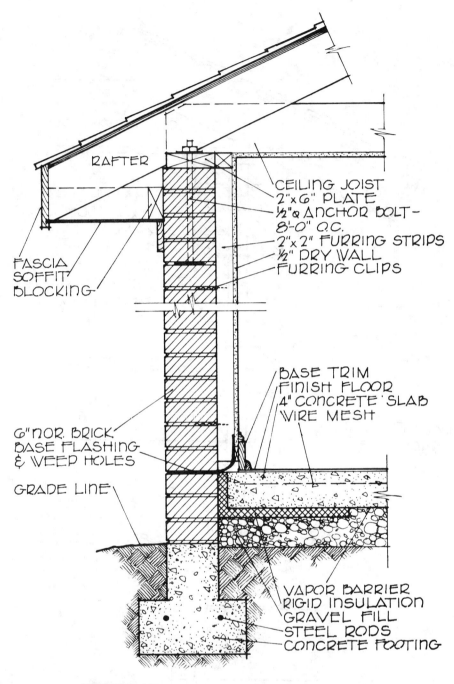

FIGURE 3–48

Typical wall section using 6" Norwegian brick.

lengths or widths of solid masonry must have these joints to allow for expansion and contraction during temperature changes.

The grade-line indication on wall sections is only approximate and may vary somewhat at different points around the building. Similarly, if the footing depth varies because of stepped footings below the wall, the typical wall section shows a break symbol to make it applicable to any of the various depths.

Above the foundation, the *Sill Detail* reveals the arrangement of the sill members and the manner by which the floor structure is attached and bears on the foundation. Good construction shows a metal termite shield between the foundation and any wood members. In

FIGURE 3-49

The 6" Norwegian brick is sometimes used in small commercial buildings of only one-story height.

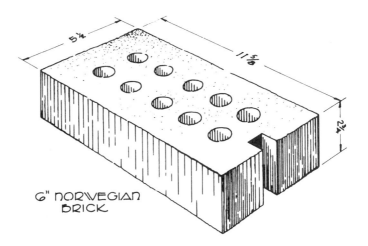

nontermite areas, this shield may not be needed; however, wood is vulnerable to moisture in moist regions when it is attached to masonry near the ground. Chemically treated lumber is generally specified at this critical point to prevent destruction. To level the foundation before the sill plate is attached, $\frac{1}{2}''$ of cement grout is often indicated on some details. Anchor bolts, which secure the wood frame to the foundation, are spaced 6' or 8' apart and are anchored well into the masonry. At least two bolts must secure each sill piece.

In platform-frame construction, the arrangement of the sill members is known as a *box sill*. It shows the floor joists resting on the sill plate with a header attached to the ends of the joists, and the subflooring fastened over the joists and headers (see Figure 3–32). Notice that the wall framing goes over the subfloor. At least 3" of bearing under the ends of the joists on the plate must be provided to carry the weight of the floor and framing above. Various details show varied sill-plate sizes.

In balloon-frame construction (Figure 3–33), the wall studs rest directly on the sill plate, and the floor joists lap against each stud—both resting on the plate. Although not as commonly used as the box sill, this arrangement reduces vertical shrinkage of wood walls by having less cross-grain wood within the height of the walls. (Cross-grain wood shrinks more than end-grain.)

Sill details also show the finish-floor material, interior wall covering and base trim, and outside wall sidings. If the wood floor is built over a crawl space or basement, vertical dimensions of these spaces are given.

Some details show slab-on-ground construction. These floor slabs are usually 4" thick with wire-mesh reinforcement within and a waterproof membrane directly below. Gravel or cinder layers below the membrane prevent continual moisture contact with the underside of the slab, as well as provide a leveling bed. Insulation or expansion-joint material is placed around the periphery of the slab if it abuts the foundation. In colder climates or if heating ducts are placed in the slab (see the Contemporary Home wall section), thicker insulation is necessary at the edges and several feet under the edges to prevent heat loss. No insulation is needed below the slab throughout its interior, since negligible heat is lost to the ground itself. Slabs can be "floating," with no foundation support below, or they can be shown resting on a portion of the foundation. Either is satisfactory construction as long as one type is used consistently around each slab. In termite areas, metal shields should be cast into the slab and foundation to seal this important junction from the ground below. A termite shield is shown with a heavy line in the detail.

Metal flashing, also shown as heavy lines in a section, is seen frequently in masonry details. In masonry walls with air cavities, the flashing directs any condensation that may form within the cavities to the exterior rather than into any interior construction. Often weep holes are indicated in outside courses to aid in this release. If wood members are set into masonry, flashing prevents direct contact between the wood and masonry. Many other situations require flashing to control moisture in masonry.

On the upper part of a wall section, the *Cornice Detail* shows the construction at the intersection of the wall and the roof. True heights of the wall from the floor to the roof are seldom drawn in the detail. Break symbols are commonly employed, but the vertical dimensions that are given reveal full height. In wood-frame construction, a double 2″ × 4″ plate is placed above the studs. This double plate provides a rigid support for the walls, allowing convenient overlapping of the plate at wall corners and at intersecting partitions (see Figure 3–17). It also provides a solid bearing for roof rafters above. Notice that a "bird's-mouth" notch is cut from the rafters of gable roofs to provide a horizontal bearing on the plate. Ceiling joists, fastened to both the rafters and the plate at this point, help prevent outward thrust of the walls from the weight of the roof. This conventional construction produces a solid, triangular roof framework. Decking and roofing are also shown on the section.

Remember that the wood frame in brick-veneer construction is the structure that carries the entire weight of the roof above and that the brick is merely an exterior wall covering. Metal ties several feet apart are inserted into every sixth course of the veneer and attached to the wood wall in order to strengthen the single brick wythe, but the wood frame provides the structure. The metal ties are usually not shown on the detail.

Solid brick or brick-cavity construction (Figure 3–47) with wood roof framing over it must have a wood plate attached with anchor bolts to secure the roof to the masonry walls. Roofs are subjected to severe wind pressures and must be securely anchored to their bearing walls regardless of the materials. On most details, rafter extensions over the outside walls provide the structural framing for the overhangs. Treatment of the eave design, both structural and trim, is shown on the cornice detail. Notice that some treatments show the extended rafters exposed on the exterior, while others show them boxed in. Others may have still different variations in keeping with the exterior appearance of the building. Usually, roof vents allow ventilation up under the roof; many soffit vents are continuous around the entire length of the overhang.

If exposed beams are intended for the roof structure, as in plank-and-beam framing, the beams are generally exposed under the overhang as well. The cornice detail of a ceiling with beams extending over the walls shows the method of blocking in the spaces between the beams to form an airtight wall. Also, it must indicate 4″ × 4″ posts in the walls beneath each beam.

Although used in many contemporary homes, plank-and-beam construction actually developed from our early building of mills, which utilized heavy posts, beams, and thick floor and roof coverings to resist fire damage. Larger and fewer wood members burn slower than more numerous thin members. Each beam supporting the roof in the construction transfers its load through posts in the walls directly to the foundation, thus leaving areas in the walls between the posts with no load over them. Fixed glass or door openings can be placed in these spaces and not require lintels above. Exposed-beam ceilings produce higher and more interesting rooms than conventional construction. The wide spacing of the beams requires at least 2″-thick tongue-and-groove decking. Two methods of beam placement are commonly used. In one, *transverse* roof beams are placed across the shorter dimension of the building, meeting at the ridge of the gable and extending down the slope to the wall plate (see Figure 3–35A). The decking is attached perpendicularly to the beams. In the other, *longitudinal* beams are placed parallel with the longer dimension of the building, usually extending out under the overhang of the gable ends (see Figure 3–35B). The decking then must run down the slope of the roof. Larger beams are necessary with the latter arrangement even though they may rest on partitions or posts throughout the interior. Interior partitions will have different heights if plank-and-beam sloping roofs are used over them, although some room ceilings can be furred down to 8′ heights if desired. Furthermore, plank-and-beam framing is sometimes combined with conventional framing. Some plans may have regular framing throughout and a plank-and-beam ceiling in a living room or den. The plan would then show the beam layout with heavy lines and a note indicating their size in the one room only.

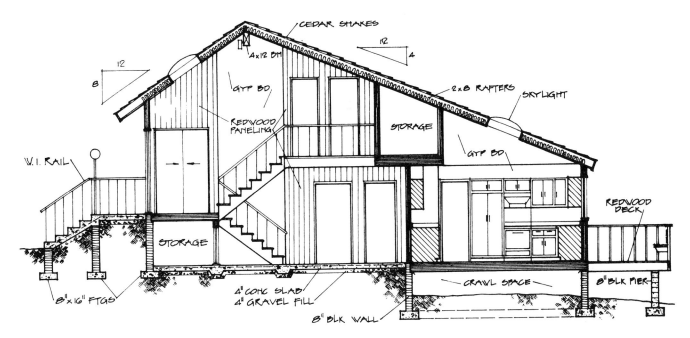

FIGURE 3-50

A transverse section.

A *fascia board* is the finish member attached to the ends of the overhanging rafters or beams. To provide a drip for water running off the roof, the fascia extends down $\frac{3}{4}''$ or so below any soffit material covering the underside of the overhang (see Figure 3–22). Trim members, roof decking, roofing materials, roof slope indication, and gutter arrangement complete the standard cornice detail. Notice that notes are used generously on typical wall details. Also, members in details are scaled to actual rather than nominal size when drawn so that relationships of the members are accurate. The actual size of a 2 × 4, for example, is $1\frac{1}{2}'' \times 3\frac{1}{2}''$.

2. TRANSVERSE AND LONGITUDINAL FULL SECTIONS

Understandably, many houses with complex framework within, possibly split-level houses with offset floor levels, may require *full sections* completely through the building (Figure 3–50). Sections through the narrow dimension are called *transverse* sections; those through the length of the building are known as *longitudinal* sections. Either type is ordinarily drawn on the same scale as the elevations for convenience, and they show all the construction in the building that falls on the cutting plane. Even features beyond the plane are shown as they appear in the interior. This "total view" is very informative, and you will find it much more revealing than isolated section details.

3. WINDOW SECTIONS

Because so many different types of windows are available for homes and there are many methods by which they are attached to the framing, window details must be included in most sets of drawings. Sections through conventional portions of windows furnish the needed information for builders. If only a vertical cutting plane were used in exposing the details of a window, only the head and sill would be seen in the section. However, information about the side (jamb) is also necessary. Therefore, window sections taken through the head, jamb, and sill are conventionally shown and placed in line vertically (see Figure 3–51), even though the jamb section is cut on a horizontal cutting plane. Only one jamb is necessary; a section of a

FIGURE 3-51

Double-hung window in a frame wall, typical window section, and rough opening.

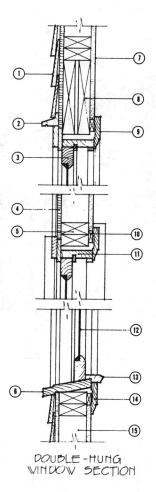

1. BEVEL SIDING
2. WATER TABLE
3. SASH
4. SHEATHING
5. DOUBLE STUD
6. WOOD SILL
7. GYP. BD.
8. HEADER
9. CASING
10. PLASTER GROUND
11. JAMB
12. GLAZING
13. STOOL
14. APRON
15. STUD

DOUBLE-HUNG WINDOW SECTION

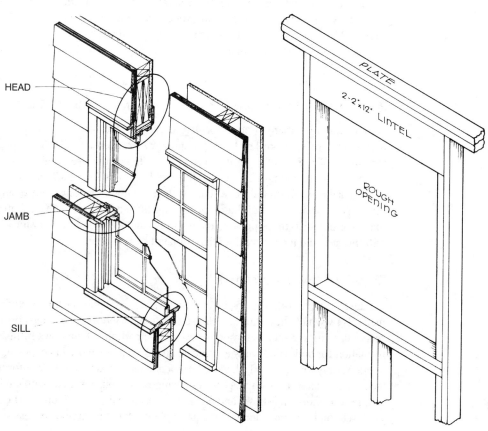

double-hung window may also include the meeting rail as well as a muntin. The vertical alignment of each part of the window helps relate corresponding features. Elevation detail views of a window are seldom needed if we understand this conventional method of showing window information. How the window is to be placed in the rough opening and how it is to be trimmed out can easily be obtained from the detail. Dimensions of windows vary greatly with types and manufacturers, as do differences between rough opening sizes and window-unit sizes. Fixed glass, which is affixed to the rough framing with wood surrounds and stops, is shown with similar section details.

4. DOOR DETAILS

Door details on drawings are limited mainly to head, jamb, and sill sections, and show the rough framing around the door and how the door-surrounds and trim are attached (Figures 3–52 and 3–53). Stock doors, which seldom require a section through the door itself, are used largely in residences. Complex front door units may be shown in full section, however, to ensure their being installed and trimmed out correctly. Door details appear similar to window details in many respects except that the door itself seems unattached in the detail, for it must be free to swing away from the stops, and clearance is provided around it.

5. FIREPLACE DETAILS

Fireplaces, if they are to operate satisfactorily, require careful construction by masons. Although the face of the fireplace may have an individualized appearance to fit the decor of the home, the interior proportions must follow rigid standards, based on experience, to make them draw properly.

Some fireplace details show the use of a prefabricated metal liner for the fire area (Figure 3–54). This steel box eliminates the need for careful details of the interior, inasmuch as it is merely set in place and the masonry built up around it. A note on the drawing then states the manufacturer of the box and its catalog size and number. Conventional fireplace details usually include a plan view, an elevation view, and a vertical section through the center of the fireplace opening (see Figure 3–55). Often the views are orthographically related for better cross-reference.

Fireplaces and chimneys in wood-frame buildings rest on isolated footings below grade, which are sized to support their concentrated weight. Framing members must not be in contact with the masonry unit if the building is to comply with fire codes. Double trimmers and headers at floor and roof levels form openings 2″ wider on all sides than the masonry surface, thus allowing clearance around the unit for fire prevention. Thinner decking and subflooring materials are placed up to the masonry surface, however. Another requirement is the masonry hearth, usually 18″ wide, in front of the fireplace opening. The hearth does not always have a full masonry foundation below it, as shown in the Colonial Home detail; instead, it is generally supported by a reinforced slab cantilevered from below the fireplace floor. Other traditional details may show a *trimmer arch* of brick, which is supported by the main masonry unit, below the hearth. The combustion chamber is lined with refractory brick, shown with a darker symbol than the other masonry. Sometimes an ashpit below the fireplace floor is shown to allow removal of ashes from a basement or an exterior ash door. Above the fireplace opening, a steel angle-iron lintel is necessary to support the brick. Usually, a metal damper to control the draft is specified above the throat. The flue size should be given on the detail. This vitrified clay lining continues from above the dome to the top of the chimney, and it must be at least one-tenth in cross-sectional area to that of the fireplace opening for proper draw. Each fireplace in a building must operate with its own separate flue. Any special construction adjacent to the masonry may also be shown as part of the fireplace detail.

Small, portable fireplaces may be installed in homes without masonry chimneys, if they rest on a fireproof pad and have a safe, fireproof flue through the roof.

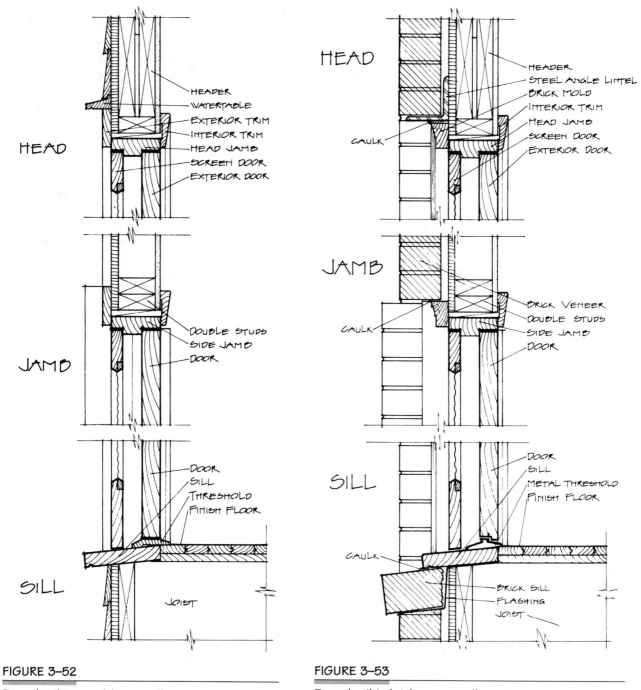

FIGURE 3-52

Door detail in wood-frame wall.

FIGURE 3-53

Door detail in brick-veneer wall.

Looking at the fireplace detail of the Colonial Home (Figure 3–30), we see that the cabinets and bookshelves are shown on both sides of the used-brick fireplace. Both the brick and the bookshelves extend to the ceiling of the den. The raised hearth brings the fire up nearer to eye level for better viewing and provides a brick seat one foot above the finished floor. We see that the hearth is to be covered with brick and is supported by a block foundation that is to be filled and tamped. The steel lintel above the fireplace opening is a $4'' \times 4''$ angle iron. A $13'' \times 13''$ flue lining is needed. All the interior proportions are clearly shown on the detail.

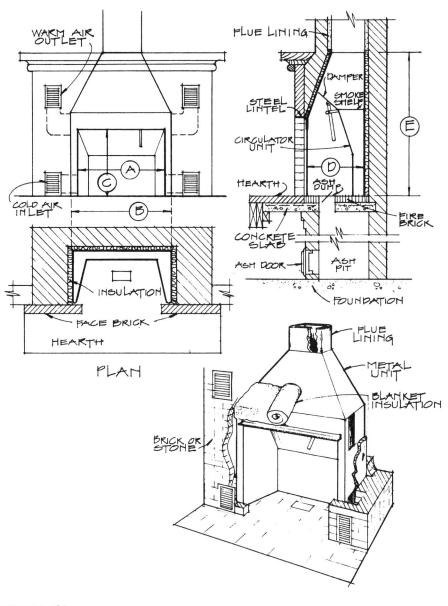

FIGURE 3–54

Prefabricated fireplace detail.

The following terms are commonly associated with residential fireplaces.

FIREPLACE OPENING: the opening on the face of the brick. It may be a single-face, two-face, or three-face opening, depending on the design.

HEARTH: fireproof masonry floor in front of the opening.

MANTLE: an extension from the face of the fireplace above the opening.

COMBUSTION CHAMBER: the area within the opening that is lined with firebrick.

THROAT: the narrow area just above the combustion chamber.

DAMPER: a steel or cast-iron door directly over the throat to control the draft through the fireplace opening.

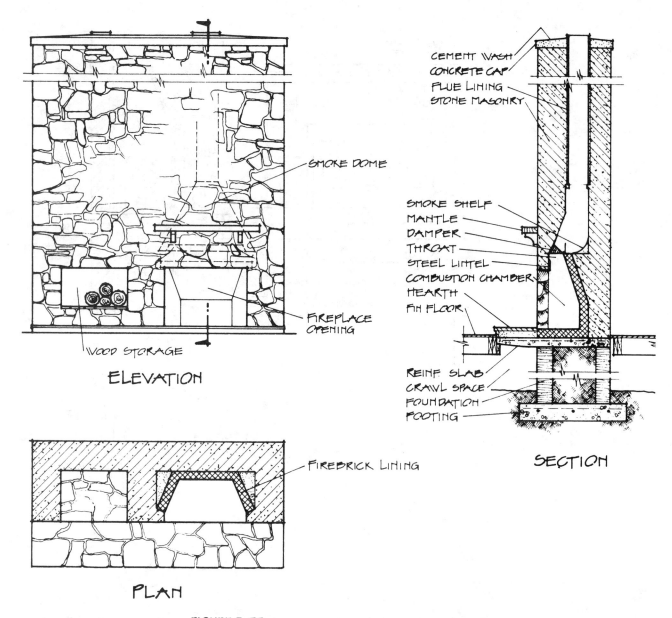

FIGURE 3–55

Residential fireplace detail.

SMOKE SHELF: the horizontal projection just above the inclined back wall of the chamber. It prevents downdrafts through the flue from putting out the fire.

SMOKE DOME: the tapered cavity above the smoke shelf and below the flue.

FLUE LINING: vitrified terra-cotta units lining the flue.

CHIMNEY POTS: extensions of the flue lining above the top of the chimney.

CEMENT CAP: concrete cap at the top of the chimney for the purpose of waterproofing the masonry. Its sloping top surface sheds water; sometimes known as a "cement wash."

ASH DUMP: small door in the floor of the fireplace for removal of ashes.

ASHPIT: cavity below the fireplace from which ashes can be removed through a basement or an exterior door.

FIGURE 3-56

Symbol for superimposed stairs.

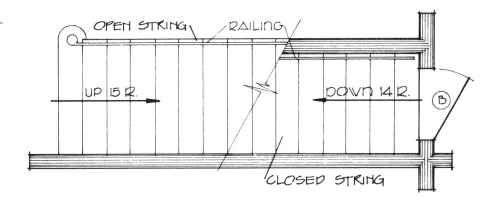

FIGURE 3-57

Stair terms.

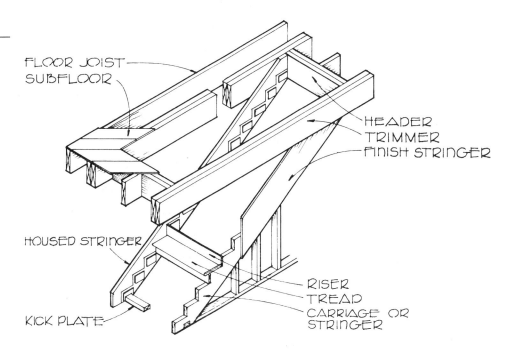

HOOD: metal covering (usually copper) above the fireplace opening, attached to the face of the brick and forming a smoke dome.

6. STAIR DETAILS

Stairs are a part of the system of hallways connecting one level of a building with another. Because they rise diagonally from one floor to the next and must be attached to each floor, their symbol is often only partially shown on each floor plan. In reality, half of the stair flight belongs to one floor plan and half to the other. General information about the stairs is found on the floor plans; more specific information is reserved for the stair detail (Figure 3–56). Stairs have their own terminology; the most commonly encountered terms are defined next (see also Figure 3–57).

BALUSTERS: the thin vertical supports of the handrail of open stairs.

BULLNOSE: the first step, which has been extended out, forming a semicircle and often receiving the newel post.

CARRIAGE: the rough structural support (usually 2″ × 12″ for treads and for risers of wood stairs, sometimes called a string or stringer.

CLOSED STRINGER: the visible member of a stairs that abuts the risers and treads and that is not cut to show the profile of the stairs.

HANDRAIL: the round or decorative member of a railing that is grasped with the hand during ascent or descent.

HEADROOM: the narrowest distance between the surface of a tread and any ceiling or header above.

HOUSED STRINGER: the stringer that has been grooved to receive the riser and tread ends.

LANDING: the floor at either the top or bottom of a flight of stairs.

NEWEL: the main post of the railing at the bottom of a stairs or at changes in direction of the railing.

NOSING: the round projection of the tread beyond the face of the riser.

OPEN STRINGER: a stringer that has been cut to fit the profile of the stairs; the riser cut is mitered, and the tread cut is square.

PLATFORM: the intermediate landing between various parts of the stair flight.

RAILING: the handrail and balusters forming the protection on open stairs.

RISE: the total floor-to-floor vertical height of a stairs.

RISER: the vertical face of the step.

RUN: the total horizontal length of a stairs, including any platform.

STAIRWELL: the enclosed chamber into which the stairs are built.

STEP: the combination of one riser and one tread.

STRINGER: the inclined member supporting the risers and treads; sometimes a visible trim member next to the profile of the stairs.

TREAD: the horizontal surface member of each step, usually hardwood.

WINDER: the radiating or wedge-shaped treads at turns of stairs.

Riser heights and tread widths are planned graphically on the stair detail so that the stairs will have uniform steps throughout the flight, thereby making it comfortable to ascend or descend from one level to another. On long flights, or if the stairs must turn to be accommodated into the plan layout, a *platform* is introduced, usually midway in the flight. This breaks the monotony of the climb, and the platform is considered as another tread. The number of risers is always shown in the stair data, as well as the total horizontal run of the unit. For safety, stairs must have railings, either along their open side or attached to the wall if the stair is enclosed between walls. Headroom above the stairs is also computed graphically on the stair detail. This should be a minimum of 7′ throughout the flight. Information about the details of the railing and all the trim to be attached is shown on the drawing. Remember that there will be one more riser than treads in the stair data. Typical residential stair dimensions are risers—$7\frac{1}{2}″$, treads—10″, and width of stairs—3′ 0″.

Stairs can be built on the job, or they can be mill-made and set in place by the builders. Both methods are in use, but the mill-made stairs produce better construction. Stairs with their profile exposed at the side are known as *open string stairs;* if they are placed between walls and their profile is not visible, they are known as *closed string* stairs. Diagonal spaces below stairs are frequently used for closets, storage, powder rooms, and so on, provided that no stairs are needed to a lower floor. Otherwise, stairs are superimposed above one another for the most efficiency. The rough framing of the stairwell opening at floor levels requires double trimmers and headers around the opening (see Figure 3–57).

Circular stairs and winders are more difficult to climb than straight flights, for their treads are wider at the ends than at the center, around which the treads radiate. Most circular stairs are supported with a steel column; the metal treads are then attached to the column in their correct place. Disappearing stairs are often indicated in upper floors for access to attic spaces. Other novelty stairs will be encountered on drawings. Some have various methods of support. Others have custom-designed railings and trim, but all stairs must have the proper riser-and-tread proportion so that their use will be comfortable and safe. Remember that, as the height of a riser is increased, the width of the tread must be decreased.

The stair detail in the Colonial Home drawings (Figure 3–30) indicates that a mill-made stair is required. It is a traditional style with balusters, bullnose, and newel post, in keeping with the other features of the home. The stair parts indicated in the notes are stock items provided by the manufacturer. On the First-Floor Plan, a storage room utilizes the diagonal space below the stairs; if a basement were to be added, this space would be utilized for the stairs.

7. TRIM DETAILS

The word *trim* refers to the finish members and moldings necessary to finish out the interiors and exteriors of a building. Trim profiles are custom designed in first-class construction, requiring mill-made moldings to order. More modest homes are usually trimmed out with stock trim carried by most lumber supply dealers.

Trim details are mainly section profiles shown full size to facilitate the reproduction of their true shape at the mill. Many trim pieces are *plowed* (see Figure 3–58) on their reverse sides to allow them to seat better on slightly uneven surfaces. *Base* trim is the finish member and molding along the lower walls at the floor intersection; it is nailed in place after the finish floor is down. *Casing* trim is used around windows and doors to cover the intersection of their surrounds (jambs) and the wall finish. Many types of trim are available for covering the intersections of dissimilar materials and for giving a finished appearance to the construction. Many traditional moldings include a reverse curve in their profiles, commonly named *ogee trim*. Contemporary interiors usually show simpler, less ornate trim. Some of the standard stock patterns are shown in Figure 3–58.

On complex trim details showing a number of moldings in combination, especially on traditional buildings, a heavy contour line is sometimes superimposed over the elevation detail of the trim to show its true profile. This line resembles the edge of a revolved section and gives a clear description of the outer surface of the molding. Sometimes hatch lines on one side of the line indicate the material side of the profile.

F. INTERIOR ELEVATIONS

Sometimes certain rooms within a house must have special wall treatment, built-in cabinetwork, or even special floor treatment (Figure 3–59). Typical rooms are bathrooms and kitchens. Other rooms may also need interior elevation views to show wall treatment information. Each wall in question is shown in elevation, with the information clearly delineated. The elevations are then oriented to the plan with identification legends, usually A, B, C, and so on. Be sure that the legend is clear when reading any of the elevation details. Subcontractors must refer to these details when bidding on the specialized work needed.

Kitchen cabinetwork sizes are standardized to fit appliances and to make work tops comfortable. Base cabinets are 36″ high and 24″ wide. Wall cabinets are usually 30″ high and 12″ deep, except over sinks, ranges, and refrigerators. On elevation views of U-shaped or L-shaped kitchens, the cabinets turn toward the viewer at one or both ends and are carefully shown in profile. Depths of the cabinets are then shown on these sectioned areas. The cabinet doors and general appearance of the cabinetwork are indicated; interior shelving and direction of door swing are shown with broken lines (see Figure 3–29). Often the high area above the wall cabinets is furred out flush with the surface of the cabinets. In addition

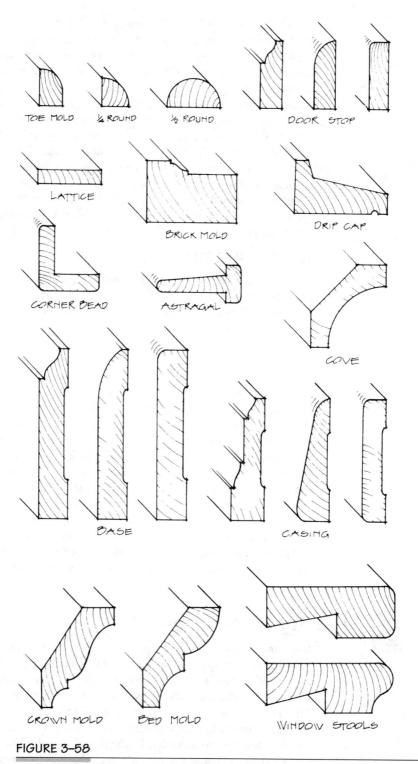

FIGURE 3-58

Typical molding profiles used in wood-frame construction.

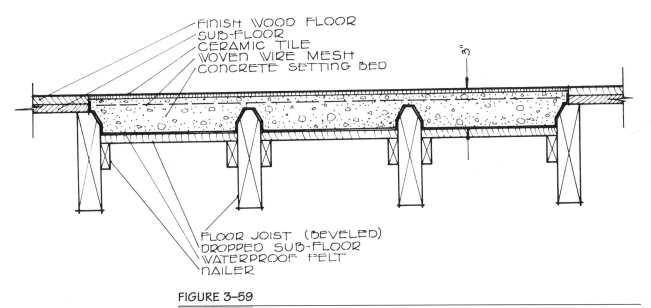

FIGURE 3-59

Detail of ceramic tile floor over wood joists.

to the countertop coverings and splash members, the views also reveal the exact placement of the major built-in appliances necessary in the kitchen.

Shop drawings, showing the details of cabinetwork construction, are the responsibility of the subcontractor, and they are not usually included in the set of residential working drawings. They are used by the cabinetwork fabricators.

In looking at the kitchen elevations of the Colonial Home (Figure 3–29), notice that Elevation A is viewed by standing in the center of the kitchen and looking toward the carport; Elevation B is the view of the window wall looking toward the rear of the house; and Elevation C is the appearance of the cabinets when looking to the right toward the utility room. The three elevations give the full picture of the cabinetwork and appliances in the kitchen. The area near the ceiling above the wall cabinets is to be furred out to the face of the cabinet work. Shelving is required in the cabinets, as shown by the broken lines, and a hood is needed above the range top. The name and number of these items should be given. Heights of the cabinets are given on the detail; widths are standardized and left to the cabinet contractor. Additional information about the type of wood, hardware, and how the cabinets are to be finished have to be found in the specifications.

Similar orientation is needed when reading bath elevations (Figure 3–60). The views show the correct size of bathroom fixtures and the extent of any ceramic tile wainscoting. Standard heights for tile wainscoting are 4' on regular walls and 6' around tub and shower enclosures; occasionally, the tile is run full height. Some baths have built-in cabinet vanities, mirrors, medicine cabinets, exhaust fans, indirect lighting, sunken tubs, and so on. The elevations are helpful to the tile contractor when bidding and ordering the correct tile fittings for the job.

Bookshelves and other special built-in cabinetwork, unless of a simple nature, commonly require elevations of the walls they are to be erected on. In the Colonial Home, bookshelves are combined with the fireplace and are therefore shown on the same detail. Other sets of drawings may have various elevation views of storage walls or shelving work.

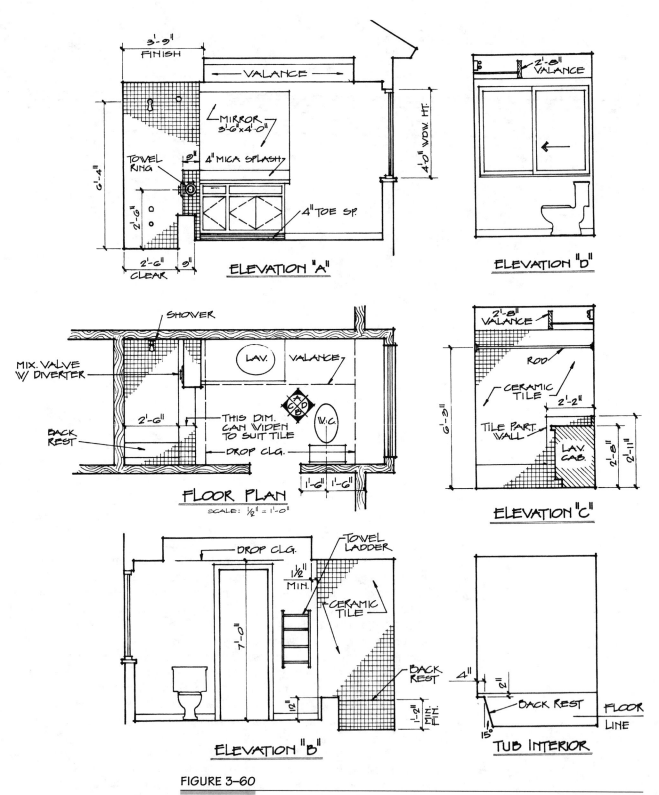

FIGURE 3-60

Residential bath details.

G. HEATING AND COOLING PLANS (HVAC)

Heating and cooling plans (Figures 3–61 and 3–62) show the information necessary for installing heating and cooling ductwork, piping, registers, furnaces, cooling units, and electrical heating equipment. On small jobs with simple systems, such as most small homes, the information is usually shown directly on the regular floor plan; no separate plan is needed. On more complex buildings, a separate plan, usually prepared by a heating and cooling engineer, is made and is included as part of the set.

Many different systems are now available for heating homes. The most widely used system has a central gas-fired furnace and blower that furnishes heated and filtered air forced through sheet-metal ducts to registers in each room. This system is known as a *forced hot-air system.* Hot air may be also forced through ductwork embedded in concrete slab floors along exterior walls; this is called a *perimeter system.*

Another system, a *hot-water system,* has a central boiler that furnishes heated water through distribution piping to radiators or convectors in each room. Return piping then brings the cooled water back to the boiler for heating and recirculation. Other variations use hot-water boilers to circulate water through base panels located along the walls in rooms or even through pipes embedded in concrete floors. This latter system is known as a *radiant heating system.* Other radiant systems utilize electricity for fuel and have resistance wiring (heat tape) embedded in either concrete floors or plastered ceilings to provide a very comfortable type of clean heat. Many variations of the systems are possible, thus providing versatility in accommodating the systems to almost any heating situation.

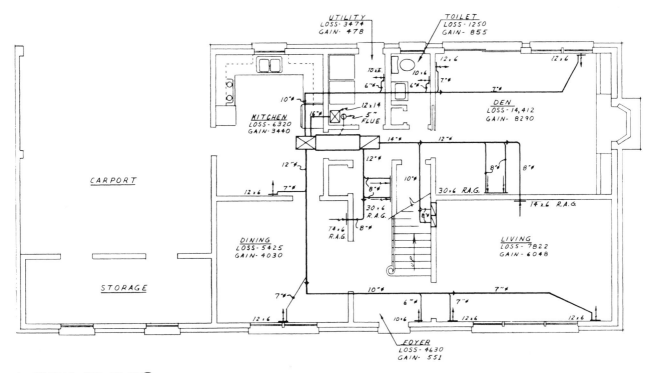

FIGURE 3–61

First-floor heating and cooling plan of Colonial Home (see Figures 3–25 to 3–30). Notice that in residential work the ductwork is often shown with a single line.

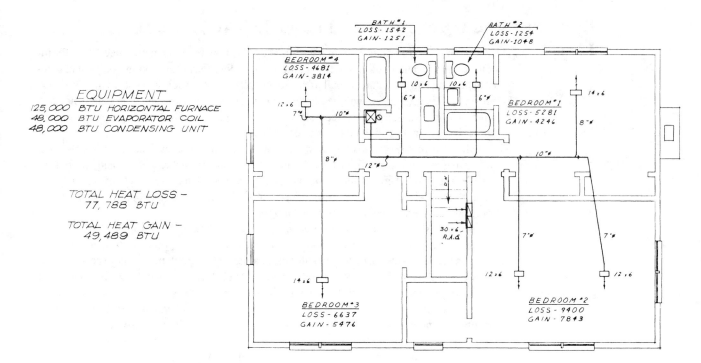

FIGURE 3-62

Second-floor heating and cooling plan of the Colonial Home.

Central heating furnaces or units are made to use any one of the various residential fuels—natural gas, L.P. gas, fuel oil, coal, or electricity. Some fuels, obviously, are more economical in one area of the country than another, which makes them more popular in those sections than in others. Some of the fuels have other advantages, such as availability, efficiency, storage, and the requirement for much simpler automatic controls. Electricity, for example, may be used in some systems that require no floor space for a central unit.

In past years, steam heat was commonly used in larger homes; today it is less popular and few residential steam systems are now installed. Like the steam engine locomotive, steam heating has almost entirely disappeared in modern homes, although steam plants are still found in industrial construction. Steam is comparatively expensive, noisy, and less sensitive to automatic controls for small homes.

Many heating or cooling units are made for basement installation; others for attics, crawl spaces, closets, and so on. Certain units require very little floor space. Some units are for slab floor systems, called counterflow furnaces, which force the air down rather than up.

Automatic controls are an important part of each type of system. Thermostats control constant room temperature and help provide comfortable heat where it is needed, with little attention or care. Even, healthful humidity levels can be maintained with a humidistat and humidifying equipment.

Many modern heating systems are now combined with cooling units to provide automatic cool air in summer and warm air in winter. *Heat pumps* are a comparatively new type of unit used for both heating and cooling. Their use is limited mainly to homes in milder climates where the temperature seldom drops below freezing. The units, operating only on electricity, take the heat from the inside and pump it outside in summer, and, conversely, pump heat from the outside air into the interior in winter. Most units have auxiliary electrical resistance heating units to supplement winter heat during extreme temperature drops (usually below 32°F). The units are placed near outside walls.

Other cooling units can be accommodated with any of the previously mentioned heating systems; some even use the same ductwork and outlet registers. Thus the air in a modern home can be maintained at a constant healthful temperature throughout all seasons in any climate and can even be cleaned by filters at the same time.

1. HOT-AIR SYSTEMS PLANS

Heating plans for hot-air systems show the furnace and ductwork layout and the register locations. The information is placed on drawings that show little of the other plan information other than the layout of walls and other features pertinent to the heating system. If the central unit is to be placed in the basement, a basement plan is used to show all ductwork leading into the unit. Sizes of trunk ducts and feeder ducts are calculated according to the amount of heated air necessary to maintain desired temperatures in the various living areas. The cross-sectional size of each duct is given on the plan. If a note 14″ × 8″ is shown, for example, the duct is to be 14″ wide horizontally and 8″ high vertically in a rectangular section. Risers (vertical ducts) are shown by standard symbols, and cold-air returns are indicated with broken-line symbols to differentiate between the two. Register outlets, which are usually located where the greatest heat loss occurs in the habitable spaces, are shown on the relating floor plans. Both the register symbol and the opening size are indicated. Standard symbols (Figure 2–27) recommended by the American Standards Association are universally used on heating and cooling drawings. Cold-air return registers are also shown. If no return-air registers are indicated on a drawing, possibly return-air circulation has been planned below doors in rooms, through central halls, and so on, back to the heating unit, with only a large cold-air register shown near the central unit.

If the house has no basement, insulated ductwork may be placed in the crawl space and shown on the foundation plan drawing; or central hall ceilings on the floor plan may show furred-down coverings over central ducts to supply heated air from a central unit placed in a closet. Information is always shown on the relating plan. Perimeter systems with ducts in the slab floor require, as we mentioned, a counterflow furnace that forces the warm air down into a plenum through ducts in the slab to register outlets near the periphery of the building. A floor plan is then needed to locate the plenum and the feeder ducts with their registers. Section details also show the construction of the slab and how it accommodates the ducts and insulation.

2. PLANS OF OTHER SYSTEMS

Heating plans having hot-water (hydronic) piping show similar information about the location and description of the equipment. Standard symbols give the location of radiators or convectors and the pipe runs. Central boilers are located on the plans and their type and size are given with a note. If the plan indicates a radiant-panel, hot-water system in the slab, a floor-plan layout of the piping is given to show the pipe size and its coil spacing within each room. Balancing and cutoff valves are also indicated with the use of symbols.

Similar floor-plan layouts are necessary when electrical heating wire is to be embedded in concrete floors or plaster ceilings. Careful layout of the installation must be shown graphically, and all electrical controls and thermostats are located and noted. Often the floor slab is constructed of insulating concrete below the embedded wires, with regular, wear-resistant concrete above. Reference must be made to the typical wall section of the slab floor for specific information about embedded wire or water-pipe placement.

Regardless of the type of heating system to be installed in a home, air conditioning systems, if desired, must include ductwork and registers for supplying cooled air into the rooms. If a separate cooling plan is needed, central units, ductwork, and registers are represented with standard symbols, similar to those on heating plans. Subcontractors prepare heating and cooling drawings after they have calculated the heat loss for winter design temperatures and the heat gain for summer temperatures. Both individual room calculations and the totals must be known. A decision is then made, based on a number of factors, as to which system and arrangement will provide the most satisfactory and economical heating and/or cooling method.

In the publication *Minimum Property Standards* (FHA No. 300), the Federal Housing Administration requires that the following information about heating and cooling be included on working drawings of homes considered for FHA financing:

HEATING SYSTEM: On a separate drawing or as a part of the floor or basement plan, show

1. Layout of system.
2. Location and size of ducts, piping, registers, and radiators.
3. Location of heating unit and room thermostat.
4. Total calculated heat loss of the dwelling, including heat loss through all vertical surfaces, ceiling, and floor. When a duct or piped distribution system is used, calculate heat loss of each heated space.

COOLING SYSTEM: On a separate drawing or as part of the floor or basement plan, show

1. Layout of system.
2. Location and size of ducts, registers, compressors, coils, and so on.
3. Model number and Btu capacity of equipment or units in accordance with applicable ARI (Air Conditioning and Refrigeration Institute) or ASRE (American Society of Refrigerating Engineers) standards.
4. Heat gain calculations, including estimated heat gain for each space conditioned.
5. Btu capacity and total kilowatt (KW) input at stated local design conditions.
6. If room or zone conditioners are used, provide the location, size, and installation details.

H. PLUMBING PLANS

Plumbing plans are seldom included in a set of residential working drawings, although, like other mechanical and electrical information, complete plumbing layouts and details are an important part of larger, commercial building drawings. However, when reading residential drawings, we must understand the function of the plumbing system and how it is accommodated into the frame structure.

Plumbing includes all the hot- and cold-water pipes that furnish water to the fixtures and the larger piping that removes the waste water and materials to the sewage lines or to an individual septic tank. Most of the piping, except the fixtures, of course, is hidden within the framework of the building.

Rough plumbing includes the piping that must be installed immediately after the framing is completed but before wall coverings are attached. In slab-on-ground construction, rough plumbing is placed below the slab before it is poured. Fixtures are set in place after the finish walls and floors are completed. All plumbing is closely controlled by local plumbing codes for the safety of the occupants as well as the public.

Locations of plumbing fixtures, including exterior hose bibbs, floor drains, and water heaters, are shown with symbols on residential drawings. Also, vertical soil stacks requiring 6″ or 8″ walls to pass through are usually shown on the plan so that the walls (wet walls) can be made wider to accommodate them. Stacks near water closets must be of 4″ soil pipe, whereas vent stacks for kitchen sinks or tubs need be of only 2″ diameter pipe. The latter can pass through regular 2″ × 4″ stud walls and therefore they seldom need to be shown on the drawing.

Otherwise, plumbing contractors, who are familiar with local plumbing codes, install residential plumbing according to local requirements and need little information other than that shown on the floor plan to complete it. Occasionally, a layout of plumbing pipe runs may be included to show unusual arrangements. The layout then shows standard plumbing symbols (Figure 2–27) representing pipe lines, fittings, and so on, similar to those used in heating and cooling plans.

If no public sewer line is accessible, the dwelling must provide its own disposal system on the property. Septic tanks below ground, with a disposal field, are the most acceptable system. Their layout, if required, is commonly shown on the plot plan. The raw sewage from the house drain is held in the septic tank until it is decomposed by bacteria. Remaining liquids slowly seep out through loose-jointed tile in disposal lines near the surface of the ground, where they saturate the soil and finally evaporate into the air. Local codes closely control their design and installation.

I. ELECTRICAL PLANS

As we have already seen, electrical information on residential drawings, like heating and plumbing information, is generally given on the floor plans if it is not extensive. If the architect has seen fit to isolate the electrical material on a separate plan, an electrical plan is included with the set.

To install residential wiring and fixtures, electrical contractors usually have sufficient information shown on the floor plan. Mainly, they need to know the locations and types of electrical outlets. Their first concern is the location of the main circuit-control box (usually placed in the utility room) from which they plan all the local circuits to meet the requirements of the outlets shown in the various rooms. Like rough plumbing, rough electrical work must be done before finish walls and floors are completed. If metal conduit is to be used for the circuit runs, the conduit is installed first, and then the wiring is pulled through the conduit to successive outlet boxes. Sheathed cable, on the other hand, is installed with the wire conductors within. All electrical fixtures are installed after finish surfaces are completed. Special outlets requiring 220 volts need to be known, as well as dual-switching arrangements that require three-wire conductors in their hookup. Other special features are merely indicated with notes on the plan.

As you will notice, outlets and switches on electrical drawings are represented with standard symbols. Broken lines show connections between the outlets and their operating switches, and outlets may be operated from several switches if three-way switch symbols are indicated. Remember that the broken lines do not represent the actual path of the wiring, however. Much of the electrical information about each job must be taken from the set of specifications that accompanies the drawings.

J. FRAMING PLANS

Unusual conditions in framing complex floors and roofs sometimes present difficult problems for carpenters. The problems can be simplified if a carefully conceived *Framing Plan* (Figure 3–63) is included with the drawings. Conventional dwellings with uncomplicated framing structures seldom require separate plans of this type; the standard notes on the regular plan, telling the size, spacing, and direction of floor or ceiling joists, are sufficient. On the other hand, if difficult framing problems are anticipated, the architect or drafter can better work out the solution graphically on a plan, thus relieving the builder of any complications. Generally, the floors and roof are the most difficult to frame, and if a framing plan is included, these are the parts of the building that are given this special consideration.

Framing plans, sometimes called *diagrams* because only single lines are mainly used to represent the framing members, are *top views* of the framing looking down from above. The plans show bearing-wall and partition outlines lightly drawn (often with broken lines), and the correctly placed framing members are shown with heavier lines. Frequently, the members are numbered or given identification for easy reference to further notes on the

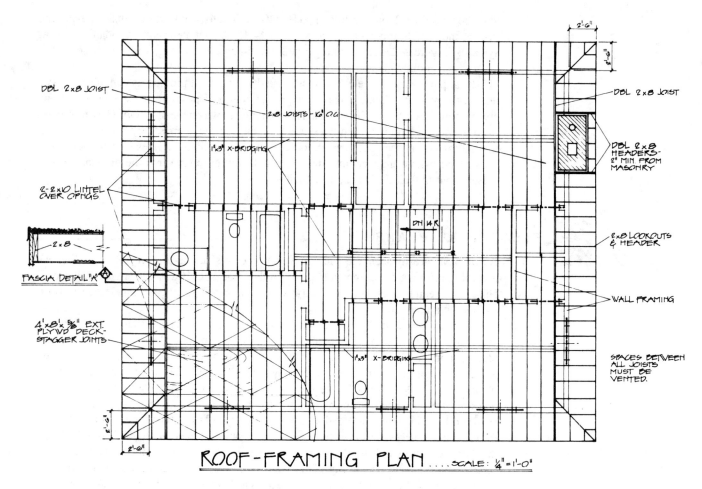

FIGURE 3–63

Roof framing diagram.

drawing. All columns, piers, and supports of any type are shown in their relationship to the framing. Also shown are masonry chimneys, stairs, vertical chases, and so on, around which trimmers and headers must be built to make the framing self-supporting under the loads it will be subjected to. Sizes of all joists, beams, and girders are given in the form of notes (callouts). Critical points may even require additional notes. If cantilever construction is used, provision for sufficient anchoring of the members is important. Remember that conventional wood framing, both walls and floors, is spaced 16″ on center; wood-trussed rafters are usually spaced 2′ on center.

Roof-framing plans of gable roofs show the supporting walls and ridges, valleys, dormer openings, overhangs, and any masonry chimneys that may penetrate the roof. Rafters are aligned so that they bear against opposites at the ridge. Hip and valley rafters on hip roofs with the same pitch on all slopes form 45-degree angles on the plan.

Flat or shed-roof framing members become the supports for any ceiling material below and are sized accordingly. Provision is also made on these roofs for sufficient ventilation of the air spaces between the framing members. Overhangs beyond the walls that are parallel to the joist direction are supported by *lookouts* cantilevered over the walls and anchored with double joists (see Figure 3–64).

Plans showing roofs framed with prebuilt trusses usually include a roof-framing plan to show definite layout of the trusses. A detail drawing of the typical truss showing its dimensions may also accompany the layout (Figure 3–65). On the plan, each truss is num-

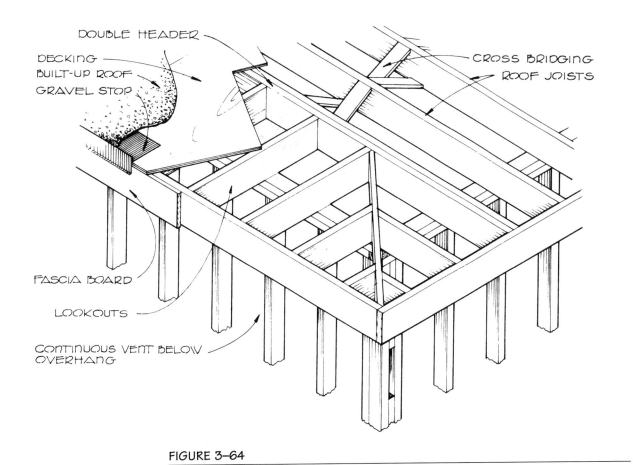

FIGURE 3–64

Flat-roof framing at corner.

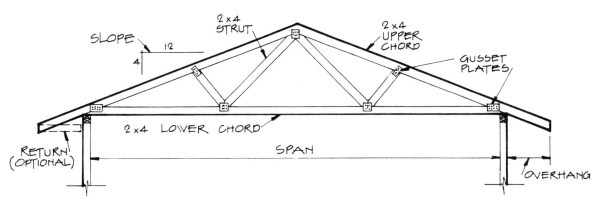

FIGURE 3–65

Typical wood-trussed rafter used in light construction.

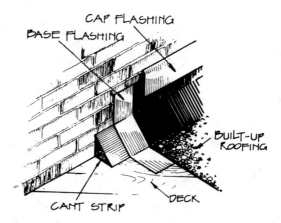

FIGURE 3-66

Two-piece roof flashing.

bered, the typical spacing is noted, and variations in the regular spacing continuity, if necessary, are also noted. It is common practice for conventional rafter framing to be combined with trusses to frame valley intersections and other changes of the truss profile. Separate framing plans may sometimes be necessary for post-and-beam construction (see Figure 3–36). Figure 3–66 shows flashing details of a possible roof/wall connection. Figure 3–67 shows typical framing conditions when using plywood I beams in light frame construction.

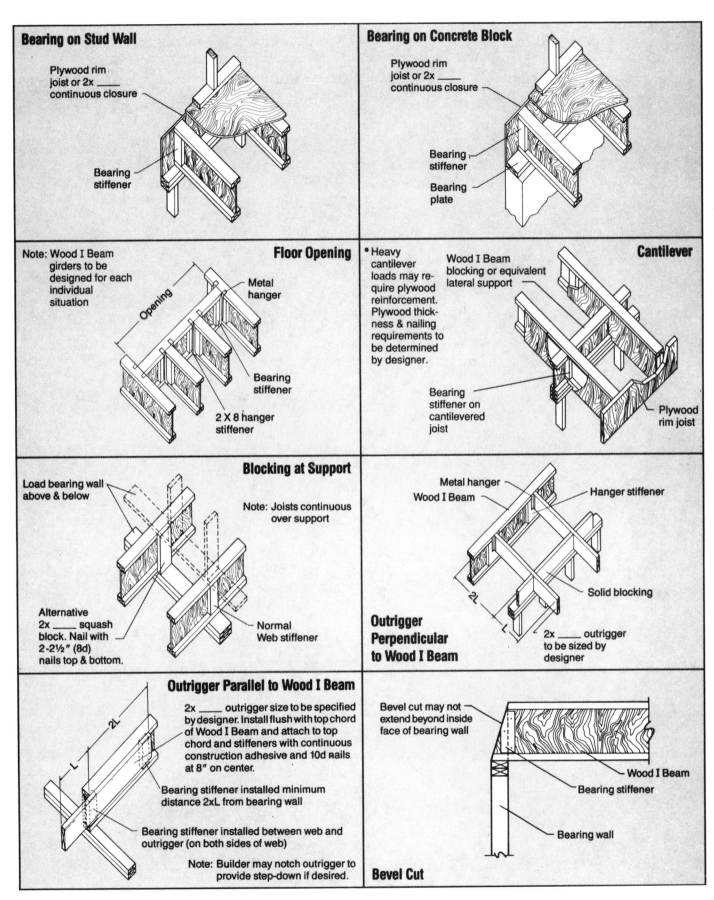

FIGURE 3–67

Use of plywood I beams in residential floor construction.

Chapter 4

SPECIFICATIONS

It is common practice in the construction industry to include a set of written documents called the Project Manual, which includes specifications, with each set of working drawings. Many other industries follow the same practice when a proposal or bid is submitted for the manufacture of a machine or other complex product.

Some specifications for a building may be only several pages in length; others, depending on the size and complexity of the structure, may include four or five (or even more) lengthy volumes.

A. THEIR PURPOSE

Numerous aspects of a project, whether in construction or otherwise, cannot be fully explained graphically on drawings. For example, how would the quality of work be shown on a drawing? Or how would one indicate on whom responsibility is placed for the various points that arise during the construction of a building? These are only a few of the points that must be clarified before construction begins and that must be explained with written words. If all the clarifying notes were put on the drawings by drafters, the result would be cluttered work, and the hand-lettered notes would make the drawings too time-consuming and expensive to produce.

However, different designers have different ideas about how much information, in the form of notes, should be placed on drawings and how much should be reserved for the specifications, or specs. Some designers advocate placing much of the spec data on the drawing itself to eliminate time-consuming cross-reference. Others place only a minimum amount of spec data on the drawing and let the specs explain the many points that must be elaborated upon.

Specifications serve three general purposes:

1. They act as a guide for contractors and bidders who must make proposals of both material and labor before a contract is signed.
2. They supply the technical descriptions of materials and construction techniques necessary to fabricate the building, as well as provide a means of checking the materials and workmanship during actual construction.
3. They serve as a legal document in case litigation must be resorted to in resolving any misinterpretations arising during construction.

Specs are also useful in appraising a proposed building for the purpose of financing. Loan companies use the specs to determine if qualifications are met before allowing requested amounts on a loan. Occasionally, specs are requested even on application for remodeling loans after the building has been in use for some time.

209

The specs are prepared by the architect or engineer during the development of the working drawings in the design stage. Some offices have a separate department for writing specs. Others have only one person; still others require the drafters to complete them. The writers must be well acquainted with building materials, supervision procedures, and building codes. By necessity, the preparation of the specs is coordinated closely with the preparation of the drawings. Not until the working drawings are near completion is it possible to write the finished specs; hence, to meet a deadline, their completion sometimes becomes a "hurry up" task. Their final form is customarily typed on $8\frac{1}{2}'' \times 11''$ paper. If a job requires a number of bidders, 25 copies of the specs are often needed. Small jobs may require only 4 or 5 copies; large jobs, on the other hand, may need 50 or 100 sets. Specs are then bound into book form and distributed to those concerned.

Almost every project requires a tailor-made set of specifications. Although sections of technical specs may be similar to those of other jobs, the complete set is almost always a combination of differing sections or items within the sections. The spec writer gets the needed information from various sources, such as old sets of specs on file, Sweet's Catalogue File, trade association literature, and manufacturers' representatives.

In short, specifications are the written description of the construction prepared by the architect or engineer. Together with the working drawings, they become the Construction Documents, upon which a basis for agreement is made between the owner and the contractor when a building contract is consummated.

B. HOW SPECIFICATIONS RELATE TO DRAWINGS

Specifications are meant to amplify, yet be compatible with, the information found on the drawings. Items are located and quantified on the drawings; and more specific information about kind and quality is located in the specs. While drawings show placement, arrangement, dimensions, and general types of materials, the specs give the supplementary and precise information about materials and the exact methods of fabrication. For example, the drawing may call for concrete on a wall detail; the specs, under the Concrete Section, would then state the type, aggregate size, ultimate strength, and so on, of the concrete, as well as how the concrete is to be formed, placed, cured, and finished. Ideally, the information in the specs is meant to start where the drawings leave off and should contain very little overlap; yet some does occur, and conflicts arise sometimes. For this reason, interpretation of specs requires the contractor to consult with the architect to determine the correct intent.

C. HOW INFORMATION IS ARRANGED IN SPECIFICATIONS

Basically, specs are divided into two major parts, General Conditions and Technical Sections. The *General Conditions* deal with the preliminary aspects of the job, such as terms of agreement, insurance, responsibility for permits, payment for the work, supervision, and temporary utilities. Although these items may seem only minor, they are often the most controversial points and therefore are carefully spelled out in the General Conditions.

The American Institute of Architects (A.I.A.) has available for architects several preprinted forms of General Conditions. A copy of one of the forms is often inserted into and included as part of many sets of specs as standard procedure, thus saving much of the repetition for spec writers. The A.I.A. Standard Form of General Conditions (Document A-201) contains articles dealing with most of the aspects of a construction project. For smaller jobs, such as residences, an A.I.A. Short Form of General Conditions is available and is frequently included in specs of less complex nature. Of course, some architects write their own General Conditions to satisfy particular requirements when it seems feasible.

When additions or deletions of items in the standard General Conditions are necessary on a job, a separate section called *Supplementary General Conditions* is prepared and placed after the General Conditions. These modifications are unique in that they refer to the one job only.

Although not considered part of the actual specs, *Bidding Requirements* are usually included with the General Conditions and are regarded as part of the legal documents, similar to the specs themselves. Listed next are some of the typical Bidding Requirements documents and their meanings:

INVITATION TO BID: a standardized form or an individualized document prepared by the architect that states briefly the type of project, place, time of completion, and other information about the project. It is usually mailed to qualified contractors or advertised in local papers or trade journals.

INSTRUCTIONS TO BIDDERS: This form gives further information about the project: bidding procedures, availability of construction documents, site examination, basis and preparation of bids, and information about the awarding of the contract.

BID FORMS: forms on which the bid itself is submitted. It spells out the commitment on the part of the contractor, the price of the project, alternates, guarantee of the work, time of completion, and complete address, as well as the signature of the bidder. On small jobs a bid may be submitted on the contractor's letterhead, and no bid form is used. *Unit Prices* are sometimes used in lieu of alternates in cases of work that cannot be estimated accurately prior to the actual construction. For example, a "rock clause" may contain a unit price for removal of rock, since it is often impossible to ascertain the amount of rock to be encountered in some excavation work. The contractor then quotes a unit price for removing the rock on a dollar per cubic yard basis. Payment is made after the rock is removed and the volume is measured.

BID BOND: a guarantee that the bidder will not rescind the bid, should the contract be offered.

PERFORMANCE BOND: a guarantee backed by a surety that the contractor will fulfill the terms of the contract, if awarded.

OTHER GUARANTEE DOCUMENTS: forms that state the length of time various parts of the construction, such as roofing, are to be guaranteed from defects.

LABOR AND MATERIAL BONDS: guarantee that workers and subcontractors, as well as material suppliers, will be paid.

ADDENDA: changes that occur before the contract has been awarded. They may describe *Alternates* in the construction or material changes that may be necessary to meet construction budgets. Or they may include other changes or clarifications that come about during the period of bidding, which must be stated for bidders.

CHANGE ORDERS: written changes that become necessary after the contract has been awarded. The exact nature of the changes, any variations in the cost of the changes, along with signatures of approval of all principals involved and the dates must be shown on the orders.

The *Technical Sections* deal with the actual construction of the building and the tradework involved. Technical information is placed in these sections so that each type of work done by one subcontractor will be found in one section. Also, it is conventional to arrange the sequence of technical sections in the same sequence as the conduct of the work on the job. For example, the Concrete Section would be listed before the Painting and Decorating Section, for concrete work is usually done early in the construction, whereas painting is done near the end. However, because each job is somewhat different, slight variations in sequence may be encountered when reading different sets of specs. On small jobs such as homes, the following spec arrangement is commonly found:

1. General Conditions
2. Excavation
3. Concrete and Masonry

4. Waterproofing
5. Miscellaneous Metals
6. Rough Carpentry
7. Finish Carpentry and Millwork
8. Roofing and Sheetmetal
9. Weatherstripping, Insulation, Caulking, and Glazing
10. Plastering or Drywall Construction
11. Ceramic Tile
12. Flooring
13. Painting and Decorating
14. Plumbing
15. Mechanical Equipment
16. Electrical
17. Landscaping

Specifications for small, project-type houses in which only stock materials and standard construction are used may be even more simplified. They are referred to as *Outline Form* specs and merely list the major work to be done and the equipment to be included. Other specs may be standard preprinted forms having fill-in blanks to be completed for each job. An example is a standard form used by the Federal Housing Administration (FHA). (See pages 245–248.) This *Description of Materials* form is required when residential loans are to be insured by the FHA. Notice that the form only itemizes the materials and equipment proposed in the dwelling. No alternates and "or equal" material designations are given, making the form precise as to what is to be included as materials only. No further information about general conditions nor instructions about *how* components are to be constructed or installed are used in this government form. The reproduced form is filled in to represent the conditions of the two-story Colonial Home described in the previous chapter (Figures 3–25 to 3–30).

D. THE CSI FORMAT

Since the introduction of their standardized specifications format in 1963, the Construction Specifications Institute (CSI), with the aid of feedback from the construction industry, has continually developed the format to make it now almost universally accepted throughout this country and Canada. Nearly all major offices use it in their spec writing. Its latest edition still retains the 16 original divisions, with only minor modifications, and the same sequence, this being one of the chief reasons for its universal acceptance (see Figure 4–1).

The set of sample specifications for a small home (pp. 221–236) follows this format.

While every professional specifier must have some latitude in judgment when dealing with local codes, trade practice variations, and client obligations, the standardization of the format when widely used has many advantages. Only the arrangement and key words are restrictive so that a set of specs, like our alphabet or numbering system, will be universally useful because of its consistency. Manufacturers, architects, contractors, and builders, who must clearly communicate in a common language, have found many advantages in the concept. Even though some jobs do not require specifications covering every division shown in the format, the division number and title are nevertheless retained so that the sequence remains uniform, and only a note, "Not Used," is placed below it. Grouping of related sections usually results in more specific information with less lengthy sections.

One of the modifications in the latest format editions is the five-digit numbering system (see Figure 4–1). It is based on the 16 fixed divisions, which are grouped by standard tradework, and five-digit labeled sections called *Broadscope Sections*. The Broadscope Sections are subdivided into more flexible *Narrowscope Sections* that have intermediary five-digit numbers. A spec writer only selects those sections that apply to a particular job.

MasterFormat™

1995 MasterFormat™ Level Two Numbers and Titles

MasterFormat™ is a master list of numbers and titles for organizing information about construction requirements, products and activities into a standard sequence. Construction projects use many different delivery methods, products and installation. Successful completion of projects requires effective communication among the people involved on a project. Information retrieval is nearly impossible without a standard filing system familiar to each user. *MasterFormat* facilitates standard filing and retrieval schemes throughout the construction industry. *MasterFormat* is a uniform system for organizing information in project manuals, organizing cost data, filing product information and other technical data, identifying drawing objects and for presenting construction market data.

The 1995 edition of *MasterFormat* replaces the 1988 edition. It is produced jointly by The Construction Specifications Institute (CSI); (800) 689-2900, and Construction Specifications Canada (CSC).; (416) 777-2198. This index of level two numbers and titles is only a highlight of the 1995 edition of *MasterFormat*. For a more detailed listing of numbers and titles and an in-depth explanation of MasterFormat and its use in the construction industry contact CSI Member/Customer Service at (800) 689-2900 or visit CSI's website at www.csinet.org.

Introductory Information
- 00001 Project Title Page
- 00005 Certifications Page
- 00007 Seals Page
- 00010 Table of Contents
- 00015 List of Drawings
- 00020 List of Schedules

Bidding Requirements
- 00100 Bid Solicitation
- 00200 Instructions to Bidders
- 00300 Information Available to Bidders
- 00400 Bid Forms & Supplements
- 00490 Bidding Addenda

Contracting Requirements
- 00500 Agreement
- 00600 Bonds & Certificates
- 00700 General Conditions
- 00800 Supplementary Conditions
- 00900 Addenda & Modifications

Division 1 - General Requirements
- 01100 Summary
- 01200 Price & Payment Procedures
- 01300 Administrative Requirements
- 01400 Quality Requirements
- 01500 Temporary Facilities & Controls
- 01600 Product Requirements
- 01700 Execution Requirements
- 01800 Facility Operation
- 01900 Facility Decommissioning

Division 2 - Site Construction
- 02050 Basic Site Materials & Methods
- 02100 Site Remediation
- 02200 Site Preparation
- 02300 Earthwork
- 02400 Tunneling, Boring, & Jacking
- 02450 Foundation & Load-bearing Elements
- 02500 Utility Services
- 02600 Drainage & Containment
- 02700 Bases, Ballasts, Pavements, and Appurtenances
- 02800 Site Improvements & Amenities
- 02900 Planting
- 02950 Site Restoration & Rehabilitation

Division 3 - Concrete
- 03050 Basic Concrete Materials and Methods
- 03100 Concrete Forms & Accessories
- 03200 Concrete Reinforcement

Division 3 - Concrete, cont.
- 03300 Cast-in-place Concrete
- 03400 Precast Concrete
- 03500 Cementitious Decks and Underlayment
- 03600 Grouts
- 03700 Mass Concrete
- 03900 Concrete Restoration & Cleaning

Division 4 - Masonry
- 04050 Basic Masonry Materials and Methods
- 04200 Masonry Units
- 04400 Stone
- 04500 Refractories
- 04600 Corrosion-resistant Masonry
- 04700 Simulated Masonry
- 04800 Masonry Assemblies
- 04900 Masonry Restoration & Cleaning

Division 5 - Metals
- 05050 Basic Metal Materials & Methods
- 05100 Structural Metal Framing
- 05200 Metal Joists
- 05300 Metal Deck
- 05400 Cold-formed Metal Framing
- 05500 Metal Fabrications
- 05600 Hydraulic Fabrications
- 05650 Railroad Track & Accessories
- 05700 Ornamental Metal
- 05800 Expansion Control
- 05900 Metal Restoration & Cleaning

Division 6 - Wood & Plastics
- 06050 Basic Wood & Plastic Materials and Methods
- 06100 Rough Carpentry
- 06200 Finish Carpentry
- 06400 Architectural Woodwork
- 06500 Structural Plastics
- 06600 Plastic Fabrications
- 06900 Wood & Plastic Restoration and Cleaning

Division 7 - Thermal & Moisture Protection
- 07050 Basic Thermal & Moisture Protection Materials & Methods
- 07100 Dampproofing & Waterproofing
- 07200 Thermal Protection
- 07300 Shingles, Roof Tiles, & Roof Coverings
- 07400 Roofing & Siding Panels
- 07500 Membrane Roofing

Division 7 - Thermal & Moisture Protection, cont.
- 07600 Flashing & Sheet Metal
- 07700 Roof Specialties & Accessories
- 07800 Fire & Smoke Protection
- 07900 Joint Sealers

Division 8 - Doors & Windows
- 08050 Basic Door & Window Materials and Methods
- 08100 Metal Doors & Frames
- 08200 Wood & Plastic Doors
- 08300 Specialty Doors
- 08400 Entrances & Storefronts
- 08500 Windows
- 08600 Skylights
- 08700 Hardware
- 08800 Glazing
- 08900 Glazed Curtain Wall

Division 9 - Finishes
- 09050 Basic Finish Materials & Methods
- 09100 Metal Support Assemblies
- 09200 Plaster & Gypsum Board
- 09300 Tile
- 09400 Terrazzo
- 09500 Ceilings
- 09600 Flooring
- 09700 Wall Finishes
- 09800 Acoustical Treatment
- 09900 Paints & Coatings

Division 10 - Specialties
- 10100 Visual Display Boards
- 10150 Compartments & Cubicles
- 10200 Louvers & Vents
- 10240 Grilles & Screens
- 10250 Service Walls
- 10260 Wall & Corner Guards
- 10270 Access Flooring
- 10290 Pest Control
- 10300 Fireplaces & Stoves
- 10340 Manufactured Exterior Specialties
- 10350 Flagpoles
- 10400 Identification Devices
- 10450 Pedestrian Control Devices
- 10500 Lockers
- 10520 Fire Protection Specialties
- 10530 Protective Covers
- 10550 Postal Specialties
- 10600 Partitions
- 10670 Storage Shelving
- 10700 Exterior Protection
- 10750 Telephone Specialties

FIGURE 4–1

The CSI MasterFormat™ Index, Level Two, Numbers and Titles.

Division 10 - Specialties, cont.		13150	Swimming Pools
10800	Toilet, Bath, & Laundry Accessories	13160	Aquariums
10880	Scales	13165	Aquatic Park Facilities
10900	Wardrobe & Closet Specialties	13170	Tubs & Pools
		13175	Ice Rinks
Division 11 - Equipment		13185	Kennels & Animal Shelters
11010	Maintenance Equipment	13190	Site-constructed Incinerators
11020	Security & Vault Equipment	13200	Storage Tanks
11030	Teller & Service Equipment	13220	Filter Underdrains & Media
11040	Ecclesiastical Equipment	13230	Digester Covers & Appurtenances
11050	Library Equipment	13240	Oxygenation Systems
11060	Theater & Stage Equipment	13260	Sludge Conditioning Systems
11070	Instrumental Equipment	13280	Hazardous Material Remediation
11080	Registration Equipment	13400	Measurement & Control Instrumentation
11090	Checkroom Equipment		
11100	Mercantile Equipment	13500	Recording Instrumentation
11110	Commercial Laundry & Dry Cleaning Equipment	13550	Transportation Control Instrumentation
		13600	Solar & Wind Energy Equipment
11120	Vending Equipment	13700	Security Access & Surveillance
11130	Audio-visual Equipment	13800	Building Automation & Control
11140	Vehicle Service Equipment	13850	Detection & Alarm
11150	Parking Control Equipment	13900	Fire Suppression
11160	Loading Dock Equipment		
11170	Solid Waste Handling Equipment	**Division 14 - Conveying Systems**	
11190	Detention Equipment	14100	Dumbwaiters
11200	Water Supply & Treatment Equipment	14200	Elevators
		14300	Escalators & Moving Walks
11280	Hydraulic Gates & Valves	14400	Lifts
11300	Fluid Waste Treatment & Disposal Equipment	14500	Material Handling
		14600	Hoists & Cranes
11400	Food Service Equipment	14700	Turntables
11450	Residential Equipment	14800	Scaffolding
11460	Unit Kitchens	14900	Transportation
11470	Darkroom Equipment		
11480	Athletic, Recreational, and Therapeutic Equipment	**Division 15 - Mechanical**	
		15050	Basic Mechanical Materials and Methods
11500	Industrial & Process Equipment		
11600	Laboratory Equipment	15100	Building Services Piping
11650	Planetarium Equipment	15200	Process Piping
11660	Observatory Equipment	15300	Fire Protection Piping
11680	Office Equipment	15400	Plumbing Fixtures & Equipment
11700	Medical Equipment	15500	Heat-generation Equipment
11780	Mortuary Equipment	15600	Refrigeration Equipment
11850	Navigation Equipment	15700	Heating, Ventilating, & Air Conditioning Equipment
11870	Agricultural Equipment		
11900	Exhibit Equipment	15800	Air Distribution
		15900	HVAC Instrumentation & Controls
Division 12 - Furnishings		15950	Testing, Adjusting, & Balancing
12050	Fabrics		
12100	Art	**Division 16 - Electrical**	
12300	Manufactured Casework	16050	Basic Electrical Materials and Methods
12400	Furnishings & Accessories		
12500	Furniture	16100	Wiring Methods
12600	Multiple Seating	16200	Electrical Power
12700	Systems Furniture	16300	Transmission & Distribution
12800	Interior Plants & Planters	16400	Low-voltage Distribution
12900	Furnishings Repair & Restoration	16500	Lighting
		16700	Communications
Division 13 - Special Construction		16800	Sound & Video
13010	Air-supported Structures		
13020	Building Modules		
13030	Special Purpose Rooms		
13080	Sound, Vibration, & Seismic Control		
13090	Radiation Protection		
13100	Lightning Protection		
13110	Cathodic Protection		
13120	Pre-engineered Structures		

The 1995 MasterFormat™ is published by CSI. Copyright 1996 U.S. & Canada. All rights reserved, including World rights and Electronic rights. U.S. copyright held by The Construction Specifications Institute, 601 Madison Street, Alexandria, VA 22314-1791; (800) 689-2900; csimail@csinet.org. Canadian copyright held by Construction Specifications Canada, 100 Lombard Street, Suite 200, Toronto, Ontario M5C 1M3; (416)777-2198. No part of this may be reproduced or transmitted in any form or by any means, electronic or mechanical including photocopying, recording, or by any information storage and retrieval system, without permission in writing from the publisher.

FIGURE 4–1

Continued.

VA Form VB4-1852
FHA Form 2005
Jan. 1955

For accurate register of carbon copies, form may be separated along above fold. Staple completed sheets together in original order.

Form approved.
Budget Bureau No. 63-R055.10.

☒ Proposed Construction
☐ Under Construction

DESCRIPTION OF MATERIALS

No. ....
(To be inserted by FHA or VA)

Property address 1230 Pinedale Avenue   City Atlanta   State Georgia
Mortgagor or Sponsor First National Bank   Atlanta, Georgia
                         (Name)   (Address)
Contractor or Builder Ace Construction Company   Atlanta, Georgia
                         (Name)   (Address)

INSTRUCTIONS

1. For additional information on how this form is to be submitted, number of copies, etc., see the instructions applicable to the FHA Application for Mortgage Insurance or VA Request for Determination of Reasonable Value, as the case may be.
2. Describe all materials and equipment to be used, whether or not shown on the drawings, by marking an X in each appropriate check-box and entering the information called for in each space. If space is inadequate, enter "See misc." and describe under item 27 or on an attached sheet.
3. Work not specifically described or shown will not be considered unless required, when the minimum acceptable will be assumed. Work exceeding minimum requirements cannot be considered unless specifically described.
4. Include no alternates, "or equal" phrases, or contradictory items. (Consideration of a request for acceptance of substitute materials or equipment is not thereby precluded.)
5. Include signatures required at the end of this form.
6. The construction shall be completed in compliance with the related drawings and specifications, as amended during processing. The specifications include this Description of Materials and the applicable Minimum Construction Requirements.

1. EXCAVATION:
Bearing soil, type Sand Clay

2. FOUNDATIONS:
Footings: Concrete mix (1:3:5) 2500#   Reinforcing 2 - #4 bars
Foundation wall: Material Concrete Block   Reinforcing
Interior foundation wall: Material   Party foundation wall
Columns: Material and size   Piers: Material and reinforcing Concrete Block
Girders: Material and sizes SYP 6" x 10"   Sills: Material SYP 2" x 6" - Celcure
Basement entrance areaway   Window areaways
Waterproofing   Footing drains
Termite protection G.I. Termite Shields, Soil Treatment
Basementless space: Ground cover Polyethylene Membrane   Insulation 4" Rock Wool   Foundation vents 4-8"x16" Cast Iron
Special foundations

3. CHIMNEYS:
Material Brick   Prefabricated (make and size)
Flue lining: Material Vit. Terra Cotta   Heater flue size 4" Dia.   Fireplace flue size 13" x 13"
Vents (material and size): Gas or oil heater Transite 4" Dia.   Water heater Transite 3" Dia.

4. FIREPLACES:
Type: ☒ Solid fuel; ☐ gas-burning; ☐ circulator (make and size)   Ash dump and clean-out None
Fireplace: Facing Brick ; lining Fire brick ; hearth Brick ; mantel None

5. EXTERIOR WALLS:
Wood frame: Grade and species Studs: Douglas Fir #2
                               Other: SYP #2 DIM.   ☐ Corner bracing. Building paper or felt
  Sheathing Plywood ; thickness 1/2" ; width 4'-0" ; ☒ solid; ☐ spaced " o. c.; ☐ diagonal;
  Siding Redwood ; grade Heart ; type Bevel ; size 10" ; exposure 9" ; fastening Alum. Nails
  Shingles   ; grade   ; type   ; size   ; exposure   ; fastening
  Stucco   ; thickness   ". Lath   ; weight   lb.
  Masonry veneer Brick, used   Sills Brick   Lintels ¼" x 3" x 4" L
Masonry: Facing   ; backup   ; thickness ". Bonding
  Door sills   Window sills   Lintels
  Interior surfaces: Dampproofing,   coats of   ; furring
Exterior painting: Material Exterior Latex   ; number of coats 3
Gable wall construction: ☒ Same as main walls; ☐ other

6. FLOOR FRAMING:
Joists: Wood, grade and species #2 SYP ; other   ; bridging #2 SYP ; anchors ½" Bolts-6'-0" O.C.
Concrete slab: ☐ Basement floor; ☒ first floor; ☒ ground supported; ☐ self-supporting; mix 1:2:4 2500# ; thickness 4"
  reinforcing 6 6/10 10 wire mesh ; insulation   ; membrane Polyethylene
Fill under slab: Material Gravel   ; thickness 6".

7. SUBFLOORING: (Describe underflooring for special floors under item 21.)
Material: Grade and species Plywood, C-D grade   ; size 4'-0" x 8'-0" ; type Interior
Laid: ☒ First floor; ☒ second floor; ☐ attic   sq. ft.; ☐ diagonal; ☒ right angles.

8. FINISH FLOORING: (Wood only. Describe other finish flooring under item 21.)

Location	Rooms	Grade	Species	Thickness	Width	Bldg. Paper	Finish
First floor	LR, Den, DR, Hall	Select	Oak	25/32"	2¼	Yes	Fill, seal, varnish
Second floor	BR.1,2,3,4,Hall	"	Oak	25/32"	2¼	"	"
Attic floor					sq. ft.		

VA Form VB4-1852
FHA Form 2005

DESCRIPTION OF MATERIALS

DESCRIPTION OF MATERIALS

9. PARTITION FRAMING:
Studs: Wood, grade and species __#2, D. Fir__ Size and spacing __2x4, 16" O.C.__ Other _____

10. CEILING FRAMING:
Joists: Wood, grade and species __# SYP__ Other _____ Bridging __X-Type__

11. ROOF FRAMING:
Rafters: Wood, grade and species __#SYP__ Roof trusses (see detail): Grade and species _____

12. ROOFING:
Sheathing: Grade and species __"Plyscore" -D. Fir__; size __3/8"__; type __Ext. C-D__ ☒ solid; ☐ spaced ____" o.c.
Roofing __Asphalt Shingles__; grade __B__; weight or thickness __260#__; size __12"x36"__; fastening __G.I. Nails__
Stain or paint _____ Underlay __#15 Felt__
Built-up roofing _____; number of plies _____; surfacing material _____
Flashing: Material __G.I.__; gage or weight __26 Gu.__; ☐ gravel stops; ☐ snow guards

13. GUTTERS AND DOWNSPOUTS:
Gutters: Material __G.I.__; gage or weight __24__; size __4"__; shape __Ogee__
Downspouts: Material __G.I.__; gage or weight __24__; size __3__; shape __Rectangular__; number __4__
Downspouts connected to: ☐ Storm sewer; ☐ sanitary sewer; ☐ dry-well. ☒ Splash blocks: Material and size __Conc. 12" x 18"__

14. LATH AND PLASTER:
Lath ☐ walls, ☐ ceilings: Material _____; weight or thickness _____ Plaster: Coats _____; finish _____
Dry-wall ☒ walls, ☒ ceilings: Material __Gypsum__; thickness __½"__; finish __Latex__; joint treatment __3 coat Tape & Cem.__
Metal corner beads

15. DECORATING: (Paint, wallpaper, etc.)

Rooms	Wall Finish Material and Application	Ceiling Finish Material and Application
Kitchen	Alkyd semi-gloss, 3 coats	same
Bath s	Alkyd semi-gloss, 3 coats	same
Others	Latex Base, 2 coats	same

16. INTERIOR DOORS AND TRIM:
Doors: Type __Panel, Sliding__; material __Wood, Pine__; thickness __1-3/8"__
Door trim: Type __Ogee__; material __Fir__ Base: Type __Ogee__; material __Fir__; size __3/4"x4½"__
Finish: Doors __Paint - Alkyd Semi-gloss__; trim __Same as walls__
Other trim (item, type and location) __Book shelves, cabinet in den, stain and two coats flat varnish__

17. WINDOWS:
Windows: Type __D. H.__; make __Anderson__; material __Wood, W. Pine__; sash thickness __1-3/8"__
Glass: Grade __SS__; ☐ sash weights; ☐ balances, type __Spring__; head flashing __G. I.__
Trim: Type __Ogee__; material __Fir__ Paint __Same as walls__; number coats __3__
Weatherstripping: Type __Friction__; material __Brass__ Storm sash, number _____
Screens: ☒ Full; ☐ half; type __Aluminum__; number __24__; screen cloth material __Aluminum__
Basement windows: Type _____; material _____; ☐ screens, number _____; ☐ Storm sash, number _____
Special windows _____

18. ENTRANCES AND EXTERIOR DETAIL: Curtis #1719
Main entrance door: Material __W. Pine__; width __3'-0"__; thickness __1-3/4"__ Frame: Material __Pine__; thickness __5/4"__
Other entrance doors: Material __Fir, Alum.__; width __2'-8__; thickness __1-3/4"__ Frame: Material __Fir__; thickness __5/4"__
Head flashing __G.I.__ Weatherstripping: Type __Brass__; saddles __Oak, Alum.__
Screen doors: Thickness __5/4__"; number __1__; screen cloth material __Alum.__ Storm doors: Thickness ____"; number ____
Combination storm and screen doors: Thickness ____"; number ____; screen cloth material _____
Shutters: ☐ Hinged; ☒ fixed. Railings _____ Louvers __Wood - Triangular__
Exterior millwork: Grade and species __B.D. Fir__ /See Elev. Paint __Ext. Latex__; number coats __3__
Sliding Glass Door, 8" - 0" wide/tempered glass

19. CABINETS AND INTERIOR DETAIL:
Kitchen cabinets, wall units: Material __Ponderosa Pine__; lineal feet of shelves __34'-0"__; shelf width __12"__
Base units: Material __Ponderosa Pine__; counter top __Formica__; edging __Formica__
Back and end splash __Formica__ Finish of cabinets __Stain and Flat Varnish__; number coats __3__
Medicine cabinets: Make __Miami-Carey 14" x 20"__; model __White__
Other cabinets and built-in furniture __Shelves and Cabinets in Den__

20. STAIRS:

Stair	Treads		Risers		Strings		Handrail		Balusters	
	Material	Thickness	Material	Thickness	Material	Size	Material	Size	Material	Size
Basement										
Main	Oak	5/4"	SYP	3/4"	SYP	2 x 12	Birch	3"	Birch	1½x1½x28"
Attic										

Disappearing: Make and model number __"Precision" Deluxe, No. 89 - 30" x 54" R.O.__

21. SPECIAL FLOORS AND WAINSCOT:

FLOORS

Location	Material, Color, Border, Sizes, Gage, Etc.	Threshold	Base	Underfloor
Kitchen	Vinyl Tile 3/32" x 9" x 9"	Alum.	W.P. Underlay	Plywood Ext
Bath	Ceramic Tile	Marble	Cer. Tile	Concrete

WAINSCOT

Location	Material, Color, Border, Cap, Sizes, Gage, Etc.	Height	Height at Tub	Height at Shower
Bath	Ceramic Tile	4'-0"	6'-0"	

Bathroom accessories: ☐ Recessed; material _____; number ____; ☒ Attached; material Ceramic; number 6

22. PLUMBING:

Fixture	Number	Location	Make	Mfr's Fixture Identification No.	Size	Color
Sink	1	Kitchen	Elkay	LWR-2522-R	25" x22"	S. Steel
Lavatory	2	Baths	Rheem	G-105	20" x 20"	White
Water closet	3	Baths	Rheem	G-3000	27-1/8"x21-7/8"	White
Bathtub	2	Baths	Rheem	1034 Regency	30" x 60"	White
Shower over tub*	2	Baths	Rheem	D71060		Chrome
Stall shower**						
Laundry trays						

*☐ Curtain rod **☐ Door ☐ Curtain rod

Water supply: ☒ Public; ☐ community system; ☐ individual (private) system. ★
Sewage disposal: ☒ Public; ☐ community system; ☐ individual (private) system. ★
★ Show and describe individual system in complete detail in separate drawings and specifications according to requirements.
House drain (inside): ☒ Cast iron; ☐ tile; ☐ other _____ House sewer (outside): ☒ Cast iron; ☐ tile; ☐ other _____
Water piping: ☐ Galvanized steel; ☒ copper tubing; ☐ other _____ Sill cocks, number 3
Domestic water heater: Type Gas; make and model Coleman #33384
  recovery 35.3 gph. 100° rise. Storage tank: Material Glass lined; capacity 40 gallons.
Gas service: ☒ Utility company; ☐ liq. pet. gas; ☐ other _____ Gas piping: ☒ Cooking; ☒ house heating.
Footing drains connected to: ☐ Storm sewer; ☐ sanitary sewer; ☐ dry well. Sump pump _____

23. HEATING:

☐ Hot water. ☐ Steam. ☐ Vapor. ☐ One-pipe system. ☐ Two-pipe system.
  ☐ Radiators. ☐ Convectors. ☐ Baseboard radiation. Make and model _____
  Radiant panel: ☐ Floor; ☐ wall; ☐ ceiling. Panel coil: Material _____
  ☐ Circulator. ☐ Return pump. Make and model _____ ; capacity _____ gpm.
  Boiler: Make and model _____ Output _____ Btuh.; net rating _____ Btuh.

Warm air: ☐ Gravity. ☒ Forced. Type of system N. Gas
  Duct material: Supply G.I.; return G.I. Insulation _____, thickness _____ ☒ Outside air intake.
  Furnace: Make and model Carrier - 58G160VG Input 135,000 Btuh.; output 108,000 Btuh.
  1180 CFM - 85° Rise 24-1/8" x 28" x 56" high. 1/3 H.P. Motor

☐ Space heater; ☐ floor furnace; ☐ wall heater. Input _____ Btuh.; output _____ Btuh.; number units _____
  Make, model _____

Controls: Make and types Minneapolis-Honeywell, fully automatic

Fuel: ☐ Coal; ☐ oil; ☒ gas; ☐ liq. pet. gas; ☐ electric; ☐ other _____ ; storage capacity _____

Firing equipment furnished separately: ☐ Gas burner, conversion type. Stoker: ☐ Hopper feed; ☐ bin feed.
  Oil burner: ☐ Pressure atomizing; ☐ vaporizing
  Make and model _____ Control _____

Electric heating system: Type _____ Input _____ watts; @ _____ volts; output _____ Btuh.

Ventilating equipment: Attic fan, make and model Hunter EA 16 21" x 21"; capacity 1750 cfm.
  Kitchen exhaust fan, make and model Thermador #H70, 36" Hood
Other heating, ventilating, or cooling equipment _____

24. ELECTRIC WIRING:

Service: ☒ Overhead; ☐ underground. Panel: ☐ Fuse box; ☐ circuit-breaker 150 Amp. Service Number circuits 10
Wiring: ☐ Conduit; ☐ armored cable; ☒ nonmetallic cable; ☐ knob and tube; ☐ other _____
Special outlets: ☒ Range; ☐ water heater; ☐ other _____
☒ Doorbell. ☐ Chimes. Push-button locations Front door, rear door

25. LIGHTING FIXTURES:

Total number of fixtures 16 Total allowance for fixtures, typical installation, $400.00
Nontypical installation _____

DESCRIPTION OF MATERIALS

DESCRIPTION OF MATERIALS

26. INSULATION:

Location	Thickness	Material, Type, and Method of Installation	Vapor Barrier
Roof			
Ceiling	4"	Rock wool, blanket. Stapled between frame	Foil
Wall	3"	Rock wool, blanket. Stapled between frame	Foil
Floor	3"	Rock wool, blanket. Stapled between frame	Foil

27. MISCELLANEOUS:
(Describe any main dwelling materials, equipment, or construction items not shown elsewhere):
Mill made stairs: Morgan, Rail #720, Balaster #777, Newel Post #895, Turnout #723, Starting Step #779 T-R.
Fireplace: Donley Damper #243
Copper weather vane vent above carport

HARDWARE: *(Make, material, and finish)* Schlage - "A" series, Plymouth type - Bronze
Baths -- Chrome - Bronze A40S, Interior - A40S
Front and Rear Exit -- A50PD

SPECIAL EQUIPMENT: *(State material or make and model.)*
Venetian blinds .... Number ....   Automatic washer ....
Kitchen range   Range top - Hotpoint #RRG (V) 822   Clothes drier ....
Refrigerator ....   Other ....
Dishwasher   Hotpoint - SRV - 80   Oven - Whirlpool, 24" - #RRG (V) 155
Garbage disposal unit ....

PORCHES:

TERRACES: 7'- 6" x 9' - 4" Rear.   Conc Base, Brick Surface

GARAGES: Carport - Conc. slab floor, ground supported, pierced brick screen wall w/ 8" x 8" solid brick pilaster, drywall ceiling

WALKS AND DRIVEWAYS:
Driveway: Width 10"-0"   Base material gravel ; thickness 6". Surfacing material concrete ; thickness 4"
Front walk: Width 4! Material Conc. ; thickness 4". Service walk: Width 3 Material Conc.; thickness 4"
Steps: Material .... ; treads ...."; risers ...." . Cheek walls ....

OTHER ONSITE IMPROVEMENTS:
(Specify all exterior onsite improvements not described elsewhere, including items such as unusual grading, drainage structures, retaining walls, fence, railings, and accessory structures.)

LANDSCAPING, PLANTING, AND FINISH GRADING:
Topsoil 2" thick; ☒ Front yard; ☒ side yards; ☒ rear yard to 30' - 0" feet behind main building.
Lawns *(seeded, sodded, or sprigged)*: ☒ Front yard seeded ; ☒ side yards seeded ; ☒ rear yard seeded
Planting: ☐ As specified and shown on drawings; ☒ as follows:

	Shade trees, deciduous, ..." caliper.		Evergreen trees, ...' to ....', B & B.
3	Low flowering trees, deciduous, 4' to 6'	18	Evergreen shrubs, 1' to 2', B & B.
	High-growing shrubs, deciduous, ...' to ....'		Vines, 2-year
	Medium-growing shrubs, deciduous, ...' to ....'		
	Low-growing shrubs, deciduous, ...' to ....'		

IDENTIFICATION.—This exhibit shall be identified by the signature of the builder, or sponsor, and/or the proposed mortgagor if the latter is known at the time of application.

Date ....   Signature ....

Signature ....

VA Form VB4–1852
FHA Form 2005

Yet the numbering system is open enough to allow unlimited sections to be introduced as the need arises without affecting the sequence of the suggested format. Thus the framework is a numbering system that can be internationally consistent, while lending itself to convenient data retrieval and the newest automated printing techniques.

Another reason for the widespread and continual use of the format is its adaptation for use other than spec writing. With feedback and cooperation from various trade and manufacturing associations, the format is now used as a *Standard Filing System and Alphabetical Guide* for the gargantuan amount of information that is available to the industry. Organization and retrieval of this information are now simplified. The format has also been adapted as a *Suggested Guide for Field Cost Accounting,* which organizes the areas of work into an orderly sequence for the Contractor or Project Supervisor on construction projects. It is especially helpful in computer-aided techniques for project organization.

E. HOW TO FIND INFORMATION QUICKLY IN SPECIFICATIONS

A necessary part of reading working drawings is the ability to find related information in the specifications. Of course, it is helpful in reading specs, as in reading drawings, to have had experience in construction and to know the characteristics and terminology of building materials commonly used in the industry. Similarly, the more experience you acquire in reading specs, the easier they will be to interpret. It is worth mentioning that you should continually refer to the specs to verify points on the drawings that are not entirely clear.

First, look over the Table of Contents to see what general areas are covered and where their position is in the set. If you have the time, read the entire specs quickly to learn the system the writer has used to organize the material and how the information is stated. Then look in the section most closely associated with the information desired and read carefully and deliberately. For instance, if information concerning wood framing is desired, consult the Carpentry Section (in the CSI Format, the Wood and Plastics Division); for concrete finishing, consult the Concrete Section, and so forth. Try to associate the information you are searching for with the trade that commonly conducts that particular work, and look first in the section dealing with the corresponding trade. Sometimes tradework overlaps and additional searching may be necessary to gather everything you need. At times, the spec reader must, in fact, play the role of a detective and use deduction to track down the desired data. Furthermore, be persistent. Always make a mental note of the items observed during previous searches.

To find information pertaining to the terms of the contract, payments, supervision, permits, and general points affecting all tradework, look in the General Conditions, keeping in mind that the General Conditions section (as well as the Supplementary General Conditions section) applies to all the technical tradework.

A typical technical section begins with a paragraph that tells in an abbreviated manner *what is to be done.* Work included and *work not included* often follow. Remember that work not included may be the responsibility of another trade, and, consequently, further information on the subject may be found in another section. The next paragraph generally covers materials to be supplied. This deals with basic definitions of the materials and may be further elaborated on in later *installation* and *workmanship* paragraphs, which state how the work is to be done. As a rule, the remainder of the section is devoted to tests, samples, guarantees, shop drawings, and so on.

Notice that the word *shall* is commonly used to command the contractor to furnish something or do something. On the other hand, the word *will,* being less strong, is used in connection with the owner furnishing something; for example, "the owner *will* furnish temporary water," and so on.

The general contractor, the superintendent, the REI, and the architect are among the few concerned with entire specs on the job. Material suppliers and subcontractors are only concerned with the technical sections relating to materials and work in their trades, including items in the General Conditions that affect them.

Material standards, often indicated by ASTM (American Society of Testing and Materials) numbers and other standard numbers, are used to establish minimum qualities of materials. The numbers are significant to subcontractors because they deal with them daily in their special trades. But the numbers may mean little to the student or layperson; usually, he or she passes over them without reading unless there is doubt about the material specified. In that case, the society's standard publication should be checked, or the material in question could be tested if necessary. However, the use of the numbers eliminates unnecessary descriptive wording in specifying quality standards in specs.

During the preparation of a bid, the general contractor should check the technical specs against the Table of Contents to see that none of the pages have been accidentally omitted.

Small House
Sample Set of
Specifications
Using the C.S.I Format

Table of Contents

General Conditions . . . . .ii–iii

Specifications

DIVISION 1– GENERAL REQUIREMENTS 1–4

DIVISION 2– SITE WORK 4–5

DIVISION 3– CONCRETE 5–7

DIVISION 4– MASONRY 7–8

DIVISION 5– METALS (Not used)

DIVISION 6– WOOD & PLASTICS 8–9

DIVISION 7– THERMAL & MOISTURE PROTECTION 10–12

DIVISION 8– DOORS & WINDOWS 12–14

DIVISION 9– FINISHES 14–16

DIVISION 10– SPECIALTIES (Not used)

DIVISION 11– EQUIPMENT (Not used)

DIVISION 12– FURNISHINGS (Not used)

DIVISION 13– SPECIAL CONSTRUCTION (Not used)

DIVISION 14– CONVEYING SYSTEMS (Not used)

DIVISION 15– MECHANICAL 17–18

DIVISION 16– ELECTRICAL 18–19

GENERAL CONDITIONS

1. <u>A.I.A. General Conditions</u>: The printed form A-201, "General Conditions of the Contract for the Construction of Buildings," latest edition, issued by the American Institute of Architects, is part of these specifications as if written in full herein. In case of conflict, these specifications take precedence over and modify the aforesaid printed form.

2. <u>Subcontractors</u>: This section of the specifications shall apply in full to subcontractors as well as to the General Contractor, as if written in full in each specific section to follow.

3. <u>Award of Contract</u>: If made, will be on standard A.I.A. Contract Form, and work shall commence with adequate force within ten days of date thereof, thence to proceed with dispatch until fully completed.
    a. <u>Completion date</u>: The work must be completed within 100 consecutive calendar days after award of the contract, lawns and landscaping excluded.
    b. <u>Liquidated damages</u>: If the contractor should fail, refuse, or neglect to complete the work within the time allowed, the Owner shall retain from amount to be paid the Contractor the sum of one hundred dollars ($100.00) per day as liquidated damages for each day the work remains incomplete.

4. <u>Performance and Payment Bonds</u>:
    a. A <u>Performance Bond</u> equal to 100% of the contract price must be furnished within ten days of date of contract. Bond shall be on form provided, written by an approved surety company authorized to do business in this state, and executed by an authorized resident agent of this state. Power-of-attorney of the signing official shall be attached to each copy thereof.

    Performance bond shall extend as a maintenance bond for one year after acceptance of work, as a guarantee that any defects which may develop during that time will be remedied.
    b. <u>Labor and Materials Bond</u> equal to 50% of the contract price must be furnished under the terms above stated.

5. <u>Owner, Designer, and Supervision</u>
    a. <u>Owner</u>: The term "Owner" used herein refers to Mr. and Mrs. B. W. Burker. Unless otherwise stated, all papers required to be delivered to the Owner shall be forwarded through the Designer.
    b. <u>Supervision</u>: If awarded, work will be supervised by the Designer's inspector.
    c. <u>Inspection of work</u>: Afford Designer and public authorities every facility for inspecting work.

6. <u>Insurance</u>
    a. <u>General</u>: The General Contractor shall carry the following insurance and shall give evidence that it is in effect. Extent of coverage required does not limit the Contractor's responsibility but rather the extent to which he will be required to protect himself.
    b. <u>Fire and Lightning</u>, including Extended Coverage and Vandalism, in the full insurable amount of the Contract, payable to the Owner, the Contractor, and all subcontractors as their interests may appear.

c. <u>General Liability</u> in the following minimum amounts:

Bodily injury	$500,000/1,000,000
Property damage	$500,000/1,000,000
Automobile	$500,000

d. <u>Worker's Compensation</u> as required by State Law.

e. Subcontractors shall carry Worker's Compensation and adequate liability insurance, and, if required, shall produce evidence thereof.

7. <u>Plans and Specifications</u> are complementary. Whatever is shown or reasonably inferred from either shall be as if required by both. If drawings and specifications conflict or require clarification which was not obtained prior to bidding, the Designer's interpretation of the true intent shall govern. Otherwise, the order of the precedence shall be:

   (a) General Conditions,   (b) Detail Specifications,
   (c) Large-scale drawings, (d) Small-scale drawings

<u>No deviation</u> shall be made from plans or Specifications except upon written order from the Designer or Owner.

8. <u>Construction Sets</u>: Contractor will be entitled to <u>10 sets</u> of plans and specifications. Additional sets may be obtained for cost of printing. One complete set must be kept on the job at all times. All drawings, specifications, and copies furnished by the Designer are his or her property and are not to be used on other work.

****

DIVISION 1—GENERAL REQUIREMENTS

<u>01010 SUMMARY OF WORK</u>

The General Contractor shall be held responsible for the execution of the work in accordance with the true intent of the drawings and specifications, which is to effect a completed first-class job and to furnish all labor and materials required therefor, whether or not each and every item is specifically mentioned.

<u>01020 APPLICABLE LAWS AND REGULATIONS</u>

All applicable Federal, State, County, and Municipal laws and regulations shall be complied with in constructing the building. In case of conflict, the most stringent requirement shall apply. Furnish certificates of inspection from all authorities having jurisdiction.

<u>01050 SURVEYS AND LEVELS</u>

1. <u>Establish lines and levels</u> from Site Plan and other drawings, and be responsible for their correctness.

2. <u>Designer's approval</u> of layout required before proceeding with excavation. Contractor shall employ at his expense an approved independent Registered Civil Engineer to certify correctness of all dimensions and level before any work is started.

3. <u>Bench Marks</u> shall be established and maintained at widely separated points until completion of work.

01300 SUBMITTALS

1. <u>Shop Drawings</u>: General Contractor shall submit four copies of all shop drawings to the Designer for approval <u>after thoroughly checking each one and affixing his signature of approval</u> thereto. The Designer will approve drawings for design only, and will assume no responsibility for dimensions or quantities.

2. <u>In requiring shop drawings</u>, the Designer is endeavoring to guard the Contractor against error, but approval thereof will in no way relieve the Contractor from his responsibility for executing the work in accordance with the contract, unless changes introduced are specifically mentioned in the letter of transmittal.

01350 PAYMENTS

1. <u>A Schedule of Cost</u> of the various divisions of work shall be submitted immediately after award of contract and upon approval shall form the basis for progress payments.

2. <u>Partial payments</u> on account less 10% will be made monthly, within ten days of receipt of application, for work completed and materials stored on the site during the preceding month. Each application for payment shall be made in tabulated form, indicating original schedule of costs and the amount earned in each division up until the last day of the preceding month. Applications shall be submitted in triplicate to the Designer and are subject to his approval.

At least 90% will be paid before the building is occupied by the Owner.

3. <u>Final Payment</u> of the remaining 10% will be made not sooner than 30 days after completion of the work and its acceptance by the Owner, provided that satisfactory evidence has been presented that all indebtedness connected with the work has been paid.

4. <u>Delay of Completion</u>: If, after substantial completion, full completion is materially delayed through no fault of Contractor and the Designer so certifies, the Owner will, without termination of the contract, make payment of the balance due for the portion of the work that is completed and accepted.

01360 EXCAVATION PRICE AGREEMENT

The <u>Contractor</u> shall submit a unit price for the following items which will be used to form a unit price agreement for extra or deleted work:

Footing Excavation	Per Yard
Rock Excavation	Per Yard
Foundation Wall	Per Sq. Ft.

01400 QUALITY CONTROL

1. <u>Testing</u> of materials will be required as specified or if, in the opinion of the Designer, portions of the work fail to meet requirements of plans or specifications or may prove insufficient for purpose intended. Testing shall be done by an agent or laboratory acceptable to the Designer with all costs paid by the Contractor.

2. <u>All field tests</u> shall be made in the presence of the Designer.

01420 WARRANTIES, GUARANTEES, ETC.

1. <u>General</u>: All warranties, guarantees, etc., shall bear the date of final acceptance by the Owner and shall be delivered on completion of

the work. The Contractor shall also furnish diagrams, maintenance manuals, parts lists, and other information pertaining to all equipment installed, together with names of nearest distributor and maintenance representative.

2. <u>General Warranty</u>: The General Contractor shall furnish a written guarantee of workmanship and materials for a period of one year, during which time he agrees to remedy all defects without charge.

3. <u>Roofing</u>, related sheet metal work, roof drains, etc., shall be covered under the above guarantee for a period of three years.

01500 TEMPORARY FACILITIES

1. <u>All permits, taxes, licenses, royalties, fees, etc.</u>, required for completion of work shall be provided.

2. <u>Water, power, and gas</u> required for construction shall be furnished with all costs paid.

3. <u>Adequate safety precautions</u> shall be taken, including barricades, signal lights, etc., to ensure protection of all workers and the public during construction.

4. <u>Job Office</u> and adequate facilities for storage of materials and equipment shall be provided.

5. <u>Provide telephone</u> in job office. Toll calls shall be paid for by party making call.

6. <u>Provide temporary toilets</u>, acceptable to local health authorities. Before completion of job, toilets shall be removed, all holes filled, and site left in clean sanitary condition. As soon as possible, locate a temporary toilet in building and connect to sewer.

01550 USE OF PREMISES

The Contractor shall confine his apparatus and operations to limits defined by the Owner. Enforce the Owner's instructions regarding signs, fires, parking, smoking, etc.

01600 MATERIALS AND WORKMANSHIP

1. <u>All material</u> furnished shall be new and without any indication of damage or breakage. If usually packaged, it shall be brought to job in original unbroken, labeled containers.

2. <u>Brand names</u> mentioned together with the phrase "or equal" indicates that other makes of equal quality or suitability may be used subject to <u>written approval</u> of the Designer. The Designer reserves the sole right to determine equality of materials.

3. <u>Materials not specified</u> but required shall be the best adapted to the purpose.

4. <u>Installation</u> of any material or product shall be in accordance with the manufacturer's directions and specifications.

5. <u>Workmanship</u> shall be equal to best standard practice, with work performed by expert, skilled craftsmen only. A competent superintendent shall be in charge of the work at all times and may not be replaced unless approval is obtained, or unless he ceases to be in the employ of the Contractor.

01650 CLEAN UP

1. Contractor shall <u>protect</u> all work during construction and repair or replace all damaged portions.

2. <u>Clean all metal</u>, exposed finished surfaces, etc., and clean and polish all glass inside and outside of building, leaving building in first-class condition. Remove all rubbish and surplus material from property.

****

DIVISION 2—SITE WORK

<u>02100 CLEARING</u>

1. <u>Trees and shrubs</u>: Remove from within building perimeter and as otherwise noted on plans, and carry from site; large trees to be removed shall be topped, trimmed and felled in such manner as to avoid damaging other trees. Grub stumps to depth of three feet. Protect all remaining trees and plants from damage; provide barricades if within 25 feet from any construction activity. Limbs shall be cut or trimmed at the direction of the Owner.

2. <u>Stripping</u>: Clear all vegetation matter and top soil from within area of building and five feet beyond, and from area for drive, parking area, and terrace. Stockpile all top soil for use in final grading, as noted below.

<u>02200 EXCAVATION</u>

1. <u>Excavation</u> shall be to exact elevations and dimensions as shown on plans. Excess out must be filled with concrete; no earth backfill will be allowed.

2. <u>If rock or latent soil is encountered</u>, the contract price will be adjusted in accordance with the unit price agreement. Rock is defined as a ledge or boulder over one-half cubic yard in volume, which cannot be broken or removed without explosives or drills. Volume shall be restricted to vertical projection of neat footing size, or to next wall or room size plus one foot working space on each side.

<u>02250 BACKFILLING</u>

Any necessary fill under concrete floor slabs shall be deposited in thin layers, slushed, tamped, and compacted; also, the concrete shall not be poured until all settlement has occurred and earth has been inspected and approved by the Designer.

Any addition required for fill under slabs shall be crushed stone or gravel four inches thick.

<u>02260 GRADING</u>

Provide such cutting, grading, and filling as is necessary to transform existing grades to finish grades as indicated on the drawings; all exterior grades shall be re-formed so as to drain away from the building on all sides.

<u>02800 LANDSCAPING</u>

The <u>Contractor</u> shall include in his bid a lump sum of $1500.00 to cover the cost of landscape work including finish grading and planting of trees, grass, and shrubs. If the cost is less than the allowance, the Owner is to receive the credit, and if more, the Owner will reimburse the Contractor.

****

DIVISION 3—CONCRETE

<u>03300 CONCRETE</u>

All concrete shall be furnished by an approved ready-mix plant, and have twenty-eight day compressive strength of 2500 P.S.I. Footings are to be of nonair-entrained concrete. Certificates shall be furnished from the mixing plant, indicating the strength, quantity, and date of each day's pour.

03350 HANDLING

1. <u>Concrete shall be placed as soon as possible</u> after mixing with maximum interval of 30 minutes from ready-mix plant. Air-entrained concrete shall not be mixed longer than 2 minutes. Deposit as near as possible to final position; avoid walking or shoveling. Proceed at such pace that no surface will have attained initial set before addition concrete is placed thereon.

2. <u>No water</u> shall be added to ready-mix at job.

3. <u>Concrete</u> shall be hand-spaded or worked with mechanical vibrator to ensure that corners are filled and voids are eliminated.

4. After beginning of pouring, work must proceed to completion, or up to a previously approved construction point.

5. <u>No work</u> shall be done during or before freezing weather. Trenches must be dry and free of mud.

03360 FORMS

1. Forms to be accurately constructed of clean, sound material true to line and level, and braced to prevent any sag or deformation; wet down before placing concrete. Footings may be formed with solid, clean soil banks.

2. <u>No penetration</u> of slab by screed stakes or similar objects will be permitted.

03370 WATERPROOFING

Waterproofing shall consist of No. 6 Poly Film placed under the floor slab and terrace slab. Overlap joints of film at least 6 inches.

03380 INSULATION

Two inches of rigid insulation shall be laid around edge of floor slab as indicated on drawings.

03390 REINFORCING STEEL

1. <u>Mesh</u> shall be welded wire fabric 6 × 6 × W2.1 × W2.1.

2. Minimum concrete protection: 1" in slabs.

3. Inspection and approval by Designer required before pouring concrete.

03400 CONCRETE FINISHES

1. Concrete floor slabs shall be depressed to receive the various types of finishes to be applied so that all finish floor areas will be the same level throughout.

2. Slabs shall be screeded to an even surface by the use of straight edge and screeding strips accurately set to the proper levels.

3. <u>All floors to remain exposed concrete</u> shall be steel troweled to a true, even plane without trowel marks or other surface imperfections with a tolerance of $\frac{1}{8}$".

4. <u>All areas to receive ceramic tile</u> or flagstone shall be left rough, after screeding to proper level.

03410 CURING CONCRETE

1. All slabs shall be <u>sprayed with water</u> at least twice daily for two days after pouring.

    2. <u>Cover all slabs</u> with 2" sand or sawdust and keep moist for 28 days.

DIVISION 4—MASONRY

04100 MORTAR

1. <u>General</u>: All mortar to be accurately measured by volume, mixed in a batch mechanical mixer at least 3 minutes after all ingredients added. Retemper as required with as little water as needed to keep workable. Discard if not used within one hour. Mortar boxes and boards to be cleaned for each batch and at end of day's work.

2. <u>For typical masonry</u>: Magnolia Masons Mix, Medusa Brixser, Atlas Mortar Cement, or Brixment, light colored nonstaining, Type II, masonry cement, using 1 part cement to $2\frac{3}{4}$ parts clean best grade sand. Use only one brand throughout job, all delivered to site in unbroken bags and stored above ground under cover. Sand to be graded fine to coarse, to be passed through No. 8 sieve before using.

3. <u>Walls below grade</u>: $\frac{1}{2}$ part Portland Cement, $\frac{1}{2}$ part typical Masonry Cement, $2\frac{3}{4}$ parts sand.

04150 WORKMANSHIP

1. The <u>quality of work to be first-class</u> with all work plumb and true to line. Vertical joints in alternate courses shall align. Horizontal courses to be level.

2. <u>Frost protection</u>: Do not work when temperature is below 40 degrees or when temperature is expected to be below freezing within 24 hours. Protect against freezing for at least 48 hours after laying in any case. Cover top of walls at end of each day's work and when rain appears imminent.

3. <u>Built-in work</u>: Build in all anchors, nailing inserts, lintels, flashings, etc., as shown or required.

04200 CONCRETE BLOCK

1. Concrete block to be lightweight made in accordance with A.S.T.M. specifications.

2. Use a <u>running bond</u> with stretchers overlapping one-half in alternating courses. All joints shall be full, both horizontal and vertical. Use $\frac{3}{8}$" tooled joints. <u>Do not clean</u> either block or stone with acid solutions.

04400 STONE

1. All stone used in walls and chimney to be Indiana limestone. Color to be approved by Designer.

2. <u>Flagstone</u> in floors to be dark gray, 1" thick, placed in pattern shown on drawings with maximum mortar joints of $\frac{1}{2}$".

3. <u>Storage</u>: Stack at site under cover and protect from damage and mud stains.

****

DIVISION 5—METALS (Not used)

DIVISION 6—WOOD & PLASTICS

06100 ROUGH CARPENTRY

1. All work shall be erected plumb, true, and square at corners. Warped or defective lumber shall be discarded.

2. Studs shall be spaced 16" o.c. and doubled around all openings. Double studs shall also be placed under each roof beam bearing on stud wall. Provide 1" × 4" let-in diagonal bracing at all corners to prevent racking.

3. <u>Wall plates</u> shall be double 2" × 4" of long lengths.

4. <u>All exposed roof beams</u> and wood ceilings shall be erected with care so that no nails or hammer marks are exposed. Exposed nailing shall be done with finish nails, set and filled.

06110 LUMBER MARKINGS

Each piece of yard lumber (not boards) and plywood shall be grade marked for identification in accordance with the recommendation of the association having jurisdiction.

06120 LUMBER GRADES

1. <u>Framing lumber</u> shall be equal to one of the following:

   Southern Yellow Pine, No. 2 dimension

   Douglas Fir standard J & P, No. 2

2. <u>Roof beams</u> shall be Douglas Fir standard J & P, No. 1.

3. <u>Roof-plank decking</u> shall be 2" × 6" Southern Yellow Pine, No. 1 Dim., kiln dried.

4. <u>Plywood</u> shall be Douglas Fir conforming to U.S. Department of Commerce Commercial Standard CS-45-48 and shall be branded or stamped with type and grade.

06200 FINISH CARPENTRY

1. Layout, cut, fit, plumb, brace, and level all items of finish carpentry work covered under this division. Do all necessary cutting of work as required by other trades for the installation of their components.

2. <u>Install interior millwork and trim</u> with tight joints securely nailed, set, and filled, ready for finish. Secure interior trim with finishing nails and exterior trim with common nails, set for putty. Sand interior woodwork as necessary to remove irregularities and machine marks. Leave work free of defects and blemishes.

3. <u>Trim and moldings</u> shall be stock patterns, unless otherwise indicated; use Southern Yellow Pine, B grade; moisture content shall not exceed 12%.

4. <u>Wood shelving</u> shall be Western White Pine, Grade #2.

06210 CABINET WORK

1. Cabinets shall be manufactured by qualified and established cabinet shop in accordance with details and the best shop practices. Fabricate in sections as suitable for installation on job.

2. <u>Materials</u> to be interior grade A-D Fir plywood for ends, backs, bottoms, and shelves; grade A-A if exposed to both sides. Fronts of cabinets in kitchen to be oak to match paneling. Shop cabinet fronts to be B grade SYP and plywood.

3. <u>Counter tops</u> to be Formica of color selected, mounted on 3/4" plywood with minimum joints. Edges to be finished with face-glued matching Formica strips.

4. <u>Shop drawings</u> shall be submitted for approval.

06230 WOOD PANELING

Paneling in family room shall be Formica V.I.P. English Oak with $\frac{1}{2}$" stained reveal, using No. 39 spline. All paneling in other areas to be prefinished veneer plywood, $\frac{1}{4}$" thick with random V-grooved planks.

DIVISION 7—THERMAL & MOISTURE PROTECTION

07150 WOOD-PRESERVATIVE TREATMENT

Bottom sills and lower two feet of studs shall be pressure treated with Celcure, Wolman, or equal standard vacuum-pressure preservative treatment.

07200 INSULATION

1. <u>Work included</u>:   (a) Thermal insulation of ceilings above wood decking.

   (b) All Exterior Walls.

2. <u>Materials</u>:   (a) One-inch rigid, waterproof insulation board above deck. A vapor barrier of No. 15 felt shall be laid between roof planks and the rigid insulation.

   (b) Four-inch mineral wool batts with reflective foil vapor barrier on room side.

3. <u>Installation</u>: After all piping and wiring is in place, install batts in walls. Cut neatly and lap vapor barrier not less than $\frac{1}{2}$" on room side of wall.

07300 ROOFING

1. <u>Roofing</u> shall be "Bird" Fiberglass Shingles as manufactured by Bird & Son or equal quality, minimum weight 235 lbs. per square. Color to be selected by Owner.

2. <u>Roofing felt</u> to be first quality, asphalt-saturated rag felt, No. 15.

07310 APPLICATION OF ROOFING

1. Over a well-nailed and clean deck apply 2 layers of No. 15 felt, covering the lower 48" of the lapped felt with roofing cement to seal it over outside walls and overhang. Fasten each strip shingle with 4 galvanized nails, allowing 5" to the weather.

2. Work shall be done by a bonded contractor approved by the manufacturer.

07320 ROOFING GUARANTEE:

Furnish full-volume bond (labor and materials) with flashing endorsement, covering a period of <u>15 years</u> from acceptance of job.

07600 FLASHING

1. <u>Base flashing</u> shall extend up vertical surfaces not less than 8" and out on flat surfaces not less than 4". Base flashings shall be formed in sections not to exceed 10 feet long. Units shall be connected together with a 3" wide loose-lock slip joint filled with plastic cement before units are hooked together. Corners shall be lapped and soldered in a flat position.

2. <u>Counter flashing</u> shall be formed in lengths not to exceed 10 feet. The counter flashing shall extend in reglet or mortar joint 1", with edge turned up $\frac{1}{4}$" and down over base flashing not less than 4". Joints shall lap 4". Metal shall be formed to provide spring action against the base flashing. Fill joints or reglets with plastic cement.

07610 CHIMNEY FLASHING

Shall extend over roofing material not less than 5", and upon vertical face of chimney minimum of 8". At corners base flashing shall be connected by a lapped or soldered seam. Provide counter flashing for chimney.

07620 GUTTERS

Shall be molded or half-round, in lengths of not over 10". Ends shall be jointed by 1" lapped, riveted, and soldered seams. Lap joints in direction of flow. Provide rolled or formed beads on front side of gutters.

1. Provide for expansion at high points of run.

2. Hangers for support of gutters shall be galvanized adjustable-strap or shank-and-circle type, spaced not more than 3' apart and securely fastened to roof.

3. Slope gutters to outlets not less than 1" in 20'.

4. Outlet tubes shall be formed with locked and soldered longitudinal seams. The upper end of tube shall be flanged and soldered to gutter. Tube shall extend into leader not less than 3".

5. Basket-type strainers formed of No. 14 gage galvanized wire shall be provided at outlet tubes and shall fit snugly in the outlet tube.

07625 LEADERS

Shall be formed in 10-foot lengths. Longitudinal joints shall be locked. End joints shall telescope not less than $1\frac{1}{2}$" and be held in place with solder. Include necessary elbows and offsets. Support leaders with hooks driven into wall. Space hooks not over 10' apart.

07900 CAULKING

1. _Materials_: Pacora, Kuhla, or Dicks-Armstrong-Pontius first-quality, gun-grade, nonstaining, elastic caulking compound, meeting Federal Specification TT-C598, Grade 1, applied without thinning.

2. _Typical Application_: (Around exterior doors, windows, etc., and elsewhere shown or required) Clean outside joints and crevices surrounding wall openings. Joints shall then be filled solid with caulking compound forced into place with gun under pressure, and surface neatly tooled. On wood surfaces, apply after priming coat but before last coat of paint. Fill deeper joints with oakum before applying finish caulk.

****

DIVISION 8—DOORS & WINDOWS

08200 WOOD DOORS

1. _All exterior and interior wood doors_: "Weldwood" as manufactured by U.S. Plywood Corp. or "Roddiscraft" as manufactured by Roddis Plywood Corp. sizes and types as shown on drawings. Interior doors shall be hollow core, paint grade; exterior to be solid core, stain grade.

2. _Wood sliding glass doors_: "Pella" as manufactured by Rolscreen Co., No. OX-033, size and type as shown on drawings.

3. _Front entrance jalousie doors_: Exterior type, manufactured by U.S. Plywood Corp., sizes shown on drawing, shall be affixed to stops with no hinges or knob hardware. Provide screens for the jalousie units, and glazed panels for winter.

4. Glass in all exterior doors to be _double glazed_.

5. Sand and hand-smooth all wood doors before finishing.

08600 CASEMENT WINDOWS

1. <u>Wood casement windows</u>: "Windowwalls" as manufactured by Anderson Corp. or equal quality, sizes, and types as shown on drawings. Operating hardware and screens to be included with units.

2. Provide <u>double-glazed</u> panes in all window units.

08700 HARDWARE

1. This section includes <u>Builder's Finish Hardware</u> and related items required to complete the work specified and/or indicated on the drawings.

2. Hardware shall conform to the following Federal Specifications and the type numbers specified:

   (a) Locks and door trim FFH-106a

   (b) Hinges FF-H-116a

3. <u>All cylinder locks</u> supplied under this section shall be the product of one manufacturer.

   (a) Hubs or cylinder knob locks for exterior doors shall be brass or bronze.

   (b) Cylinder rings of wrought bronze of proper size to fit door thickness shall be supplied.

   (c) Hardware attached to flush hollow core doors where blocking is not provided shall be attached to the door by through-bolts with sleeve nuts.

08710 WEATHERSTRIPPING

1. Work includes:

   (a) Spring bronze at head and jambs of all exterior doors.

   (b) Aluminum thresholds for exterior doors.

   (c) Caulking of all thresholds under this section.

2. Equipment and materials:

   (a) Head and jamb stripping, equal to Accurate No. 600.

   (b) All stripping shall be adjusted as necessary for proper operation of doors and windows, and the entire installation shall present a continuous air barrier around the perimeter of all openings.

   (c) Thresholds shall be set on full bed of caulking compound, and secured with flat-head aluminum bolts operating in expansion shields, one being near each end and at not over 10" on centers at intermediate points.

08800 GLASS & GLAZING

1. This section includes the furnishing and setting of all glass as required to complete the work as indicated on the drawings and/or specified.

2. <u>Material</u>: <u>Glass</u> shall conform to FS DD-451a for clear Window Glass, Type B; Polished Plate Glass, Type 1, glazing quality.

3. Manufacturers labels showing strength and quality will be required on all glass.

4. Glazing compound and putty shall be equal in quality to Hamstrong, "Glaze-All."

5. <u>Glazing</u>: Surfaces shall be dry and free from dust, rust, or ice before glazing. Dirt surfaces shall be cleaned with a cloth saturated with tur-

pentine or mineral spirits before glazing. Putty shall not be applied in temperatures below 40°F or during damp or rainy weather. Do not handle windows after glazing until the putty or glazing compound has set.

****

DIVISION 9—FINISHES

09250 GYPSUM WALLBOARD

This section includes the furnishing of $\frac{1}{2}$" "Drywall" for walls and ceilings as shown on drawings.

1. <u>Materials</u>:
   (a) $\frac{1}{2}$" thick gypsum wallboard, meeting Federal Specifications SS-51a and ASTM specifications C-36.
   (b) <u>Use metal corner beads</u> on all exterior corners. Metal bead trim shall be U.S. Gypsum No. 200-A metal trim.
   (c) <u>Nails</u> shall be GWB-54, $1\frac{5}{8}$" long annular ring, meeting the requirements of ASTM C-380.

2. <u>Installation</u>:
   (a) <u>Apply Gypsum Wallboard</u> first to ceilings and then to side walls; plan wallboard lengths to minimize end joints.
   (b) <u>Space nails</u> not more than 7" apart on ceiling panels and not more than 8" apart on walls. Dimple the nail heads slightly below surface of wallboard.
   (c) <u>Finish joints</u>, nail dimples, corners, and edges using the three-coat cement and tape system. Sand each coat after it is dry to a smooth, even surface.
   (d) <u>Wallboard installation</u> shall be done by experienced drywall applicators, normally engaged in the trade.

09300 CERAMIC TILE

This section includes furnishing and installing ceramic tile finish on all surfaces as indicated on drawings.

1. <u>Material</u> to be Mosaic Tile Co., American-Olean, or Cambridge Tile Co. manufactured tile.

2. <u>Floor tile</u> to be standard grade, factory mounted unglazed ceramic tile with square edges. Sizes, patterns, and colors to be selected from Harmonitone solid color or Velvetex mottled-color group. Set in 1:3 mix Portland cement bed at least $\frac{3}{4}$" thick with uniform joints not wider than $\frac{1}{16}$", with best waterproof white Portland cement.

3. <u>Walls and Wainscot</u> to be $4\frac{1}{4} \times 4\frac{1}{4}$ standard grade cushion edge, white nonvitreous body glazed tile in color selected from Harmonitone Satin or Bright Glaze group. Edges to have spacers to provide uniformly narrow joints. Provide all necessary trim shapes including 2" cap.

4. <u>Bath accessories</u> to be recessed nonvitreous body, glazed tile in same color as wainscot. Provide the following in each bath:
   (a) 1 paper holder, 1 soap dish, 1 grab bar.
   (b) 2–20" towel bars, 1–30" towel bar.

09650 CORK TILE FLOORING

<u>Cork tile flooring</u>, placed as shown on drawings, shall be $\frac{1}{8}$" thick, 9" × 9" as manufactured by Armstrong Cork Co. Color to be selected by Owner. Rubber base, 4" cove type, to be installed in all rooms having the cork tile floors. Installation shall be in accordance with the latest edition of the

Armstrong Cork specifications and shall be by a qualified Armstrong Contractor.

<u>09680 Carpeting</u> and rubber underlay pad are to be furnished under another contract.

09900 PAINTING & DECORATING

1. <u>Work includes all surfaces</u> requiring finishing unless expressly eliminated. Work to include touching-up prime coats, prefinished items, etc., finished by other trades.

2. <u>Acceptance of all surfaces</u> shall be assumed by the fact that the painting contractor has begun work. Thereafter, he shall be responsible for all surfaces.

09910 PAINTING MATERIALS

1. The <u>same brand</u> of paint, stain, varnish, etc., shall be used throughout, except in cases where unavoidable.

2. Products shall be best grade Du Pont, Pittsburgh, Sherwin-Williams, DeVoe, Pratt and Lambert, or Ben Moore, delivered to job in original unopened containers.

3. <u>Colors</u>, where applicable, to be selected by Owner.

4. <u>All paints and finishes</u> shall be applied as follows:

   (a) <u>Galv. iron and steel</u>: After treatment with vinegar, prime with oil-base primer, two coats of latex exterior paint.

   (b) <u>Exterior wood</u>: Oil-base primer coat, two coats exterior latex paint.

   (c) <u>Painted-interior wood trim</u>: Prime with oil-base interior sealer, two coats interior latex paint.

   (d) <u>Gypsum board walls and ceilings</u>: Two coats flat latex base paint. Baths and kitchen—two coats latex semigloss paint.

   (e) <u>Kitchen cabinets</u>: Stain to match paneling; two coats varnish.

   (f) <u>Wood shelving</u>: Apply one coat of boiled linseed oil.

09920 PAINTING WORKMANSHIP

1. <u>Application</u>: shall be in accordance with manufacturer's directions. Hardware, etc., shall be removed before painting and replaced thereafter. Exterior painting shall not be done in wet weather or when temperature is below 50 degrees F. Temperature shall be maintained continuously between 65 and 75 degrees F for all interior painting.

2. <u>Surfaces</u> shall be clean, dry, smooth, and free of dust, scratches, or hammer marks. Wood surfaces shall be lightly sanded before painting. Coat all knots and pitch streaks with shellac or Stopping Varnish before priming. Fill nail and minor blemishes with putty or plastic wood.

3. Allow first coat of all latex paint to dry a minimum of 12 hours before applying the second coat.

****

DIVISION 10—SPECIALTIES (Not used)

DIVISION 11—EQUIPMENT   (Not used)

DIVISION 12—FURNISHINGS (Not used)

DIVISION 13—SPECIAL CONSTRUCTION (Not used)

DIVISION 14—CONVEYING SYSTEMS (Not used)

DIVISION 15—Mechanical

15400 PLUMBING

1. <u>Scope</u>: This section includes all plumbing work and related items required to complete the work indicated on the drawings and in the specifications. All plumbing work shall conform to the Local Plumbing Code and the National Plumbing Code. All plumbing shall be run to and connected to city stub, outlets, meters, etc.

2. <u>Materials</u>:

   (a) <u>Cast iron pipe</u> and fittings to be extra heavy weight.

   (b) <u>Water and gas pipe</u>: Underground—soft-drawn, type K, copper piping, standard weight. Above ground—hard-drawn, type L, copper pipe, standard weight.

   (c) <u>Sheet lead</u> weighing not less than 4 lbs. per sq. ft.

   (d) <u>Water pipe insulation</u>: $\frac{3}{4}$" thermal molded glass fiber.

   (e) <u>Fixtures</u>: shall be American-Standard as follows, colors to be selected by Owner:

   <u>Lavatories</u>—19" × 17" Ledgewood with p4105 fittings.

   <u>Lavatories, built-in</u>—Dresslyn with F-131A fittings.

   <u>Sink</u>—B-939 with B-876 ST MF fittings.

   <u>Water Heater</u>—Westinghouse AF-40T-2, 40 gal. with 10 yr. guarantee.

15410 INSTALLATION

1. <u>Soil pipe</u> shall be installed in a neat manner with packed oakum or hemp filled with molten lead not less than 1" deep. Lead shall be run in one pour. Slopes of pipe shall be in accordance with existing codes.

2. <u>Water pipe</u> shall be run concealed in a neat manner, all lines run at 90 degrees or parallel to others. Joints to be brass with preinserted solder rings. Capped air chambers to be not less than 8" high and full diameter of the riser, and shall be provided at the top of each hot and cold water riser. Water piping to be insulated in attic and exterior walls. Hot and cold water lines to have 6" clearance throughout.

15420 TESTING

1. <u>Soil pipes</u> and vents shall be filled with water and tested for leaks. Any leaks shall be repaired by recaulking and leading joints. Test shall be repeated until system is watertight.

2. <u>Water pipes</u> shall be tested under 100-psi hydrostatic pressure for a period of <u>one hour</u>. Leaking joints shall be reheated or replaced as required.

15600 HEATING SYSTEM

The heating system will be installed under a separate contract. The General Contractor shall perform minor carpentry work as required to assist in the proper installation. Work shall be scheduled in order to allow the heating contractor adequate time to install each stage of his work without unnecessary cutting. The heating system shall be operational in time for interior finish work requiring heat. See drawings.

****

DIVISION 16—ELECTRICAL

16010 GENERAL

This division includes all electrical work and related items required to complete the work indicated on the drawings and specifications. All work shall conform to the local electrical code and National Electrical Code. Work includes all services, mast, rack, entrance cable, meters, panels, fixtures, controls, etc., required for a complete electrical system as indicated on plans.

16100 MATERIALS

Materials shall be equal to Wesco as manufactured by the Westinghouse Electrical Supply Company:

- Entrance cable—Type RR Neophrane-jacketed underground cable
- Panel—NCAB 22-4L 100
- Sheathed cable—Durex or Romex with "R" conductors (not less than No. 12 wire)
- Wire connectors—Bryant, solderless
- Switch plates—Bryant, Ivory Bakelite
- Receptacles—(waterproof) No. 3894
- Receptacles—(grounded) No. 5282
- Switches—Bryant, mercury 15-amp., 125 volts
- Receptacles—Bryant No. 61221
- Chimes—Edwards, Cordetto II, No. 1636
- Push buttons—Ivory with "Nite-lite"
- Outlet boxes—No. 5717-$\frac{1}{2}$
- Switch and receptacle boxes—Type LCN

16500 FIXTURE ALLOWANCE

The <u>Contractor</u> shall include in his bid the sum of <u>$1000.00</u> for the purchase of lighting fixtures, including built-in surface cooking unit. If the cost is less, the Owner is to receive credit, and if more, the difference will be reimbursed to the Contractor.

16600 APPLIANCE WIRE SIZES

Appliance	No. of Wires	Min. Wire Size
Range	3	No. 6
Elec. Dryer	3	No. 8
Small motors	2	No. 12
Water heater	3	No. 8

****

Architectural Working Drawings

TEST 1
COMPLETION

TEST ON FIRST-FLOOR PLAN OF COLONIAL HOME
(Figure 3-26)

NAME _____

TOTAL SCORE _____

Value of each answer ____4 POINTS____

DIRECTIONS: *Fill in the blank with the missing word to make the statement correct. Use only the First-Floor Plan.*

1. The overall-length of the house (including the brick veneer) is _____.

2. Is a Garage door required? _____

3. The entire first floor has how many entrance doors from the exterior? _____

4. A _____ threshold is required for the front door.

5. The switch for the Carport lighting is found on the wall of the _____.

6. The Den patio is to be surfaced with _____.

7. The size and spacing of the ceiling joists over the Den are _____.

8. A _____ door is required between the Kitchen and the Dining Room.

9. The inside walls in the Living Room are to be covered with _____.

10. The stairway is to be _____ wide. (Consider the partitions to be 6″ thick.)

11. Between the front Foyer and the Living Room a _____ allows communication.

12. The stairs have _____ risers to the second floor.

13. Are wall cabinets required in the Kitchen? _____

14. The cased opening in the Dining Room is _____ wide.

Continued

237

Architectural Working Drawings

TEST 1 (continued) COMPLETION

TEST ON FIRST-FLOOR PLAN OF COLONIAL HOME
(Figure 3-26)

15. Natural lighting for the front Foyer is provided by two _____.

16. The ceiling joists over the Dining Room will span a _____ distance.

17. The number of bath fixtures in the Toilet is _____.

18. The total exterior width of the house is _____.

19. The walls in the Living Room are to be painted with _____ paint.

20. Is ceramic tile wainscot required in the Utility Room? _____

21. Ceramic tile floors are required in the _____ and _____ rooms.

22. The Foyer is to have _____ -type flooring.

23. A _____ hearth is called for in the Den.

24. The space below the stairs is to be used for _____.

25. A _____ -type door is needed between the Den and the back Hall.

Architectural Working Drawings

TEST 2
COMPLETION

TEST ON ELEVATIONS
(Sheets Nos. 4 & 5)
OF COLONIAL HOME
(Figures 3-28 & 3-29)

NAME _____

TOTAL SCORE _____

Value of each answer _____ **4** POINTS

DIRECTIONS: *Fill in the blank with the missing word to make the statement correct. Use the Elevations only.*

1. The _____ on the gable ends allow positive roof ventilation.

2. Both roofs have a slope of _____.

3. The cupola is to be made of _____.

4. All window units are to be _____ type of windows.

5. There will be _____ lights in the front entrance door unit.

6. The chimney is to be _____ feet higher than the ridge of the upper roof.

7. Where is a fixed sash required? _____

8. Why was wood selected as siding for the upper floor? _____

9. The access door of the crawl space is located on the left side of the _____ Elevation view.

10. Each sash of the upper-story windows is to have _____ lights.

11. The second-story floor-to-ceiling height is _____.

12. Board-and-batten siding is specified on one wall of the _____.

13. The grade line is lowest near the _____ corner of the house.

14. A patio sliding door of _____ size is needed in the rear of the house.

Continued

239

Architectural Working Drawings

TEST ON ELEVATIONS
(Sheets Nos. 4 & 5)
OF COLONIAL HOME
(Figures 3–28 & 3–29)

TEST 2
(continued)
COMPLETION

15. Are metal gutters and downspouts shown? _____

16. A _____ (type) lintel is necessary to support the brick veneer above the front door.

17. A _____ is shown to provide water runoff on the chimney.

18. The sills below first-floor windows are to be made of _____.

19. The finish second floor is _____ in distance above the first floor.

20. Where will bevel siding be needed on the Carport exterior walls? _____

21. The concrete floor in the Carport Storage Room is _____ higher than the Carport floor itself.

22. The front door is to be _____ high.

23. The second-floor level requires _____ double-hung windows.

24. The rough opening width for the front door and window unit is _____.

25. More complete information about the foundation and footings will be found on the _____ detail.

Architectural Working Drawings

TEST 3

TEST ON DETAILS
(Sheet No. 6)
OF COLONIAL HOME
(Figure 3-30)

COMPLETION

NAME _____

TOTAL SCORE _____

Value of each answer _____**4** POINTS_____

DIRECTIONS: *Fill in the blank with the missing word to make the statement correct. Use the Detail Sheet only.*

1. The fireplace is to be made of _____ brick.

2. Lumber of _____ size is required for the rafters.

3. The interior walls of the house are to be covered with _____ thick wallboard.

4. Is a shelf required inside the base cabinets found on each side of the fireplace? _____

5. The stairs are _____ string stairs.

6. A _____ crawl-space height is needed.

7. The stud-to-stud distance that the second floor extends out beyond the first floor of the front is _____.

8. Two No. _____ reinforcing bars are required in the footings.

9. The riser height of the stairs is _____.

10. A _____ damper is required in the fireplace.

11. The decking is to be _____.

12. The size of the fireplace opening is _____.

13. The exterior soffit under the cantilevered second floor is to be of what material? _____

Continued

Architectural Working Drawings

TEST ON DETAILS (Sheet No. 6) OF COLONIAL HOME (Figure 3-30)

TEST 3 (continued) COMPLETION

14. The bookshelf base cabinets are to be how deep? _____

15. The floor of the fireplace will be how many inches above the finish floor? _____

16. Sill plates on the foundation are to be made of what size lumber? _____

17. Give the size description of the metal lintel to be installed above the fireplace opening. _____

18. Indicate the slope of the roof. _____

19. What type of insulation is specified for all the outside walls? _____

20. Is a newel post required? _____

21. What size cap block is to be used on the foundations? _____

22. The finish floor is to be made of what type of wood? _____

23. Are headers required at the outer ends of the floor joists? _____

24. What is the crawl-space earth to be covered with? _____

25. The base trim is to be of what size? _____

Architectural Working Drawings

TEST 4
TRUE-FALSE

TEST ON COMPLETE SET OF DRAWINGS OF COLONIAL HOME
(Figures 3–25 to 3–30)
(Figures 3–61 & 3–62)

NAME _____

TOTAL SCORE _____

Value of each answer __**4** POINTS__

DIRECTIONS: *If the statement is true, circle the T; if false, circle the F.*

1. Insulation is to be installed between the first-floor joists. — T F
2. All the shelves in the Den bookshelves are to be spaced the same distance apart. — T F
3. Two-inch by four-inch lookouts in the cornice construction will support the soffit. — T F
4. The brick screen in the Carport is to be made of used brick. — T F
5. The main girder under the first floor is a steel I beam. — T F
6. The second-floor joists are to be made of 2″ × 8″ lumber. — T F
7. A step is required between the Carport and the Kitchen. — T F
8. Foundation vents are indicated for the crawl space. — T F
9. The foundation piers are to be of 12″ × 8″ × 16″ concrete blocks. — T F
10. The concrete floor in the Carport Storage Room is to be the same level as the Carport floor. — T F
11. The fireplace hearth is to be 1′-6″ higher than the first-floor level. — T F
12. A six-panel door is indicated for the rear Utility Room entrance. — T F
13. The heating unit is to be installed below the stairs. — T F
14. A 5′-6″-wide cased opening is shown between the front Foyer and the Dining Room. — T F
15. Brick veneer exterior walls are shown for the second story. — T F
16. A ceramic tile floor is specified for the Utility Room. — T F

Continued

Architectural Working Drawings

TEST 4 (continued)
TRUE-FALSE

TEST ON COMPLETE SET OF DRAWINGS OF COLONIAL HOME
(Figures 3–25 to 3–30)
(Figures 3–61 & 3–62)

17. The lights in the front Foyer are operated with three-way switches. T F

18. The ceiling joists over the Den will have to span 19'-2" T F

19. A panel door is required in the stairwell at the second-floor level. T F

20. A two-panel door is specified for the front Foyer closet. T F

21. Insulation is to be placed around the periphery of the Carport floor slab. T F

22. The maximum span for the girder supporting the first-floor framing is 8'-6". T F

23. Ten piers are needed under the first floor. T F

24. The surface of the rear patio is below the level of the Den finish floor. T F

25. A vanity cabinet is needed in Bath No. 1. T F

Architectural Working Drawings

TEST 5
SHORT ANSWER

TEST ON SPECS AND ALL DRAWINGS OF COLONIAL HOME
(Figures 3–25 to 3–30)
(Figures 3–61 & 3–62)

NAME _____

TOTAL SCORE _____

Value of each answer _____ **4** POINTS

DIRECTIONS: *Place the correct answer in the blank following each question. Use all the Colonial Home drawings and the FHA materials form (pp. 215–218).*

1. How much is to be allowed for the purchase of lighting fixtures? 1. _____

2. How wide is the front walk to be? 2. _____

3. What size wall footings are required? 3. _____

4. What is the bottom of the crawl space to be covered with? 4. _____

5. The chimney flashing is to be made of what material? 5. _____

6. What special glass is to be used in the sliding doors? 6. _____

7. Are screens to be included with the doors? What material? 7. _____

8. What roofing is specified? 8. _____

9. Is a scuttle indicated? 9. _____

10. What reinforcement is specified for the Carport slab? 10. _____

11. What material are the gable louvers to be made of? 11. _____

12. What type of wood are the Kitchen cabinets to be made of? 12. _____

13. What type of paint are the walls of Bedroom No. 2 to be painted with? 13. _____

14. How many vents are required in the crawl space? 14. _____

Continued

TEST ON SPECS AND ALL DRAWINGS OF COLONIAL HOME
(Figures 3–25 to 3–30)
(Figures 3–61 & 3–62)

Architectural Working Drawings

TEST 5
(continued)
SHORT ANSWER

15. What material is required on the Carport Cupola roof?

15. _____

16. What is the wall between the Carport and the Storage Area to be surfaced with?

16. _____

17. How deep is the Carport interior?

17. _____

18. What is the cross-sectional size of the steel plate in the flitch beam over the Carport?

18. _____

19. What material is the chimney flue to be made of?

19. _____

20. Are screens to be furnished for the windows?

20. _____

21. How high will the finish second floor be above the finish first floor?

21. _____

22. Where is the access door into the attic area over the Carport located?

22. _____

23. The floor in Bath No. 2 is to be covered with what material?

23. _____

24. What is the storage capacity of the domestic hot-water heater tank?

24. _____

25. The ceiling in the Carport is to be covered with what material?

25. _____

Architectural Working Drawings

TEST 6

IDENTIFICATION

CONSTRUCTION TERMS

NAME _____

TOTAL SCORE _____

Value of each answer __**8** POINTS__

DIRECTIONS: *Using the following detail drawing, identify each labeled part and place the correct term in the appropriate blank.*

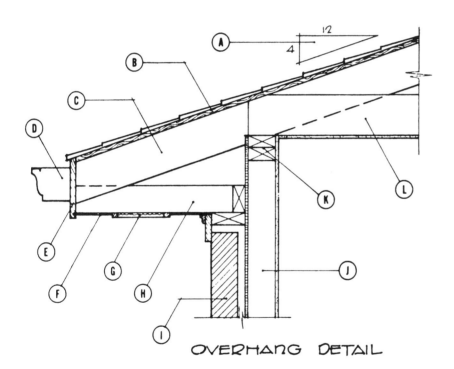

OVERHANG DETAIL

A. _____

B. _____

C. _____

D. _____

E. _____

F. _____

G. _____

H. _____

I. _____

J. _____

K. _____

L. _____

247

Architectural Working Drawings

TEST

IDENTIFICATIO

CONSTRUCTION TERMS

NAME _____

TOTAL SCORE _____

Value of each answer **6.66** POINTS

DIRECTIONS: *Using the following detail drawing, identify each labeled part and place the correct term in the appropriate blank.*

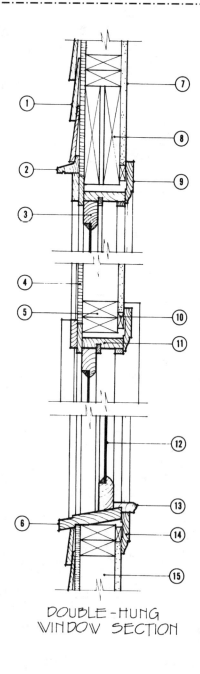

DOUBLE-HUNG WINDOW SECTION

1. _____
2. _____
3. _____
4. _____
5. _____
6. _____
7. _____
8. _____
9. _____
10. _____
11. _____
12. _____
13. _____
14. _____
15. _____

248

Architectural Working Drawings

TEST 8

IDENTIFICATION

CONSTRUCTION TERMS

NAME _____

TOTAL SCORE _____

Value of each answer ____**6.66** POINTS____

DIRECTIONS: Using the following detail drawing, identify each labeled part and place the correct term in the appropriate blank.

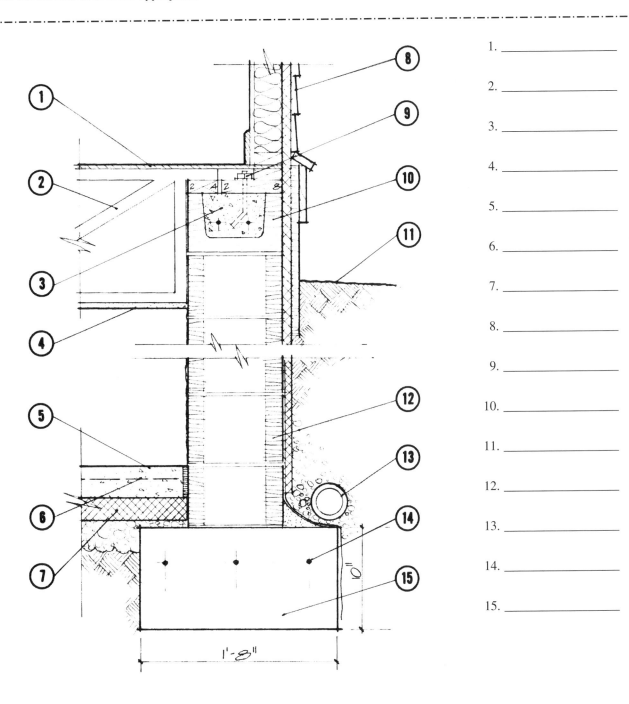

1. _____
2. _____
3. _____
4. _____
5. _____
6. _____
7. _____
8. _____
9. _____
10. _____
11. _____
12. _____
13. _____
14. _____
15. _____

249

Architectural Working Drawings

TEST 9

SHORT ANSWER

READING EXERCISE
(Figures 3–36 to 3–41)

NAME _____

TOTAL SCORE _____

Value of each answer ____**3** POINTS____

DIRECTIONS: *Using the drawings in Figures 3–36 to 3–41, place the correct answer in the blank provided.*

1. Allowing the frame walls to be 6″ thick, what are the inside dimensions of the Carport?

   1. _____

2. How many linen closets are shown?

   2. _____

3. What type of closet doors are needed in Bedrooms No. 1 and No. 2?

   3. _____

4. What is the distance between the center lines of the sliding glass doors in the Family Living Room?

   4. _____

5. How thick is the gravel fill below the floor slab?

   5. _____

6. What type of exterior door is shown for the rear entrance?

   6. _____

7. Are the roof beams to be transverse or longitudinal?

   7. _____

8. How far apart are the roof beams to be spaced?

   8. _____

9. What is the material in the foundation wall?

   9. _____

10. Where is the roof overhang the widest?

    10. _____

11. Is a built-in dishwasher to be included?

    11. _____

12. How many waterproof thresholds are required?

    12. _____

13. List the two different types of windows shown in the Family/Dining/Living space.

    13. _____

14. What type of exterior siding is predominantly used throughout the exterior walls?

    14. _____

15. The roof over the Family/Dining/Living space has what slope?

    15. _____

16. What kind of roofing is to be used?

    16. _____

Continued

Architectural Working Drawings

TEST 9
(continued)
SHORT ANSWER

READING EXERCISE
(Figures 3–36 to 3–41)

17. How wide is the roof overhang over the rear entrance door? 17. _____

18. Is a saddle required? Where? 18. _____

19. What steel reinforcement is specified in the floor slab? 19. _____

20. Is the outdoor screen wall to be near the front or rear of the house? 20. _____

21. What is the finish floor material to be in the Utility Room? 21. _____

22. Are there to be windows in the Carport? 22. _____

23. What is the cross-sectional *actual size* of the treated sill plate? 23. _____

24. What steel reinforcing is required in footings? 24. _____

25. Where in the frame walls are 4″ × 4″ posts to be placed? 25. _____

26. Is the roof decking exposed to the rooms below? 26. _____

27. How thick is the fascia board? 27. _____

28. What size are the headers to be? 28. _____

29. How many opening sashes does the window in Bedroom No. 1 have? 29. _____

30. What type of outlets are needed in the Front Entrance? 30. _____

31. How many cars will the Carport accommodate? 31. _____

32. How high from the finish floor are the exterior walls on which roof beams rest? 32. _____

33. What is used to separate Bedrooms No. 2 and No. 3? 33. _____

Architectural Working Drawings

TEST 10

TEST ON SYCAMORE WALK FLOOR PLAN
(Figure 3-70)
(Figure 3-70 is located in back of the book)

MULTIPLE CHOICE

NAME _____

TOTAL SCORE _____

Value of each answer _____ **10** POINTS

DIRECTIONS: *Place an X in the corresponding blank that identifies each correct answer.*

---

                                                                                                         a   b   c

1. How many square feet are there in the first floor? . . . . 1. ( )   ( )   ( )
      a. 1875   b. 2850   c. 2150

2. What is the ceiling height in the Dining Room? . . . . 2. ( )   ( )   ( )
      a. 12'-0"   b. 8'-6"   c. 9'-0"

3. What is the length of the house (excluding brick veneer)? . . . . 3. ( )   ( )   ( )
      a. 67'-6"   b. 63'-0"   c. 51'-6"

4. How many water closets are required on the first floor? . . . . 4. ( )   ( )   ( )
      a. 5   b. 3   c. 1

5. How many columns are shown on the Rear Porch? . . . . 5. ( )   ( )   ( )
      a. 5   b. 3   c. 4

6. What is the depth of the 2-car garage? . . . . 6. ( )   ( )   ( )
      a. 21'-10"   b. 20'-0"   c. 22'-10"

7. What size are the windows in the Kitchen? . . . . 7. ( )   ( )   ( )
      a. 2'-8" × 4'-2"   b. 2'-8" × 6'-8"   c. 2'-4" × 4'-6"

8. What size door is required in the Master Bedroom walk-in closet? . . . . 8. ( )   ( )   ( )
      a. 2'-6" × 7'-0"   b. 2'-8" × 6'-8"   c. 2'-4" × 6'-8"

9. What is the ceiling height in Bedroom No. 3? . . . . 9. ( )   ( )   ( )
      a. 10'-0"   b. 6'-8"   c. 9'-0"

10. How wide is the brick fireplace in the Great Room? . . . . 10. ( )   ( )   ( )
       a. 5'-4"   b. 6'-4"   c. 4'-8"

Architectural Working Drawings

TEST 11
SHORT ANSWER

TEST ON SYCAMORE WALK FOUNDATION/BASEMENT PLAN
(Figure 3–69)

NAME _____

TOTAL SCORE _____

Value of each answer __**5** POINTS__

DIRECTIONS: *Place the correct answer in the blank provided. Use the Sycamore Walk Foundation/Basement Plan.*

1. What scale was used on the original drawings?

   1. _____

2. What type of light fixture is required in the Family Room?

   2. _____

3. What type of fireplace is required between the Recreation Room and the Family Room?

   3. _____

4. How many columns are needed in the Recreation/Family Room area?

   4. _____

5. What type of steel reinforcing is needed in the floor slab?

   5. _____

6. What size door is required in the walk-in closet?

   6. _____

7. Where will the furnace be located?

   7. _____

8. How many risers are shown in the stairs to the first floor?

   8. _____

9. What will be the height of the foundation wall?

   9. _____

10. What size and type of door is needed to the exterior from the Family Room?

    10. _____

11. Where is a Floor Drain required?

    11. _____

12. How many linen closets are shown in the plan?

    12. _____

13. Are all the columns in the Recreation/Family Room area uniformly spaced?

    13. _____

14. What is the total length of the foundation as observed from the front?

    14. _____

Continued

Architectural Working Drawings

TEST 1
(continued)
SHORT ANSWER

TEST ON SYCAMORE WALK FOUNDATION/BASEMENT PLAN
(Figure 3-69)

15. What is the size of the fireplace foundation as shown in the left Storage area?

15. _____

16. What size is the door to the Mechanical Room?

16. _____

17. Are the two light fixtures in the Recreation Room operated with separate switches?

17. _____

18. Are the partitions in the Basement to be made of masonry or wood?

18. _____

19. How many wash basins (lavatories) are needed in the Bedroom Bath?

19. _____

20. Why are double doors necessary to the front Storage Room?

20. _____

Architectural Working Drawings

TEST 12
TRUE-FALSE

TEST ON SYCAMORE WALK DRAWINGS AND NOTES
(Figures 3–69 to 3–77)

NAME _____

TOTAL SCORE _____

Value of each answer _____ **4** POINTS

DIRECTIONS: *If the statement is true, circle the T; if false, circle the F.*

1. All ridge lines on the roof are the same height.   T   F

2. The outer finish on the Left Elevation is to be brick veneer.   T   F

3. All first-floor ceiling framing is to be 2″ × 10″-16″ o.c.   T   F

4. Angular walls in the First-Floor Plan are 45 degrees.   T   F

5. The ceiling height of the Front Foyer is 12′-0″.   T   F

6. A ceiling fan is required in Bedroom No. 3.   T   F

7. The Foundation wall is to be 8′-0″ high.   T   F

8. The front-door columns are 12″ diameter.   T   F

9. An island work surface is shown in the Kitchen.   T   F

10. The front stoop and steps are to be made of brick.   T   F

11. The Keeping Room will have a sloping ceiling.   T   F

12. The Stoop finish floor will be 6″ lower than the first-floor level.   T   F

13. The overall width of the house will be 64′-6″.   T   F

14. The dormer on the Front Elevation will have a copper roof.   T   F

15. The finish floor of the Garage is to be 8″ lower than the house finish floor.   T   F

16. The size of the walk-in closet door in the Future Bedroom will be 2′-0″ × 6′-8″.   T   F

17. Four water-closet fixtures are needed for the entire house.   T   F

Continued

TEST ON SYCAMORE WALK DRAWINGS AND NOTES
(Figures 3–69 to 3–77)

Architectural Working Drawings

TEST 12
(continued)
TRUE-FALSE

18. Anchor bolts are not required where sill plates rest on a slab.   T   F

19. The roof overhang is to be 1'-6" from the outer surface of the perimeter studs.   T   F

20. First-level floor joists are to be 2" × 8"-16" o.c.   T   F

21. A flitch beam is required in the First Level Floor Framing Plan.   T   F

22. The Box-Bay roof is made of standing seam copper.   T   F

23. The floor in the Master Bedroom is finished with carpet.   T   F

24. The perimeter stud walls will be made of 2" × 4" studs.   T   F

25. All valley rafters are to be 2" × 10".   T   F

Architectural Working Drawings

TEST 13

SHORT ANSWER

TEST ON SYCAMORE WALK DRAWINGS AND NOTES
(Figures 3–69 to 3–77)

NAME _____

TOTAL SCORE _____

Value of each answer _____**4** POINTS_____

DIRECTIONS: *Place the correct answer in the blank provided.*

1. How many dormers are shown on the front of the house? 1. _____
2. What size and type of door is required to the Mechanical Room? 2. _____
3. The Interior Elevations for the Floor Plan are drawn at what scale? 3. _____
4. Is a seat required in the Master Bath? 4. _____
5. How many plumbing walls are needed throughout the First-Floor Plan? 5. _____
6. What is the window height of the Kitchen window? 6. _____
7. Where are jack arches shown? 7. _____
8. What type of switch will operate the Kitchen lights? 8. _____
9. How many square feet will there be in the first-floor level? 9. _____
10. What is the ceiling height of the Porch? 10. _____
11. Where does the furnace flue extend up through the first-floor level? 11. _____
12. At what scale were the Rear and Side Elevations drawn? 12. _____
13. What size joists are used for the first-floor framing? 13. _____
14. What width are the typical footings below the foundation wall? 14. _____

Continued

Architectural Working Drawings

TEST ON SYCAMORE WALK DRAWINGS AND NOTES
(Figures 3-69 to 3-77)

TEST 13
(continued)
SHORT ANSWER

15. What finish-floor material is required in the Front Foyer? 15. _____

16. What is the total design load for the roof framing? 16. _____

17. Welded wire fabric in concrete floors shall be lapped how much at seams? 17. _____

18. Where is a saddle required? 18. _____

19. What is the beam size above the front entrance columns? 19. _____

20. What is the window height above the finish floor in the Kitchen? 20. _____

21. What does the OFS abbreviation stand for? 21. _____

22. What size is the door to the Front Powder Room? 22. _____

23. Where are stationary windows required? 23. _____

24. What size footings are needed for the Interior Columns? 24. _____

25. Is insulation required in the first-floor framing? 25. _____

Architectural Working Drawings

TEST 14

TEST ON SYCAMORE WALK DRAWINGS AND NOTES
(Figures 3–69 to 3–77)

SHORT ANSWER

NAME _____

TOTAL SCORE _____

Value of each answer __**4** POINTS__

DIRECTIONS: *Place the correct answer in the blank provided.*

1. The driveway slab is how far below the intersection of the Garage slab? 1. _____

2. The finish floor of the Keeping Room is to be of what material? 2. _____

3. The joists and rafters lumber is to be of what type? 3. _____

4. Openings in masonry a maximum of 6′-0″ wide shall have what size metal lintels? 4. _____

5. The soffit is to be of what material? 5. _____

6. What are the exterior rear steps to be made of? 6. _____

7. Approximately how many pieces of 2″ × 12″ × 16′-0″ lumber are needed for the house? 7. _____

8. What material is the Porch floor? 8. _____

9. What size is the plynth below the front columns? 9. _____

10. How many walk-in closets are shown? 10. _____

11. How many treads will be in the interior stairs? 11. _____

12. Where is a Butler's Pantry shown? 12. _____

13. How many ceiling-light outlets are needed in the Kitchen? 13. _____

14. How does one go from the Garage to the interior of the house? 14. _____

15. Are any exterior hose bibbs shown? 15. _____

Continued

Architectural Working Drawings

TEST 14 (continued)
SHORT ANSWER

TEST ON SYCAMORE WALK DRAWINGS AND NOTES
(Figures 3–69 to 3–77)

16. How many ceiling-light outlets are needed in the Garage? 16. _____

17. How many bookshelves are shown by the Keeping Room fireplace? 17. _____

18. What size and type of gutter is typical around the roof? 18. _____

19. What is the typical roof overhang from the outside face of studs? 19. _____

20. What material and thickness is the typical interior wall finish? 20. _____

21. Concrete shall develop what compressive strength in 28 days? 21. _____

22. What type vapor barrier is needed below concrete slabs? 22. _____

23. What material is the dormer sill? 23. _____

24. Is a chimney cap required? 24. _____

25. Where is a whirlpool tub shown? 25. _____

Architectural Working Drawings

TEST 15

TEST ON SYCAMORE WALK DRAWINGS AND NOTES
(Figures 3–69 to 3–77)

SHORT ANSWER

NAME _____

TOTAL SCORE _____

Value of each answer _____**4** POINTS_____

DIRECTIONS: *Place the correct answer in the blank provided.*

1. Why are two switches shown for the ceiling-light outlet in the Master Bedroom?　　1. _____

2. What is the finish floor in the Keeping Room?　　2. _____

3. Which wall in the Kitchen is the washer installed on?　　3. _____

4. What is the maximum pitch of the copper flashing below the front gables?　　4. _____

5. Where is a beam (CB-5) located?　　5. _____

6. How many support columns are needed in the Garage?　　6. _____

7. Why is a double ceiling-joist needed between the Keeping Room and the Breakfast Room?　　7. _____

8. Is a cased opening shown between the Dining Room and the Keeping Room?　　8. _____

9. What two ceiling heights are shown in the Master Bedroom?　　9. _____

10. What width cased opening is shown between the Future Family Room and the lower hall?　　10. _____

11. What is the finish floor in the Future Family Room?　　11. _____

12. How many water basins are in the No. 2 and No. 3 Bedrooms related bath?　　12. _____

13. What is the total depth of the chair-rail molding?　　13. _____

14. What size and type of exterior door is needed in the Great Room?　　14. _____

Continued

Architectural Working Drawings

TEST 15 (continued)
SHORT ANSWER

TEST ON SYCAMORE WALK DRAWINGS AND NOTES
(Figures 3-69 to 3-77)

15. How high will the header be above F.F. between the Foyer and Great Room?

15. _____

16. What is the width of the air space between exterior masonry and frame construction?

16. _____

17. Will the columns on the back Porch be round or square?

17. _____

18. How are the studs placed in the dormer side-wall framing?

18. _____

19. How many switches are needed for the Foyer ceiling light?

19. _____

20. Is a wall cabinet shown in the Laundry?

20. _____

21. What is the ceiling height above the window in Bedroom No. 3?

21. _____

22. Who is to provide installation details for all trusses?

22. _____

23. What lateral spacing is required for thru-bolts in flitch plates?

23. _____

24. What vertical thickness of crushed stone is needed below concrete slabs?

24. _____

25. How are load-bearing stud walls anchored to concrete slabs?

25. _____

Architectural Working Drawings

TEST 16

MULTIPLE CHOICE

WHERE TO LOOK FOR INFORMATION

NAME _____

TOTAL SCORE _____

Value of each answer __**4** POINTS__

DIRECTIONS: *Place an X in the proper column indicating the most logical place to find the following items of construction information. Only one answer for each item.*

	PLAN	ELEV.	SPECS
1. The number of coats of paint to be applied.			
2. The finish-floor level.			
3. Number of risers in the stairs.			
4. Capacity of the hot-water tank.			
5. The size of a floor beam.			
6. Width of the wall footings.			
7. The exposure of wood shingles.			
8. The amount of roof overhang.			
9. Layout of rooms.			
10. Responsibility for workers' insurance.			
11. Height of the chimney.			
12. Type of the framing lumber.			
13. The thickness of the window glass.			
14. The slope of the roof.			
15. The spacing of the floor joists.			
16. Where a light switch is located.			
17. The heights of room ceilings.			
18. The location of a scuttle.			
19. Type of hardware to be used on doors.			
20. The location of a cased opening.			
21. Room dimensions.			
22. Exterior finish materials.			
23. Size of a terrace.			
24. The appearance of a window.			
25. The spacing of studs in a wall.			

Chapter 5
READING THE WORKING DRAWINGS OF A SMALL CHURCH

Construction with heavy wood members, which was commonplace in mills during Colonial days, still finds application in many newer buildings. The working drawings (Figures 5–5 to 5–20, fold-out drawings at the end of the text) for a small church, which utilizes laminated arches for the structure of the main hip roof, are included to further your experience in reading various types of construction drawings. Because they have been reduced in size photographically, the drawings cannot be scaled directly.

Other types of light-commercial buildings, besides churches, commonly employ laminated members for special situations within their structures (Figure 5–1). Each piece (often referred to as *glue-lam*) is made up of smaller-lumber laminations, glued under pressure with clamping devices and dressed to specific sizes at the fabricating plant before being transported to the job site. Because of the comparatively fragile nature of wood, many of the members are wrapped in waterproof paper for protection in transit. Many stadiums, field houses, and so on, where wide, open spans are needed with no intermediate support, employ structural-engineered systems with laminated members to solve the problems. Many buildings expose the structural members as well as the exposed plank ceilings, resulting in interesting interiors.

Heavy wood members used in buildings require numerous types of fasteners and steel connectors to fasten and anchor them securely. A few of the metal connectors are shown in Figure 5–2. Others are detailed in the church Shop Drawings. Bolts, with hex heads and washers, and lag screws are commonly used as fasteners.

The laminated members are custom fabricated for each project and require the use of Shop Drawings, prepared by the fabricator, to ensure proper fit and installation. The sizes and details found on shop drawings are based on the sizes and conditions given in the architectural drawings.

Southern yellow pine and Douglas fir, both in abundant supply, are the two main types of structural lumber used for laminated members and other timbers. The controlled grading of laminated lumber and timbers provides designers with consistent assurance of load-carrying requirements. Heavy timber is generally considered to be structural members of minimum 6″ nominal dimension.

Laminated members are compatible with conventional light-frame construction; the wood can be worked with ordinary hand tools and power equipment, similar carpenters and workers are used, and the wood has good structural characteristics. Usually, large structural members are employed when open spans are needed to satisfy the intent of the building.

Although wood will readily burn, it can be considered "fire resistant" if heavy, massive members are used. Extremely severe fires are necessary before large members burn through and fail under load. Not only are chemical fire retardants available for treating wood members, but when heavy members are exposed to fire, the outside surface first chars to form an

FIGURE 5-1

Wood is an important structural material in many light-commercial buildings. (A) Erecting glued laminated wood arches. (B) Laminated wood beams with prefabricated purlins and decking being erected for a warehouse. (Courtesy American Institute of Timber Construction.)

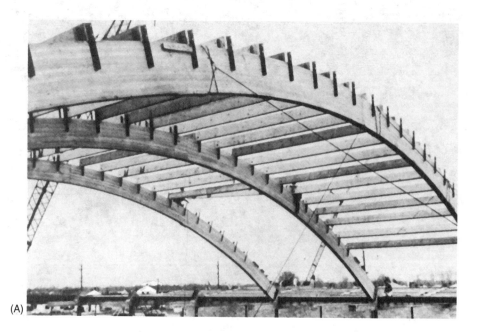

(A)

(B)

insulating barrier to protect the interior section of the member. Thinner wood pieces, of course, would quickly burn through and collapse. But heavy members have been shown to sustain their loads during exposure to high fire temperatures, even when steel members were found to be unable to retain their original shapes.

Prefabricated, light-wood trusses (often called *trussed rafters*), (Figure 3–65) are now used in the roof framing of many residences and small commercial buildings. They are commonly spaced 24" o.c. and save about one-third on material and construction costs, as compared to conventional rafter-and-ceiling-joist construction of gable framing. In reading a set of working drawings, look for evidence of their use on Roof Framing Diagrams and Typical Wall Sections. Often the diagram or Framing Plan will show each truss with a heavy line and a number so that estimators will be able to count the exact number of trusses required in the building. Other drawings may show the number of trusses needed in a series, with a note indicating the number and type.

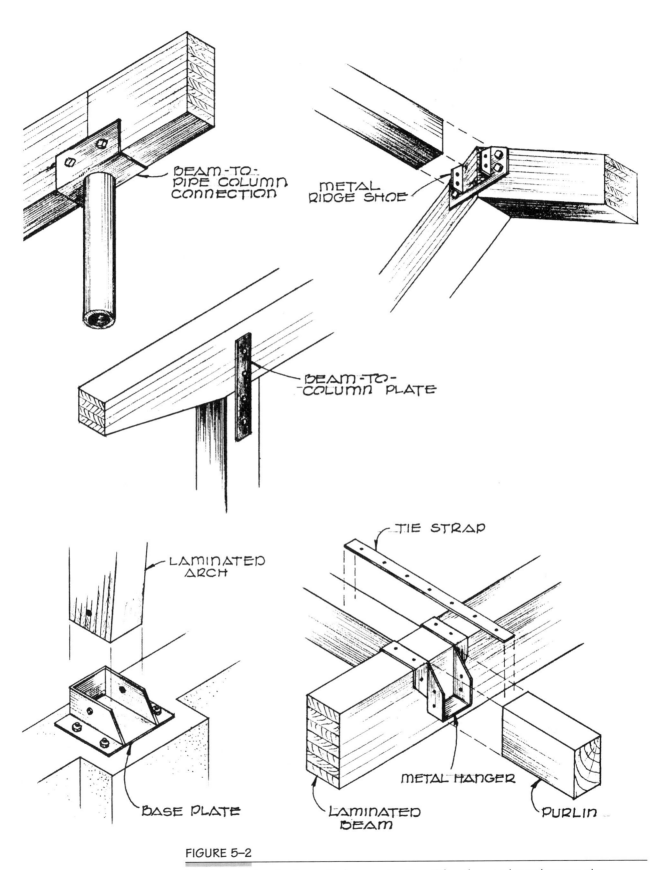

FIGURE 5–2

Metal connectors for heavy timber construction. Often the wood members are glue-laminated (glue-lam).

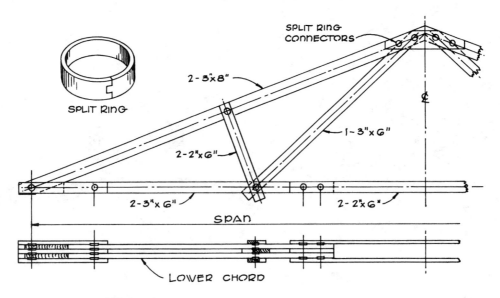

FIGURE 5–3

Engineered wood truss.

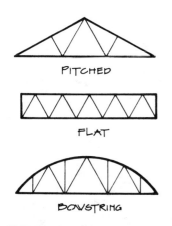

FIGURE 5–4

Basic types of truss shapes.

Another advantage in the use of prefabricated trusses is the elimination of any need for intermediate support within the building; the entire load of the roof is transferred to the load-bearing exterior walls. This allows more flexibility in the placement of partitions in the plan. Also, the roof can be framed quicker, using power lifting equipment, so the building can be "dried in" in the minimum amount of time.

In the fabricating plant, light trusses are assembled after the members are accurately cut, and engineered connector plates are fastened at the joints. Various types of connector plates may be used; economical light-wood trusses are now assembled using *pressure plates* that quickly squeeze together using power-pressure equipment. Heavier, engineered trusses may utilize *split rings* and bolted connections (see Figure 5–3). Other bolted connectors, as well as plywood gusset plates, may also be used. Many lumber dealers now stock light-wood trusses having standard spans and roof slopes, which can be quickly delivered to the job site (Figure 5–4).

START WITH A GENERAL CONCEPT OF THE BUILDING STRUCTURE

Begin by looking over the exterior views of a building to have a mental idea of the general shapes, roof type, and exterior material shown. This mental image of the building often gives the reader clues as to the structural system used to support the building. Then relate the exterior elevations to the Plan views, possibly by relating a prominent feature that is identifiable on both drawings. Next look for the *Typical Wall Section,* which will reveal further information about the general construction. Investigation of other relating drawings may be necessary before a more positive idea of the designer's intention is verified. But students reading a set of drawings must understand the general concept of the building structure at the very outset.

B. READING THE SMALL CHURCH DRAWINGS

In looking at the *Roof Framing Plan* we see that the structural members are wood and that heavy laminated arches are employed to support the roof and the heavy steeple in the central section; prefabricated wood trusses form the structure in the other wings. Further information about the arches and their relating beams is shown in the set of condensed shop drawings, which were reduced from the original 15 sheets for the sake of brevity in this manual. Other drawings from the original architectural drawings had to be omitted as well. Yet the principal drawings and details are included to provide a complete understanding of the church building.

The block-and-brick veneer exterior walls support the roof trusses, similar to many light-construction commercial buildings. Notice that the floor throughout is a slab-on-grade, and many of the details are similar to those found in residential construction. Gypsum wallboard is the interior finish material, and the wallboard is attached directly to the inside of block walls to provide similarity throughout. The drop ceiling, which is hung from the arches in the Multipurpose Room, results in a somewhat contemporary styling within; yet the brick-veneer exterior with a hip roof and arched heads over door and window openings provides a French Provincial exterior styling.

Churchgoers can enter the building through the Carport Entrance or the Front Entrance at the left of the Multipurpose Room. Flexibility in seating for the assembly area is gained by the use of movable partitions shown between the Multipurpose Room and the Classrooms. This is a flexible plan, so necessary for a small church building. Study the plan carefully to understand traffic patterns and the relative rooms within. It is particularly important that you understand at this point how features within the building relate to features on the exterior, as mentioned earlier.

In looking at the *Foundation Plan,* we see that steel tie rods are placed in the concrete slab to counteract the outward thrust at the bases of the laminated arches when the weight of the roof and the steeple is applied. Details of this feature are found on the same sheet. One major problem in the design of the structure was supporting the weight of the steeple, yet having an open area below.

Considerable information is given on the Foundation Plan about footings and the walls that are load bearing. Notice that a *Footing Schedule* is included providing identifying marks, size, and reinforcing for both concentrated loads and walls. The mark CLF-2, for example, indicates a footing 3'-6" 3'-6" 12" deep with #4 bars spaced 6" apart each way. The widened footings support concentrated loads that are transferred through the walls. Notice also that offsets of the stepped footings are indicated with a line and an S through it, and elevations (heights) of the *tops* of the footings are given. Check the Elevation Views to verify where these steps occur because of variations of the grade level. Be sure you can relate the Section Details shown to the plane on the Plan where they apply. For example, the Section Detail 4-A1 shows the construction of the concrete floor slab where the slab is thickened and the placement of the steel tie rods occurs. The footing construction that supports the arches is shown in Section Detail 1-A1, which was engineered and had to be carefully detailed. Notice that considerable steel reinforcing had to be used throughout the various footings.

Because the *frost line* is rather shallow in a warm climate, some of the footings are combined with the floor slab and poured simultaneously.

The *Foundation Plan* (Sheet A-1) is actually a horizontal section, cut through the lower-wall materials. Therefore, the outline of the footings, below the plane of the section, are shown in broken lines. Occasionally, as you read the Foundation Plan, refer to the Floor Plan to relate features; it may even be necessary to refer to the Roof Framing Plan as well to understand why concentrated loads occur and where the variations are needed in the foundation construction.

The *Floor Plan* (Sheet A-3) shows the layout of rooms, their dimensions, and the materials used in the walls, similar to a residential plan. The location dimensions of walls and partitions are given to their surfaces; however, the windows in outside walls are located to their center lines. A *Reflected Ceiling Plan* of the Multipurpose Room is included and shown in broken lines. Study this plan to see how it relates to the roof structure. Notice from Section 1-A7, Sheet A-7, that the pyramidal ceiling is offset from the hip roof structure above.

Broken lines also are used to locate features in the Kitchen and Chancel layout. In this case they indicate features that are not in the contract (N.I.C.)

Ceiling heights are shown in rooms where horizontal ceilings are needed. Further information about the Chancel platform is found on Sheet A-7. Doors are labeled with a number within a circle so their types can be found on the *Door Schedule* (Sheet A-4). Keep in mind the large, central square shape of the plan that will have the laminated arches anchored in the corners. We see several *Alternate Notes,* which indicate that options were available in case it was necessary to reduce the total cost of the building.

A *Finish Schedule* is shown on the same sheet to indicate the finish materials to be used in each room; the relating mark on the plan is surrounded with a hexagon. Further notes about construction and *Finish Trim Details* are also shown on the sheet. Take the time at this point to study the *Transverse Section,* 1-A7, while reviewing information shown on the Floor Plan.

Be sure you understand the system the drafter has used to relate sections to the place where they have been taken from in the other views. This set is well detailed, and the reader must understand where the detail relates. A small circle with a letter or number above and the sheet number where the section is found below is used throughout the set.

The *Elevation Views* (Sheets A-5 and A-6) reveal the exterior appearance of the church, exterior trim materials, and vertical heights. Relate the elevations to the plan at the outset so that you have a mental image of the entire building. Notice that roof slopes on all hips are similar, resulting in a higher ridge over the left wing because of the span. The details of the steeple indicate that it will be prefabricated and anchored on the center of the main hip. We see by a note that the steeple is one of the *Alternates.* Extended, simulated brick quoins are shown on the major front corners for appearance. Notice that carefully detailed sections show the construction at critical points in the concrete block-and-brick load-bearing walls. As mentioned earlier, the stepped footings below grade are shown with broken lines.

It is common practice when drafting symmetrical features, such as the *Typical Window Elevation* (Sheet A-6), to draw only half of the feature, thus saving drafting costs. The window shutters are another option, as shown by the note.

Exterior masonry walls are insulated with 1 1/2″ rigid insulation between the block and brick veneer. Insulation is also required around the periphery of the concrete floor slab. The block exterior wall is visible on the end of the back wing and will be insulated with loose fill in the cavities.

Study the details shown to see where special construction within the concrete block cavities occurs. Notice that the block adjacent to window and door openings is filled with grout and a #5 vertical bar is used for reinforcement. Block is also used to support the brick veneer below grade, and the window units are backed with 8″ block up to sill height. In fact, the entire block bearing wall is capped with a U-block-grout-#5 bar bond beam for stability and resistance to cracking. Lintel blocks with reinforcing are also required above window and door openings. Steel beam lintels are also required above the brick arches in the Carport walls.

Interior Elevations found throughout the architectural sheets show special interior treatment. Elevations 2-A7, Sheet A-7, show how the wall behind the Chancel and Choir Area will appear. Interior Elevations F, G, H, and J on Sheet A-4 show the wall layouts in toilets where plumbing fixtures are to be installed. Locate these elevations on the Floor Plan to further understand their location and purpose.

The *Plumbing Plan* (Sheet P-1), prepared by the plumbing engineers, is a typical floor plan of the building with both water and drainage piping located and sized. Notice the legend that indicates the symbols for piping, fittings, and fixtures. Some of the piping will be in the concrete floor slab, and some will be installed above the ceilings. Here, again, relate the plumbing layout to the Floor Plan (Sheet A-3) to understand the system used by the plumbing engineer to represent each feature.

Two domestic water heaters are used to avoid long hot-water runs. It is common practice on plumbing drawings to use pictorial isometric-type drawings to more clearly show offsets, risers, and so on, in piping diagrams.

In referring to the *Site Plan* we see that water and sewer utilities come from the public mains on Henderson Drive and enter at the rear of the building.

One *Mechanical Plan* (Sheet M-1) is included with the set. This shows the typical method of locating the heating and air conditioning units and the distribution ductwork necessary to carry the conditioned air throughout the building. Instead of one large central unit to process the air, a number of smaller heat pumps placed in strategic places was the best solution to the problem. This provides quicker start-up time, more flexibility, and more uniform air distribution. Each register is located, and its size and capacity are shown on the arrow symbol. Ducts are carefully laid out on the plan and their cross-sectional sizes shown. Notice that the largest cross-sectional size of the runs is nearest the units so that required volumes of air are delivered to the outer registers with similar velocities. Provision for disposal of the liquid condensate from the units is shown. In looking at the *Roof Framing Plan* (Sheet A-2), details are given to show the construction of the mechanical-unit platforms.

Two *Electrical Plans* (Sheets E-1 and E-2) are included, one for the lighting circuits and one for power circuits. The Lighting Plan locates each fixture and outlet to be used for lighting and switching, and a *Lighting Fixture Schedule* provides further information about the fixtures. Notice that both incandescent- and fluorescent-type fixtures are specified. Uniformly loaded circuits are located and shown with curved lines between the outlets. Marks through the circuit lines indicate the number of conductors within the line. A detail also shows that spotlights are to be installed around the base of the steeple. The main transformer is to be located at the rear of the building, and the main panel (1200-A main breaker) is located in the rear-right storage room. Three subpanels are needed.

The *Electrical Power Plan* (Sheet E-2) locates outlets that supply the various power equipment throughout the building. Provision is also made for safety equipment and future circuits.

Because the heating and cooling units are placed on platforms above the ceiling, as earlier mentioned, an *Attic Plan Detail* is included to provide that information. An enlarged *Kitchen Plan* shows power outlets more clearly than the smaller layout.

Only a *condensed* set of shop drawings of the laminated arches and main-hip structural members is included in this material. Typically, shop drawings are not considered as part of the set of architectural drawings. Yet they are included here to show students the type of drawings that are required to fabricate and install heavier members that are custom fabricated for buildings. Only the principal views are included to conserve space in the manual. In smaller buildings, *Shop Drawings* (as included with the Small Church set) usually replace *Structural Drawings,* which are commonly associated with large commercial buildings. The *Framing Plan* (Sheet 2) is the major layout, which labels each member and connection. A schedule of the members and connectors is shown on Sheet 1. The typical elevation view of the arches is shown on Sheet 2. Various beams and purlins are shown throughout the sheets. Study the views carefully to understand the labeling system used and the method of identifying each joint in the structure. The metal connectors also are fabricated from the shop drawings, and, of course, the structure was erected with the help of the shop drawings.

The author is grateful to the Koppers Co. for permission to use the original set of shop drawings, and to Cunningham, Forehand, Stringer, Architects, of Atlanta, Georgia, for permission to use the set of architectural drawings.

Architectural Working Drawings

TEST 1

SHORT-ANSWER TEST ON THE SMALL CHURCH BUILDING

NAME _____

TOTAL SCORE _____

Value of each answer _____**4** POINTS_____

DIRECTIONS: *Use FLOOR PLAN only (Sheet A-3).*

1. What two major materials are used in the outside walls?
   1. _____

2. What are the inside dimensions of the Multipurpose Room?
   2. _____

3. How many doors are required for entrance to the Kitchen?
   3. _____

4. What system is used to expand the Multipurpose Room?
   4. _____

5. Where is the Chair-and-Table Storage Room located?
   5. _____

6. How wide is the Corridor in the left wing?
   6. _____

7. What is the finish floor material in the Pastor's Study?
   7. _____

8. What is the overall length of the church?
   8. _____

9. Is a Janitor's Room provided?
   9. _____

10. What materials is the partition between the 1st and 2nd Grade Room and Kindergarten Room to be made from?
    10. _____

11. How deep is the Vestibule in the outside entrance of the left wing?
    11. _____

12. What are the ceiling heights of the Classrooms?
    12. _____

13. How wide are the arched openings in the Carport?
    13. _____

Architectural Working Drawings

TEST 1 (continued)

SHORT-ANSWER TEST ON THE SMALL CHURCH BUILDING

14. Where is the entrance to the Ladies' Toilet? 14. _____

15. How many square feet are in the total plan (use exterior dimensions, except Carport)? 15. _____

16. How wide is the Pass-Through Opening (between Kitchen and Classroom Area)? 16. _____

17. How wide is the paved area leading to the Carport? 17. _____

18. How high is the base trim in the carpeted rooms? 18. _____

19. Are the kitchen cabinets and appliances to be furnished by the Contractor? 19. _____

20. How high is the ceiling to be in the Choir and Assembly Room? 20. _____

21. What is the thickness of the exterior veneer wall? 21. _____

22. How many individual classrooms can be made with the use of the folding partitions? 22. _____

23. Where will the Choir Platform be located? 23. _____

24. Can cars be driven through both entrances of the Carport in front of the building? 24. _____

25. Where is the Fire Alarm panel located? 25. _____

Architectural Working Drawings

TEST 2

TRUE-FALSE

TRUE-FALSE TEST ON THE SMALL CHURCH BUILDING

NAME _____

TOTAL SCORE _____

Value of each answer _____ **4** POINTS

DIRECTIONS: Use ELEVATIONS only (Sheets A-5 and A-6). Circle the correct letter.

1. Aluminum gutters and downspouts are required.	T	F
2. The roof deck is to be 3/4" plywood.	T	F
3. Steel lintels are used to strengthen the brick archways in the Carport.	T	F
4. All the roofs on the church have the same slope.	T	F
5. Three downspouts are needed on the facade.	T	F
6. The balustrade railing around the spire is to be made of wood.	T	F
7. All door and window openings on the Front Elevation have arched heads.	T	F
8. Brick quoins are shown on all corners of the church.	T	F
9. The sloping portion of the spire is 16'-0" tall.	T	F
10. The finish ceiling in the Carport is 10'-0" above the concrete floor.	T	F
11. The radius of the Carport arches will be 8'-0".	T	F
12. The roof is to be covered with asphalt shingles.	T	F
13. The Canopy over the rear entrance door is 8'-0" wide.	T	F
14. Casement-type windows are required throughout the church.	T	F
15. The windows are to be double-glazed.	T	F

Architectural Working Drawings

TEST 2 (continued)
TRUE-FALSE

TRUE-FALSE TEST ON THE SMALL CHURCH BUILDING

16. A running bond is used with the exposed block on the rear elevation.   T   F

17. The downspouts are to be 3" × 4" in size.   T   F

18. On the interior surface of exterior walls, the window stools will be 6" above the finish floor.   T   F

19. The brick quoins will be 2'-8" wide.   T   F

20. An ALTERNATE is shown that eliminates the Carport.   T   F

21. The fascia is to be painted wood.   T   F

22. Only one entrance is shown on the Front Elevation.   T   F

23. All exposed exterior trim is to be painted.   T   F

24. Face brick is specified for exterior veneer.   T   F

25. The Front Entrance has the same detail treatment as the side entrance.   T   F

Architectural Working Drawings

TEST 3
IDENTIFICATION

TEST ON THE SMALL CHURCH DRAWINGS

NAME _____

TOTAL SCORE _____

Value of each answer __**8.3** POINTS__

DIRECTIONS: Use the following SECTION DETAIL, write the correct names in the spaces provided, as identified by the numbers.

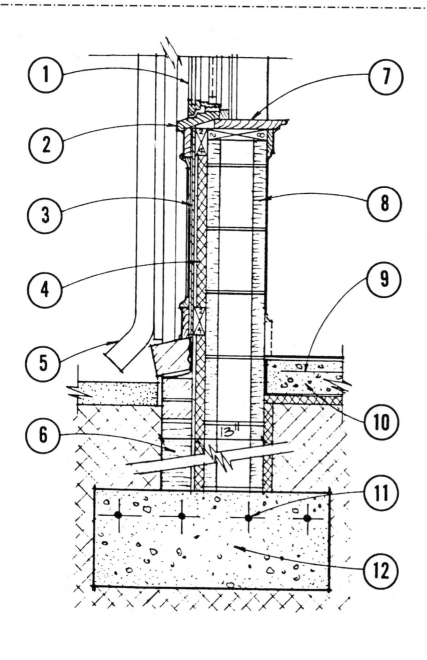

1. _____
2. _____
3. _____
4. _____
5. _____
6. _____
7. _____
8. _____
9. _____
10. _____
11. _____
12. _____

Architectural Working Drawings

TEST 4 — IDENTIFICATION

TEST ON THE SMALL CHURCH DRAWINGS

NAME _____

TOTAL SCORE _____

Value of each answer __**8.3** POINTS__

DIRECTIONS: Using the following SECTION DETAIL, write the correct names in the spaces provided (next page), as identified by the numbers.

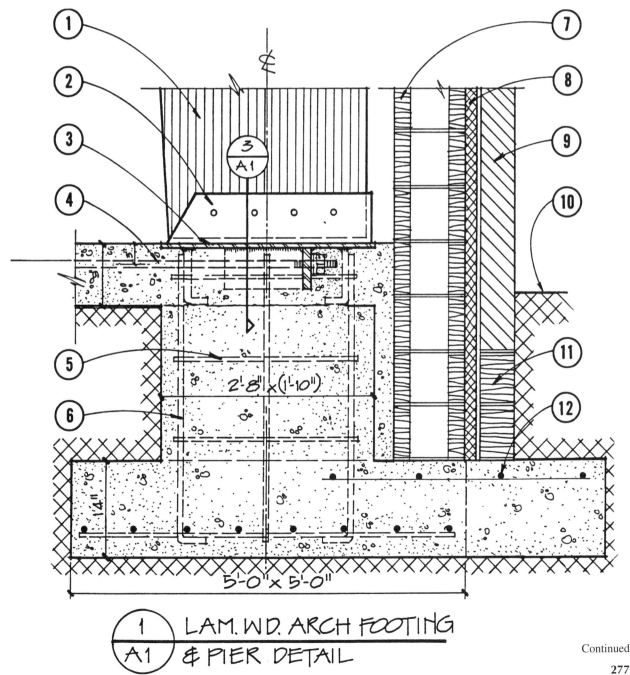

3/A1 — LAM. WD. ARCH FOOTING & PIER DETAIL

Continued

Architectural Working Drawings

TEST 4
(continued)
IDENTIFICATION

TEST ON THE SMALL CHURCH DRAWINGS

1. _____   2. _____   3. _____

4. _____   5. _____   6. _____

7. _____   8. _____   9. _____

10. _____   11. _____   12. _____

Architectural Working Drawings

SHORT-ANSWER TEST ON THE SMALL CHURCH BUILDING

NAME _____

TOTAL SCORE _____

Value of each answer _____ **10** POINTS

DIRECTIONS: *Use the FOUNDATION PLAN (Sheet A-1) only.*

1. How thick is the concrete floor slab? 1. _____

2. What size is the central main square in the plan? 2. _____

3. How deep is the footing that supports the laminated arches? 3. _____

4. Where are 4″ steel pipe columns to be installed? 4. _____

5. What reinforcing is to be used in the turned-down slabs? 5. _____

6. What size are the typical exterior wall footings? 6. _____

7. What strength concrete is to be used in the footings? 7. _____

8. What is the finish-floor elevation? 8. _____

9. What is the exterior width of the left-wing walls? 9. _____

10. What size (#) are the steel tie rods? 10. _____

Architectural Working Drawings

TEST 6

NAME _____

TOTAL SCORE _____

Value of each answer _____ **10** POINTS

DIRECTIONS: *Use the FRAMING PLAN (Sheet A-2) only.*

1. Why are 2 × 10s used in the center of the lower chord of A-type wood trusses?   1. _____

2. What is the span of C-type wood trusses?   2. _____

3. What is the thickness of the laminated arches?   3. _____

4. What size steel lintels are required above the Carport archways?   4. _____

5. What size valley rafter is needed at the junction of the right wing and the main hip?   5. _____

6. What is the slope of the hip roofs?   6. _____

7. What ceiling framing is needed to support the folding partitions?   7. _____

8. How high from the finish floor are the bearing walls in the left wing?   8. _____

9. Where are the type-B wood trusses to be installed?   9. _____

10. What size are the jack rafters in the right-wing roof?   10. _____

APPENDIX

1. REVIEW OF PLANE GEOMETRY CALCULATIONS

A. SQUARE: a rectangle having four equal sides and four right angles (Figure A-1).

$$\text{Area} = a^2$$
$$\text{Perimeter} = 4a$$
$$\text{Length of diagonal} = a\sqrt{2}$$
$$\approx 1.4a$$

B. RECTANGLE: a parallelogram with right angles (Figure A-1).

$$\text{Area} = L \times W$$
$$\text{Perimeter} = 2L + 2W$$
$$\text{Length of diagnoal} = \sqrt{L^2 + W^2}$$

C. PARALLELOGRAM: a four-sided figure with opposite sides parallel and of equal length (Figure A-2).

$$\text{Area} = bh$$

D. QUADRILATERALS: figures with four sides and various angles. A *trapezoid* (Figure A-3) is a four-sided figure with two sides parallel.

$$\text{Area} = \tfrac{1}{2}h \times (b_1 + b_2)$$

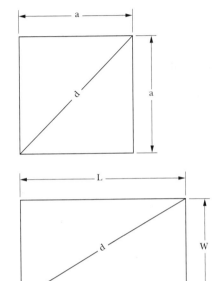

FIGURE A–1

Square and rectangle.

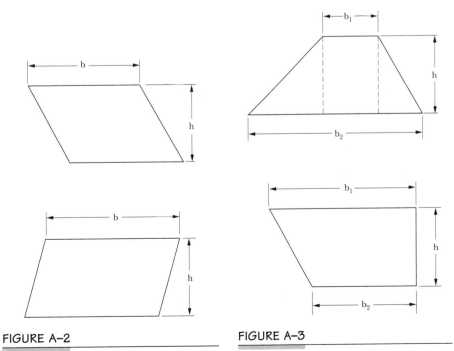

FIGURE A–2

Parallelograms.

FIGURE A–3

Trapezoids.

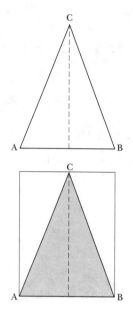

FIGURE A-4

Areas of triangles.

E. TRIANGLES: figures enclosed by three straight sides. The sum of the three angles of any triangle is 180 degrees. The area of any triangle is found by taking one-half the base (*b*) times the altitude (*h*), measured perpendicular to the base (see Figures A-4 and A-5). If the altitude is unknown, use the following formula to find the area of any triangle, with *a, b,* and *c* as the lengths of the sides:

$$\text{Area} = \sqrt{S(S-a)(S-b)(S-c)} \quad [\text{where } S = \tfrac{1}{2}(a+b+c)]$$

For areas of equilateral triangles, use

$$A = 0.433c^2 \text{ (where a side is } c)$$

SCALENE: a triangle with sides of different lengths (see Figure A-6).
ISOSCELES: a triangle with two sides of equal length. In Figure A-7, side *a* equals side *b*. Also, the angles (*A, B*) opposite the equal sides are equal.
EQUILATERAL: a triangle with all three sides of equal length. Also, all angles are equal (see Figure A-8). Side a = side b = side c, and $<A = <B = <C = 60°$.

$$\text{Area} = 0.433c^2$$

RIGHT TRIANGLE: a triangle having one right angle (90°) (Figures A-9 and A-10). *The square of the hypotenuse of a right triangle equals the sum of the squares of the two sides* (the Pythagorean theorem) (see Figure A-11). To verify

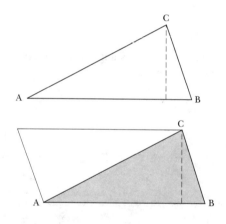

FIGURE A-5

Areas of triangles.

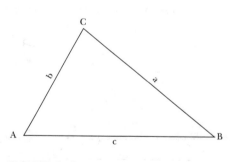

FIGURE A-6

Scalene triangle.

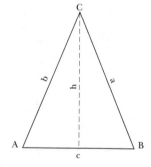

FIGURE A-7

Isosceles triangle.

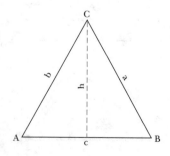

FIGURE A-8

Equilateral triangle.

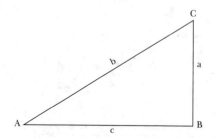

FIGURE A-9

Right triangle.

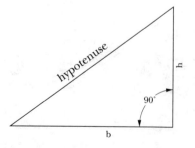

FIGURE A-10

Right triangle.

FIGURE A–11

The right triangle verifies the Pythagorean theorem.

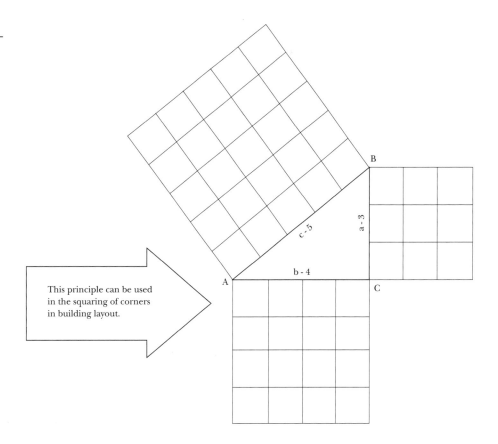

This principle can be used in the squaring of corners in building layout.

this theorem, let us look at the right triangle ABC (Figure A-11) with $\angle C$ a right angle. We find that the length of the unknown hypotenuse is as follows:

$$c^2 = a^2 + b^2 \text{ (Theorem)}$$

$$\begin{aligned} c &= \sqrt{a^2 + b^2} \\ &= \sqrt{9 + 16} \\ &= \sqrt{25} \\ &= \textbf{5 Ans.} \end{aligned}$$

Another example of the use of the theorem is to find one of the sides rather than the hypotenuse: If side $c = 13$ and side $b = 5$, to find side a we rearrange the theorem:

$$\begin{aligned} a^2 &= c^2 - b^2 \\ a &= \sqrt{c^2 - b^2} \\ &= \sqrt{13^2 - 5^2} = \sqrt{169 - 25} \\ &= \sqrt{144} = \textbf{12 Ans.} \end{aligned}$$

F. OTHER REGULAR POLYGONS

Pentagon (5 sides) Area = $1.72c^2$
Hexagon (6 sides) Area = $2.598c^2$
Octagon (8 sides) Area = $4.828c^2$

(c is the length of one side.)

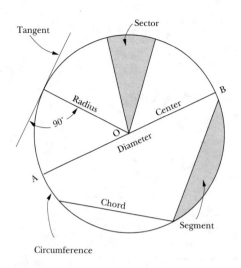

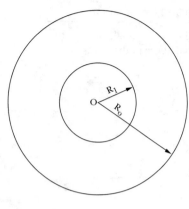

FIGURE A-13

Area of an annulus (ring).

FIGURE A-12

Circle terms.

G. CIRCLE: a uniform, closed curve on which any point is the same distance from a fixed center. Listed next are the major terms and characteristics associated with the circle, regardless of size (see Figure A-12).

1. There are 360 degrees in a circle around the fixed point.

2. *Diameter:* a straight-line segment passing through the center and having its two ends on the curve.

3. *Radius:* a straight-line segment extending from the center to the curve. $R = \tfrac{1}{2}D$.

4. *Circumference:* the curve forming the circle, the length of which is called its perimeter.

5. *Pi* (π): the ratio of the circumference to the diameter of any circle. $\pi = 3.1416$. Circumference $= 2\pi \times R = \pi D$.

6. *Area of a circle* $= \pi R^2$ or $\dfrac{\pi D^2}{4}$

7. *Area of an annulus (ring)* $= \pi(R_0^2 - R_1^2)$

$$= \frac{\pi}{4}(D_0^2 - D_1^2) \text{ (Figure A-13)}$$

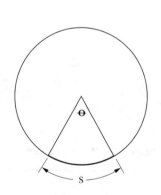

FIGURE A-14

Area of a circular sector.

8. *Area of a circular sector* (Figure A-14) $= \dfrac{\pi R^2 \theta}{360}$ (θ is in degrees)

9. *Length of arc* $S = 0.0174 \times R \times \theta$ (θ is in degrees)

H. ELLIPSE: a curved figure with major and minor axes (see Figure A-15). *Area of an ellipse* $= \pi \tfrac{1}{4} M m$, where M = major axis and m = minor axis.

$$Perimeter = \pi\left(\frac{M + m}{2}\right)$$

FIGURE A-15

Ellipse.

2. REVIEW OF SOLID GEOMETRY CALCULATIONS

A. RECTANGULAR SOLIDS AND CUBES: solids having six plane faces, straight edges, and square corners (Figure A-16). They have three dimensions: length, width, and height.

$$Volume = Length \times Width \times Height \text{ (Figure A-17)}$$

$$= L \times W \times H$$

$$= \text{cubic feet}$$

For example, if $L = 3$ ft, $W = 2$ ft, and $H = 4$ ft:
$$Volume = LWH$$
$$= 3 \times 2 \times 4$$
$$= \textbf{24 cu ft Ans.}$$

The volume of a cube having all equal dimensions (Figure A-18), such as $3 \times 3 \times 3$, is found in the same manner:

$$Volume = a^3 \text{ (}a\text{ is the length of an edge)}$$

Surface area of a rectangular figure equals the sum of its face areas.

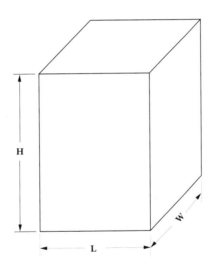

FIGURE A-16

Rectangular solid.

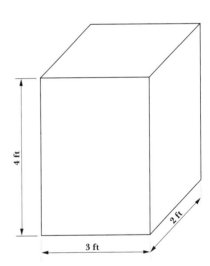

FIGURE A-17

Volume of a rectangular solid.

FIGURE A-18

The volume of a cube is a^3.

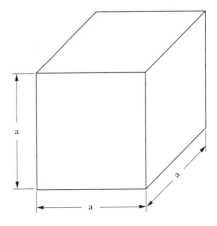

This fact is easily seen if the figure is mentally disassembled (Figure A-19). Notice that it has six faces or sides and opposite sides are equal. The front and back are equal, the right and left ends are equal, and the top and bottom are equal. Therefore, *surface area* = 2LH + 2WH + 2LW.

FIGURE A-19

Surfaces of a rectangular solid.

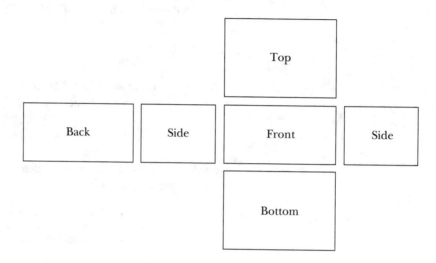

As an example, find the surface area of a box having $L = 6$ ft, $W = 4$ ft, and $H = 3$ ft (Figure A-20).

$$\text{Area} = 2(6 \times 3) + 2(4 \times 6) + 2(3 \times 4)$$
$$= 36 + 48 + 24$$
$$= \mathbf{108 \text{ sq ft Ans.}}$$

Surface area of a cube = $6a^2$ (*a* is the length of an edge)

Another method of finding the volume of a rectangular solid is by thinking of the volume as the area of one face in square units and then multiplying it by the thickness; the result is in cubic units. For example, we can calculate the volume of a solid by finding the area of its base and multiplying this area by its height as follows (see Figure A-17):

$$\text{Area of base} = 3' \times 2' = 6 \text{ sq ft}$$
$$\text{Volume} = \text{Area of Base} \times \text{Height}$$
$$= 6 \times 4 = \mathbf{24 \text{ cu ft Ans.}}$$

FIGURE A-20

Rectangular solid.

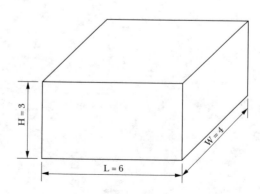

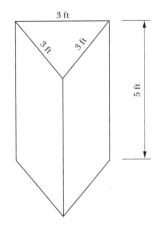

FIGURE A–21

Triangular prism.

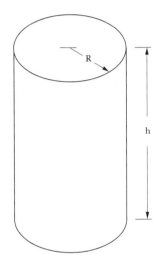

FIGURE A–22

Cylinder.

B. PRISMS: solids having various polygon ends and parallel edges, such as triangular prisms, hexagonal prisms, and octagonal prisms. *To calculate the volume of a prism, find the area of its base (polygon) and multiply by its height.* For example: Find the volume of the triangular prism in Figure A-21.

$$\text{Area of base} = 0.433c^2 \text{ (see Section 1)}$$
$$= 0.433 \times 3 \times 3 = 3.897 \text{ or } 3.9 \text{ sq ft}$$
$$\text{Volume} = 3.9 \times 5 = \textbf{19.5 cu ft Ans.}$$

Naturally, this calculation is valid regardless of the orientation or position of the prism.

SURFACE AREAS OF PRISMS: Calculate the surface area of each face and add them. Faces may be polygons or parallelograms.

C. CYLINDERS: solids having a circular base and top and enclosed with a uniformly curved surface parallel to their axes, such as a round metal rod (Figure A-22). The length of the axis is taken as the height (h).

Volume of a cylinder $= \pi R^2 h$ (area of base times height)
Surface area $=$ *area of base plus area of top plus area of lateral side.*

If the curved lateral side is flattened out on a horizontal plane (see Figure A-23), it becomes a rectangle. Its dimensions are the circumference of the base and the height of the cylinder. Therefore,

$$\textit{Surface area} = 2\pi Rh + 2\pi R^2$$

For example: find the surface area and volume of a cylinder such as the one shown in Figure A-22, with a radius of 3′ and a height of 10′.

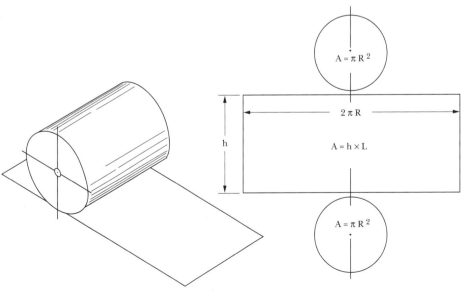

FIGURE A–23

Finding the volume and surface area of a cylinder.

$$\text{Area} = 2 \times 3.1416 \times 3 \times 10 + 2 \times 3.1416 \times 3 \times 3$$
$$= 188.5 + 56.5$$
$$= \mathbf{245 \text{ sq ft Ans.}}$$

$$\text{Volume} = \pi R^2 h$$
$$= 3.1416 \times 3 \times 3 \times 10$$
$$= \mathbf{282.78 \text{ cu ft Ans.}}$$

To calculate gallons in this volume: 282.7 × 7.48 (gallons per cubic foot) = **2114.6 gal Ans.**

D. RIGHT PYRAMIDS AND RIGHT CIRCULAR CONES: solid figures with polygon or circle bases and apexes or points falling upon axes that are right angles from the plane of their bases (see Figures A-24 through A-27).

A *regular pyramid* is a pyramid with a polygon base and isosceles triangles forming its other faces. To find the surface area of a regular pyramid, multiply the surface area of one isosceles-triangular side by the number of sides, and add the area of the base.

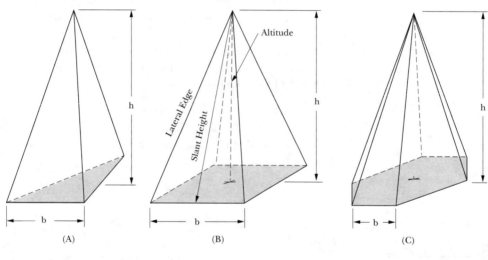

FIGURE A-24

Right regular pyramids. (A) triangular base. (B) square base. (C) hexagon base.

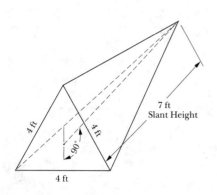

FIGURE A-25

Surface area of a regular triangular pyramid.

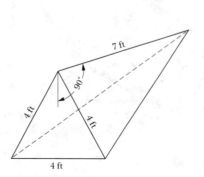

FIGURE A-26

Volume of a right-triangular pyramid.

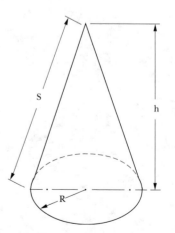

FIGURE A-27

Right circular cone.

For a *triangular-base pyramid*:

$$\text{Surface area} = 3 \times \frac{bS}{2} + 0.433b^2$$

For a *square-base pyramid*:

$$\text{Surface area} = 4 \times \frac{bS}{2} + b^2 = 2bS + b^2$$

For a *hexagon-base pyramid*:

$$\text{Surface area} = 6 \times \frac{bS}{2} + 2.598b^2$$

Find the surface area of the right triangular pyramid shown in Figure A-26 having base sides equal to 4′ and height equal to 7′.

$$\begin{aligned}
\text{Surface area} &= \frac{3(bS)}{2} + 0.433b^2 \\
&= \tfrac{3}{2}(4 \times 7) + 0.433(4 \times 4) \\
&= 42 + 6.928 \\
&= \mathbf{48.93 \text{ sq ft Ans.}}
\end{aligned}$$

To find the volume of a right pyramid, multiply the area of its base by one-third its height:

$$V = A_b \times \frac{h}{3}$$

Volume of Figure A-26 $= 6.928 \times \tfrac{7}{3}$

$$= \mathbf{16.16 \text{ cu ft Ans.}}$$

A *right circular cone* is a solid with a circle for its base and having the axis of its height perpendicular to and passing through the center of its base (see Figure A-27).

To find the surface area of a cone, add the area of its curved surface to the area of its base.

$$\text{Total surface area} = \pi RS + \pi R^2 \ (S = \text{slant height} = \sqrt{R^2 + h^2}).$$

For example, find the total surface area of a cone such as that shown in Figure A-27, with a base radius of 3′ and a height of 4′.

$$\begin{aligned}
\text{Curved surface area} &= \pi RS, \ S = \sqrt{3^2 + 4^2} = 5 \\
&= 3.1416 \times 3 \times 5 \\
&= 47.1 \text{ sq ft}
\end{aligned}$$

$$\begin{aligned}
\text{Base area} &= \pi R^2 \\
&= 3.1416 \times 9 \\
&= 28.27 \text{ sq ft}
\end{aligned}$$

$$\text{Total surface area} = 28.27 + 47.1$$

$$= \mathbf{75.37 \text{ sq ft Ans.}}$$

To find the volume of a cone, multiply the area of its base by its height and divide by 3:

$$V = \frac{\pi R^2 h}{3}$$

Volume of Figure A-27 = $1.047 \times 9 \times 4$

Volume = **37.7 cu ft Ans.**

To find the surface area of a sphere, multiply the square of its radius by 4π:

$$\text{Surface area of a sphere} = R^2 4\pi$$

$$\text{Volume of a sphere} = R^3 \frac{4\pi}{3}$$

TABLE A-1

Weights and Measures.

Linear Measure			
1 inch		= 2.54	centimeters
12 inches = 1 foot		= 0.3048	meter
3 feet = 1 yard		= 0.9144	meter
$5\frac{1}{2}$ yards or $16\frac{1}{2}$ feet = 1 rod (or pole or perch)		= 5.029	meters
40 rods = 1 furlong		= 201.17	meters
8 furlongs or 1760 yards or 5280 feet = 1 (statute) mile		= 1609.3	meters
3 miles = 1 (land) league		= 4.83	kilometers

Square Measure			
1 square inch		= 6.452	square centimeters
144 square inches = 1 square foot		= 929	square centimeters
9 square feet = 1 square yard		= 0.8361	square meter
$30\frac{1}{4}$ square yards = 1 square rod (or square pole or square perch)		= 25.29	square meters
160 square rods or 4840 square yards or 43,560 square feet = 1 acre		= 0.4047	hectare
640 acres = 1 square mile		= 259	hectares or 2.59 square kilometers

Cubic Measure		
1 cubic inch	= 16.387	cubic centimeters
1728 cubic inches = 1 cubic foot	= 0.0283	cubic meter
27 cubic feet = 1 cubic yard	= 0.7646	cubic meter
(in units for cordwood, etc.)		
16 cubic feet = 1 cord foot		
8 cord feet = 1 cord	= 3.625	cubic meters

TABLE A-1

Weights and Measures. Continued

Chain Measure

(for Gunter's, or surveyor's, chain)

7.92 inches = 1 link	= 20.12	centimeters
100 links or 66 feet = 1 chain	= 20.12	meters
10 chains = 1 furlong	= 201.17	meters
80 chains = 1 mile	= 1609.3	meters

(for engineer's chain)

1 foot = 1 link	= 0.3048	meter
100 feet = 1 chain	= 30.48	meters
52.8 chains = 1 mile	= 1609.3	meters

Surveyor's (Square) Measure

625 square links = 1 square pole	= 25.29	square meters
16 square poles = 1 square chain	= 404.7	square meters
10 square chains = 1 acre	= 0.4047	hectare
640 acres = 1 square mile or 1 section	= 259	hectares or 2.59 square kilometers
36 square miles = 1 township	= 9324.0	hectares or 93.24 square kilometers

Dry Measure

1 pint	= 33.60 cubic inches =	0.5505 liter
2 pints = 1 quart	= 67.20 cubic inches =	1.1012 liters
8 quarts = 1 peck	= 537.61 cubic inches =	8.8096 liters
4 pecks = 1 bushel	= 2150.42 cubic inches =	35.2383 liters

1 British dry quart = 1.032 U.S. dry quarts

Liquid Measure

1 gill = 4 fluid ounces =	7.219 cubic inches =	0.1183 liter
4 gills = 1 pint	= 28.875 cubic inches =	0.4732 liter
2 pints = 1 quart	= 57.75 cubic inches =	0.9463 liter
4 quarts = 1 gallon	= 231 cubic inches =	3.7853 liters

British imperial gallon (4 imperial quarts) = 277.42 cubic inches = 4.546 liters

Time Measure

60 seconds = 1 minute
60 minutes = 1 hour
24 hours = 1 day
7 days = 1 week
4 weeks (28 to 31 days) = 1 month
12 months (365 or 366 days) = 1 year
100 years = 1 century

Angular and Circular Measure

60 seconds = 1 minute
60 minutes = 1 degree
90 degrees = 1 right angle
180 degrees = 1 straight angle
360 degrees = 1 circle

TABLE A-1
Weights and Measures. *Continued*

Troy Weight

24 grains = 1 pennyweight
20 pennyweights = 1 ounce
12 ounces = 1 pound

Avoirdupois Weight

$12\frac{11}{32}$ grains = 1 dram
16 drams = 1 ounce
16 ounces = 1 pound
100 pounds = 1 short hundredweight
20 short hundredweight = 1 short ton

TABLE A-2
The Metric System.

Linear Measure

10 millimeters = 1 centimeter
10 centimeters = 1 decimeter
10 decimeters = 1 meter
10 meters = 1 decameter
10 decameters = 1 hectometer
10 hectometers = 1 kilometer

Square Measure

100 sq millimeters = 1 sq centimeter
100 sq centimeters = 1 sq decimeter
100 sq decimeters = 1 sq meter
100 sq meters = 1 sq decameter
100 sq decameters = 1 hectometer
100 sq hectometers = 1 sq kilometer

Cubic Measure

1000 cu millimeters = 1 cu centimeter
1000 cu centimeters = 1 cu decimeter
1000 cu decimeters = 1 cu meter

Liquid Measure

10 milliliters = 1 centiliter
10 centiliters = 1 deciliter
10 deciliters = 1 liter
10 liters = 1 decaliter
10 decaliters = 1 hectoliter
10 hectoliters = 1 kiloliter

Weights

10 milligrams = 1 centigram
10 centigrams = 1 decigram
10 decigrams = 1 gram
10 grams = 1 decagram
10 decagrams = 1 hectogram
10 hectograms = 1 kilogram
100 kilograms = 1 quintal
10 quintals = 1 ton

TABLE A-3
Arabic and Roman Notations.

0 = 0	15 = XV	30 = XXX	150 = CL
1 = I	16 = XVI	31 = XXXI	200 = CC
2 = II	17 = XVII	32 = XXXII	300 = CCC
3 = III	18 = XVIII	40 = XL	400 = CD
4 = IV	19 = XIX	50 = L	500 = D
5 = V	20 = XX	60 = LX	600 = DC
6 = VI	21 = XXI	70 = LXX	700 = DCC
7 = VII	22 = XXII	80 = LXXX	800 = DCCC
8 = VIII	23 = XXIII	90 = XC	900 = CM
9 = IX	24 = XXIV	100 = C	1000 = M
10 = X	25 = XXV	101 = CI	2000 = MM
11 = XI	26 = XXVI	102 = CII	10,000 = $\overline{X}$
12 = XII	27 = XXVII	103 = CIII	100,000 = $\overline{C}$
13 = XIII	28 = XXVIII	104 = CIV	1,000,000 = $\overline{M}$
14 = XIV	29 = XXIX		

TABLE A-4
Decimals of an Inch Equivalents.

$\frac{1}{64}$ = .015625	$\frac{9}{32}$ = .28125	$\frac{35}{64}$ = .546875	$\frac{13}{16}$ = .8125
$\frac{1}{32}$ = .03125	$\frac{19}{64}$ = .296875	$\frac{9}{16}$ = .5625	$\frac{53}{64}$ = .828125
$\frac{3}{64}$ = .046875	$\frac{5}{16}$ = .3125	$\frac{37}{64}$ = .578125	$\frac{27}{32}$ = .84375
$\frac{1}{16}$ = .0625	$\frac{21}{64}$ = .328125	$\frac{19}{32}$ = .59375	$\frac{55}{64}$ = .859375
$\frac{5}{64}$ = .078125	$\frac{11}{32}$ = .34375	$\frac{39}{64}$ = .609375	$\frac{7}{8}$ = .875
$\frac{3}{32}$ = .09375	$\frac{23}{64}$ = .359375	$\frac{5}{8}$ = .625	$\frac{57}{64}$ = .890625
$\frac{7}{64}$ = .109375	$\frac{3}{8}$ = .375	$\frac{41}{64}$ = .640625	$\frac{29}{32}$ = .90625
$\frac{1}{8}$ = .125	$\frac{25}{64}$ = .390625	$\frac{21}{32}$ = .65625	$\frac{59}{64}$ = .921875
$\frac{9}{64}$ = .140625	$\frac{13}{32}$ = .40625	$\frac{43}{64}$ = .671875	$\frac{15}{16}$ = .9375
$\frac{5}{32}$ = .15625	$\frac{27}{64}$ = .421875	$\frac{11}{16}$ = .6875	$\frac{61}{64}$ = .953125
$\frac{11}{64}$ = .171875	$\frac{7}{16}$ = .4375	$\frac{45}{64}$ = .703125	$\frac{31}{32}$ = .96875
$\frac{3}{16}$ = .1875	$\frac{29}{64}$ = .453125	$\frac{23}{32}$ = .71875	$\frac{63}{64}$ = .984375
$\frac{13}{64}$ = .203125	$\frac{15}{32}$ = .46875	$\frac{47}{64}$ = .734375	1 = 1.000
$\frac{7}{32}$ = .21875	$\frac{31}{64}$ = .484375	$\frac{3}{4}$ = .750	
$\frac{15}{64}$ = .234375	$\frac{1}{2}$ = .500	$\frac{49}{64}$ = .765625	
$\frac{1}{4}$ = .250	$\frac{33}{64}$ = .515625	$\frac{25}{32}$ = .78125	
$\frac{17}{64}$ = .265625	$\frac{17}{32}$ = .53125	$\frac{51}{64}$ = .796875	

TABLE A-5
Numbers and Powers.

Number	Square	Square Root	Number	Square	Square Root
1	1	1.000	51	2601	7.141
2	4	1.414	52	2704	7.211
3	9	1.732	53	2809	7.280
4	16	2.000	54	2916	7.348
5	25	2.236	55	3025	7.416
6	36	2.449	56	3136	7.483
7	49	2.646	57	3249	7.550
8	64	2.828	58	3364	7.616
9	81	3.000	59	3481	7.681
10	100	3.162	60	3600	7.746
11	121	3.317	61	3721	7.810
12	144	3.464	62	3844	7.874
13	169	3.606	63	3969	7.937
14	196	3.742	64	4096	8.000
15	225	3.873	65	4225	8.062

TABLE A-5
Numbers and Powers. Continued

Number	Square	Square Root	Number	Square	Square Root
16	256	4.000	66	4356	8.124
17	289	4.123	67	4489	8.185
18	324	4.243	68	4624	8.246
19	361	4.359	69	4761	8.307
20	400	4.472	70	4900	8.367
21	441	4.583	71	5041	8.426
22	484	4.690	72	5184	8.485
23	529	4.796	73	5329	8.544
24	576	4.899	74	5476	8.602
25	625	5.000	75	5625	8.660
26	676	5.099	76	5776	8.718
27	729	5.196	77	5929	8.775
28	784	5.292	78	6084	8.832
29	841	5.385	79	6241	8.888
30	900	5.477	80	6400	8.944
31	961	5.568	81	6561	9.000
32	1024	5.657	82	6724	9.055
33	1089	5.745	83	6889	9.110
34	1156	5.831	84	7056	9.165
35	1225	5.916	85	7225	9.220
36	1296	6.000	86	7396	9.274
37	1369	6.083	87	7569	9.327
38	1444	6.164	88	7744	9.381
39	1521	6.245	89	7921	9.434
40	1600	6.325	90	8100	9.487
41	1681	6.403	91	8281	9.539
42	1764	6.481	92	8464	9.592
43	1849	6.557	93	8649	9.644
44	1936	6.633	94	8836	9.695
45	2025	6.708	95	9025	9.747
46	2116	6.782	96	9216	9.798
47	2209	6.856	97	9409	9.849
48	2304	6.928	98	9604	9.899
49	2401	7.000	99	9801	9.950
50	2500	7.071	100	10000	10.000

INDEX

A

Abbreviations, and architectural drawings, 65–73
Abutment, 58
Actual size, 44, 81
Addenda, and bidding requirements, 211
Addition; *See* Construction mathematics
Adobe construction, 58
A-frame, 58
Aggregate, 58
A.I.A; *See* American Institute of Architects (A.I.A.)
A.I.A. Short Form of General Conditions (Document A-201), 210
A.I.A. Standard Form of General Conditions (Document A-201), 210
Air conditioning; *See* HVAC
Alternate notes, and small church, 270
American Institute of Architects (A.I.A.), 210
American Metric Construction Handbook, 83–84
American National Metric Council, 83
Anchor bolt, 58
Angular lines, 89
Apron, 58
Arch, 58
Architect-designed homes, 11
Architect's scale, 73–76
Architectural drawings; *See* Working drawings
Architectural materials, symbols for, 52–54
Arcs; *See* Circles and arcs
Areaway, 58
Arithmetic; *See* Construction mathematics
Arris, 58
Asbestos board, 58
Ash dump, and fireplace and chimney, 192
Ashlar, 58
Ashpit, and fireplace and chimney, 192
Asphalt, 58
Astragal, 58
Atrium, 58
Attic, 58
Auxiliary, symbols for, 55
Auxiliary views, and working drawings, 36
Awning, 58, 178
Axis, 58

B

Babylonian civilization, 1
Backfill, 58
Balloon frame, 164, 166
 and sill detail, 185
Balusters, stairs and stairwells, 193
Banister, 58
Bargeboard, 58
Base cabinets, 58
Baseboard, 58
Basement floor slabs, 183–184
Basement, test on, 253–254
Batt, 58
Batten, 58
Batter boards, 58
Bay window, 58
Beamed ceiling, 58
Beams, 58, 148
Bearing of the line, 137
Bearing plate, 58
Bearing wall, 58
Bench marks, 58, 140
Bid bonds, and bidding requirements, 211
Bidding requirements, 211
 addenda, 211
 bid bonds, 211
 bid forms, 211
 change orders, 211
 instructions to bidders, 211
 invitation to bid, 211
 labor and material bonds, 211
 other guarantee documents, 211
 performance bond, 211
Bidet, 58
Bid forms, and bidding requirements, 211
Bird's mouth, 58
Blind nailing, 58
Blocking, 59
Board feet, calculation of; *See* Construction mathematics
Board measure, 59
Bond, 59
Bond beam, 58
Bow windows; *See* Windows and doors
Box sill, 185
Bracing, 59
Break lines, 40
Break symbols, 186
Brick, 59
 6" Norwegian, 185
Brick-cavity construction, 183
Brick-veneer construction, 186
Bridging, 59
Broadscope sections, and CSI format, 212
Building codes, 8
Building construction; *See also* Working drawings
 contract documents, 7

295

Building construction; *See also* Working drawings *(Continued)*
   major parts of, 7
   working drawings, 7
   written specifications, 7
Building line, 59
Building materials; *See* Materials
Built-up roof, 59
Bullnose, 59
   stairs and stairwells, 193
Butt, 59
Buttress, 59

C

Cabinet-oblique, 94, 96
Cabinetwork, sizes, 195, 197
CADD (computer-aided design and drafting), 17–22
   desktop printers, 21
   future developments, 22
   hardware recommendations, 20–21
   inkjet plotter, 21
   Internet, 22
   introduction, 17–19
   LED plotter, 21
   operating system (OS), 19–20
   peripheral devices, 21
   and software, 20
   tools used, 18–19
Callout, 59
Cantilever, 59
Cant strip, 59
Carriage, stairs and stairwells, 193
Casing, 59
Caulking, 59
Cavity wall, 59
Ceiling beams, 168
Ceiling framing, 167, 173
Ceiling, suspended, 64
Cement cap, and fireplace and chimney, 192
Center lines, 39
Chaldean art, 1
Chamfer, 59
Chancel, 59
Change orders, and bidding requirements, 211
Chase, 59
Chimney pots, and fireplace and chimney, 192
Chord, 59
Church drawings; *See* Small church
Circles and arcs, 89–91
  isometric, 94–98
  partial, 94
Circuit, 59
Civil engineer's scale, 76–79
Clerestory, 59
Climate, and construction, 8
Closed stringer, stairs and stairwells, 194
Closet doors; *See* Windows and doors
Collar beam, 59

Colonial home
   brick-screen wall construction, 145
   cantilevered second-floor joists, 151
   elevation reading, 181–182
   elevation view, 159, 160
   exploded view, 147
   fireplace, 150
   floor plan reading, 173
   foundation construction, 144
   front evaluation, 146
   HVAC plan, 199, 200
   model of, 153
   perspective view, 146
   pier construction, 145
   second-floor plan, 158
Column, 59
Combustion chamber, and fireplace and chimney, 191
Commercial
   architect-designed homes, 11
   light construction procedures, 9, 11
   manufactured homes, 11
Communication, and working drawings, 5, 7
Computer-aided design and drafting (CADD); *See* CADD (computer-aided design and drafting)
Concrete, 59
   symbols for, 53
Concrete slabs, 143
Conduit, 59
Control joint, 59
Construction
   architect-designed homes, 11
   and building codes, 8
   and climate, 8
   loan for, 11
   manufactured homes, 11
   material availability, 8
   residential and light commercial procedures, 9, 11
   and subsoil, 8
Construction drawings; *See also* Working drawings
   preliminary, 12
   presentation, 12, 13
   shop, 15
   types of, 12, 14–15
   working, 12, 14–15
Construction loan, 11
Construction mathematics, 41–51
   angular dimension, 48, 50
   board feet calculation, 44–46
   decimal and fraction calculations, 43–44
   decimal to fraction conversion, 43
   English dimensions to metric conversion, 84–86
   exercises for, 122–131
   feet and inches to decimals conversion, 44
   feet-and-inch addition and subtraction, 41–42
   feet-and-inch multiplication, 42–43
   fraction to decimal conversion, 43
   fractional pitch indication, 48, 50
   grade slope indications, 50–51

major geometric shapes, 49
metric dimensions, 83–86
nominal size, 44
percentage of grade, 50–51
plane figures, 47
plane geometry, 47
rafter length calculation, 51
ratio and proportion, 46–47
roof pitches, 48, 49
rule of three, 47
slope-ratio diagram, 48, 50
solid figures, 47–48
symbols, 41
terminology, 41
Construction Specification Institute; *See* CSI; CSI format
Construction terminology
exercises for, 112–120
glossary of, 58–65
test on, 247–249
Construction types
adobe, 58
A-frame, 58
architect-designed homes, 11
balloon frame, 164, 166
brick-cavity, 183
brick-veneer, 186
double joist, 148
dry-wall, 60
eastern frame, 164, 166
half-timber, 61
light commercial, 9, 11
light-frame, 164, 167
light wood-frame, 164
manufactured homes, 11
plank-and-beam, 186
platform frame, 164, 165
post-and-beam, 164, 167, 168, 169
procedures, 9, 11
slab-on-ground, 185
western frame, 164, 165
Continuous X-bridging, 148
Contour interval, 140
Contour lines, 140
Contour Maps, 139
Contract documents, 7
Concrete blocks, 143
Convenience outlets, symbols for, 55
Cooling system; *See* Heating and cooling (HVAC)
Coordinate paper, and pencil sketching, 92–93
Coping, 59
Corbel, 60
Cornice detail, 60, 186
Cost per lot, 9
Cove, 60
Crawl spaces, 60, 143
Cricket, 60
Cripple, 60
Cronoflex™, 17

Crown molding, 60
CSI (Construction Specification Institute), 212; *See also* CSI format
CSI format, 212–236
broadscope sections, 212
MasterFormat™, 213–219
narrowscope sections, 212
sample set of, 221–236
Custom-design, 11
Cut stone, 60
Cutting-plane line, 30, 33, 40

D

Dado joint, 60
Damper, and fireplace and chimney, 191
Dampproofing, 60
Dashed lines, 39, 40
Datum point, 139–140
Dead load, 60
Deck, 60
Decimals; *See* Construction mathematics
Description of materials, 212
Desktop printers, and CADD, 21
Details, 182–195
basement floor slabs, 183–184
cornice detail, 186
fascia board, 187
fireplace and chimney, 189–193
foundation walls, 182–184
plank and beam construction, 186
roofing support, 186
sill detail, 184–185
stairs, 193–195
test on, 241–242
trim, 195
wall, 182–187
and working drawings, 30–35
Developer, 9
construction procedure, 9, 11
Diagrams, 203; *See also* Framing plan
Diazo prints, 16
Dimension lines, 39, 40
Dimension lumber, 60
Dimensions; *See* Scales and dimensions
Distribution panel, 60
Division; *See* Construction mathematics
Domestic hot water, 60
Door schedule, and small church, 270
Doors; *See* Windows and doors
Door stop, 60
Dormers, 60, 162
Dots per inch (DPI); *See* DPI
Double-hung windows; *See* Windows and doors
Double joist construction, 148
Double glazing, 60
Downspout, 60
DPI (dots per inch), 21

Drawings; *See* Working drawings
Dressed size, 60
Drip, 60
Dry-wall construction, 60
Duplex outlet, 60

E

Earthwork, symbols for, 53
Easement, 60
Eastern frame, 164, 166
Eave, 60
Efflorescence, 60
Electrical, symbols for, 55
Electrical plan, 203
   and small church, 271
Electrical, symbols for, 55
Electric, and floor plan, 157, 162
Elevation, 28–30, 174–182
   colonial home, 146, 159
   exercises for, 109–110
   front, 170
   grade line, 174
   interior, 195, 197
   left, 170
   reading the colonial home, 181–182
   rear, 170
   roof information, 174, 175
   schedules and details, 160
   and small church, 270
   symbols for, 52
   test for, 239–240
   windows and doors, 174, 176–182
Elevation details, 160
Elevation schedule, 160
Elevation views; *See* Elevation
Ell, 60
English-metric conversion; *See* Construction mathematics
Excavation, 60
Expansion joint, 60
Exploded, colonial home, 147
Extension lines, 40

F

Facade, 60
Face brick, 60
Fascia board, 60, 187
Fasteners, 60
Federal Housing Administration (FHA); *See* FHA
Feet-and-inch dimensions; *See* Construction mathematics
Fenestration, 60
FHA (Federal Housing Administration), 11, 212
Fiberboard, 61
Finishes, symbols for, 54
Finish schedule, 164
   and small church, 270
Finish trim details, and small church, 270
Fire cut, 61

Fireplace and chimney
   ash dump, 192
   ashpit, 192
   cement cap, 192
   chimney pots, 192
   combustion chamber, 191
   damper, 191
   details, 189–193
   and floor plan, 157
   flue lining, 192
   hearth, 191
   hood, 192
   mantle, 191
   opening, 191
   prefabricated, 191
   smoke dome, 192
   smoke shelf, 192
   throat, 191
Fire rated, 61
Fire resistant, and wood, 265–266
Fire-stop, 61
First-floor plan, 155
   test for, 237–238
Flagstone, 61
Flashing, 61
Flitch beam, 61
Floor plan, 30, 156–158, 162, 164, 167, 173
   balloon frame, 164, 166
   electrical information, 157, 162
   fireplace and chimney, 157, 161
   kitchens and baths, 157
   light-frame construction, 164, 167
   platform frame construction, 164, 165
   post-and-beam construction, 164, 167, 168, 169
   reading of colonial home, 173
   schedules, 164
   second, 158
   and small church, 269–270
   split-level, 162–164
   stairs and stairwells, 157, 161, 193–195
   test on, 252
   trim detail, 195
   windows and doors, 156
   wood-frame walls, 156–157, 161, 171
Flue, 61
Flue lining, and fireplace and chimney, 192
Footer, 61
Footing, 61
Footing schedule, and small church, 269
Footings, 143, 182
Forced hot-air system, 199; *See also* HVAC
Foundation plans, 140, 142–145, 148–149, 154
   colonial home, 144
   small church, 269
Foundation, test on, 253–254
Foundation walls, 182–183
Fractions; *See* Construction mathematics
Framing plan, 203–207
Frieze, 61

Frost line, 61, 182
    and small church, 269
Full sections, 187
Furring strips, 61

G

Gable, 61
Glazing, 61
General conditions
    A.I.A. Short Form of General Conditions, 210
    A.I.A. Standard Form of General Conditions (Document A-201), 210
    and specifications, 210–211
Geometric shapes, 49
Geometry; See Construction mathematics
Girders, 148
Glass, symbols for, 54
Glue-lam, 265
Government Lots, 139
Grade, 61
Grade line, 184
Gradient, 61
Gravel stop, 61
Grounds, 61
Grout, 61
Guarantee documents, and bidding requirements, 211
Gudea, statue of, 1, 2
Gusset, 61
Gutter, 61
Gyp board, 61

H

Half-timber, 61
Handrail, stairs and stairwells, 194
Hanger, 61
Hardware, 61
    and CADD, 20–21
Header, 61
Headroom, 61
    stairs and stairwells, 194
Hearth, 61
Hearth, and fireplace and chimney, 191
Heartwood, 62
Heating and cooling (HVAC), 199, 201–202
    colonial home plan, 199, 200
    cooling system, 201, 202
    electrical heating, 201
    forced hot-air system, 199, 201
    heat pumps, 200
    hot-water system, 199, 201
    Minimum Property Standards (FHA No. 300), 202
    perimeter system, 199
    radiant heating system, 199
Heating system; See Heating and cooling (HVAC)
Heat pumps, 62, 200
Hidden lines, 39
    elimination of, 30

Hip rafter, 62
Home plan service, 9
Hood, and fireplace and chimney, 192
Hopper, 178
Hose bibb, 62
Hot-water system, 199
House stringer, stairs and stairwells, 194
House sweeper, 62
HVAC; See Heating and cooling (HVAC)

I

Incandescent lamp, 62
Inches; See Construction mathematics
Inkjet plotter, and CADD, 21
Instructions to bidders, and bidding requirements, 211
Insulation, symbols for, 54
Interior elevation, 195, 197
    and small church, 270
Interior trim, 62
Internet, and CADD, 22
Invitation to bid, and bidding requirements, 211
Isometric arcs, and pencil sketching, 94–98
Isometric circles
    partial, 94
    and pencil sketching, 94–98
Isometric drawing, 36–37, 38
    cabinet-oblique, 94, 96
    oblique, 94, 96
    and pencil sketching, 93–98
Isometric view, and pencil sketching, 93–98

J

Jack rafter, 62
Jalousie, 178
Jamb, 62
Jointery, 62
Joists, 62, 164

K

Kiln-dried lumber, 62
Kitchens and baths
    cabinetwork sizes, 195, 197
    and floor plan, 157
    residential bath detail, 198
Knee walls, 62, 162
Knocked down, 62

L

Labor and material bonds, and bidding requirements, 211
Lally column, 62
Laminated beam, 62
Laminated members, 265
Landing, stairs and stairwells, 194
Landscape plans, 140, 141
Lap joint, 62

Lath, 62
Lattice, 62
Layouts; *See also* Plan
Leaders, 40, 62
Ledger, 62
LED plotter, and CADD, 21
Light commercial, construction procedures, 9, 11
Light-frame construction, 164, 167
Light wood-frame construction, 164
Lineal foot, 62
Lines
 angular, 89
 break, 40
 center, 39, 40
 cutting-plane, 40
 dashed, 39
 dimension, 39, 40
 extension, 40
 hidden, 39, 40
 leaders, 40
 object, 39, 40
 pencil sketching, 88–89
 section, 40
 visible, 39
 and working drawings, 39–40
Lintel, 62
Live load, 62
Load-bearing wall, 62
Longitudinal sections, 187
Lookout, 62
Lot line, 62
Louver, 62
Louvre, 1
Low-voltage switching system, 157
Luminaire, 62

M

Mantle, and fireplace and chimney, 62, 191
Manufactured homes, 11
Masonry, 62
 symbols for, 53
Mastic, 62
Materials
 availability of, 8
 symbols, 30, 52–54
Mathematics, and construction, 41–52; *See also* Construction mathematics
Matte finish, 62
Meandered water, 139
Mechanical plan, and small church, 271
Metals, symbols for, 53
Metes and Bounds system, 135
Metric dimensions, 83–86
 English dimensions conversion to, 84–86
 exercises for, 132–133
 rounding off, 86
 scales on drawings, 86
Microfilm, 17
Millwork, 62

Minimum Property Standards (FHA No. 300), 202
Miter joint, 62
Modular coordination, 79–83
Modular dimensions, reading of, 79–83
Modular masonry units, 81
Module, 62
Monolithic, 62
Mosaic, 62
Mudsill, 62
Mullion, 62, 174
Multicolor offset prints, and working drawings, 17
Multiplication; *See* Construction mathematics
Multitasking, and CADD, 20
Multiview drawing, 25
Muntin, 62
Mylar, 17

N

Narrowscope sections, and CSI format, 212
Narthex, 62
Newel, stairs and stairwells, 194
Ningirsu, temple of, 1
Nominal size, 44, 62, 81
Nonferrous metal, 62
North arrow, 28, 29
Nosing, stairs and stairwells, 194

O

Object lines, 39, 40
Oblique drawing, 37, 38, 94, 96
On center, 62
Opening, and fireplace and chimney, 191
Open stringer, stairs and stairwells, 194
Operating system (OS); *See* OS (operating system)
Orthographic projection, 25–27
OS (operating system), and CADD, 19–20
Outlet, 62
 symbols for, 55
Outline form, and Federal Housing Administration (FHA), 212
Overhang, 63
Ozalid prints, 16

P

PalmPilot, 22
Parapet, 63
Parge coat, 63
Parquet flooring, 63
Partition, symbols for, 52
Party wall, 63
Pencil, 5
Pencil sketching, 86–99
 angular lines, 89
 basics of, 86–88
 cabinet-oblique, 94, 96
 circles and arcs, 89–91
 and coordinate paper, 92–93
 dimensioning pictorial drawings, 98–99

exercises for, 100–106
isometric circles and arcs, 94–98
isometric drawing, 93–98
isometric view, 93–98
line weights, 99
oblique, 94, 96
partial circles, 94
and pictorial drawings, 93–94, 98–99
practice strokes, 88–89
proportion, 91
vertical lines, 88–89
Penny, 63
Percentage of grade, 50–51
Performance bond, and bidding requirements, 211
Pergola, 63
Perimeter, 47
Perimeter system, 199
Peripheral devices, and CADD, 21
Periphery, 63
Perspective view, 146
Photographic reproduction, and working drawings, 17
Pictorial drawings, 36–39
isometric drawing, 36–37, 38
oblique drawing, 37, 38
and pencil sketching, 93–94, 98–99
perspective drawings, 35, 38–39
vanishing drawings, 38
Pier, 63
Pilaster, 63
Pitch, 63
Plain-paper copies, and working drawings, 17
Plan, 5, 7
floor, 5, 6, 7
roof, 5
and symbols, 7
Plane geometry, 47; *See also* Construction mathematics
Plank, 63
Plank-and-beam construction, 186
Plate, 63
Plat map, 137
Platform frame construction, 164, 165
sill detail, 185
Platform, stairs and stairwells, 194
Plot plan, 135–140, 172; *See also* Site plan
Plumb, 63
Plumbing and heating, symbols for, 56
Plumbing plans, 202–203
rough, 202
and small church, 271
Post-and-beam construction, 164, 167, 168, 169
Prefabricated, and fireplace and chimney, 191
Preliminary drawings, and construction process, 12
Pre-sale, 11
Presentation drawings
and construction process, 12
as selling tools, 12
Prime coat, 63
Prints
diazo, 16
photographic reproduction, 17

plain-paper copies, 17
multicolor offset, 17
Proportion; *See* Ratio and proportion
Punch List software, 22
Purlin, 63

Q
Quarry tile, 63
Quarter round, 63
Quarter sawed, 63
Quoins, 63

R
Rabbet, 63
Racking action, 164, 167
Radiant heating system, 199
Rafter length calculation, 51; *See also* Construction mathematics
Rafters, 63, 164
Railing, stairs and stairwells, 194
Rake, 63
Random rubble, 63
Ratio and proportion, 46–47
and pencil sketching, 91
proportion defined, 46
ratio defined, 46
rule of three, 47
Rebar, 63
Reflected ceiling plan, and small church, 270
Regions, and working drawings, 7–8
Residential
architect-designed homes, 11
construction procedures, 9, 11
manufactured homes, 11
Residential terms, 57
Residential working drawings, 135–207
foundation plans, 140, 142–145, 148–149, 154
site plan, 135–140
Reveal, 63
Rezoned, 9
Ribbon, 63
Ridgeboard, 63
Riprap, 63
Riser, stairs and stairwells, 194
Rise, stairs and stairwells, 63, 194
Roofing
and elevation view, 174, 175
framing diagram, 204–206
prefabricated trusses, 266, 268
and small church, 271
small church framing plan, 269
support of, 186
trussed rafters, 266
trusses, 266, 268
Roof overhang
construction of, 152
corner construction, 153
Roof pitches, 48, 50; *See also* Construction mathematics

Roof plan, 30
Roofs, types, 175
Rough hardware, 63
Rough opening, 63
Rough plumbing, 202
Rowlock, 63
Rule of three, 47
Run, stairs and stairwells, 63, 194

S

Saddle, 63
Sash, 63
Scales and dimensions, 73–79
    architect's scale, 73–76
    civil engineer's scale, 76–79
    exercises for, 111
    fully divided, 74
    metric, 83–86
    open divided, 74
    and pictorial drawings, 98–99
    reading modular, 79–83
Scarf joint, 63
Schedules, and floor plan, 63, 164
Scuttle, 63, 162
Second-floor plan, 162
Sectioning lining, 40
Sections, 182–187
    horizontal, 30, 34
    longitudinal, 30, 34, 187
    transverse, 30, 34, 187
    windows, 187–189
    of working drawings, 30–35
Seismic code, 63
Selling tools, and presentation drawings, 12
Setback, 63
Shake, 63
Sheathing, 63
Shoe mold, 63
Shop drawings, 15
Sill, 64
Sill detail, 184–185
    and balloon frame construction, 185
    and metal flashing, 185
    platform-frame construction, 185
    slab-on-ground construction, 185
Site plan, 10, 135–140
    plat map, 137
    plot plan, 136, 172
    and small church, 271
Sketching; *See* Pencil sketching
Skylights, 178
Slab-on-ground construction, and sill detail, 185
Sleepers, 64
Small church
    alternate notes, 270
    door schedule, 270
    electrical plan, 270
    elevation, 270
    finish schedule, 270
    finish trim details, 270
    floor plan, 269–270
    footing schedule, 269
    foundation plans, 269
    Interior elevation, 270
    mechanical plan, 271
    plumbing plans, 271
    reflected ceiling plan, 270
    roofing, 271
    site plan, 271
    test for, 272–280
    transverse sections, 270
    working drawings reading, 265–271
Smoke dome, and fireplace and chimney, 192
Smoke shelf, and fireplace and chimney, 192
Soffit, 64
Software, and CADD, 20
Soil stack, 64
Soleplate, 64
Span, 64
Specifications, 209–236; *See also* CSI format
    American Institute of Architects (A.I.A.), 210
    arrangement of information, 210–212
    bidding requirements, 211
    CSI format, 212–236
    description of materials, 212
    Federal Housing Administration (FHA), 212
    finding information, 219–220, 263
    general conditions, 210
    outline form, 212
    purpose of, 209–210
    relationship to drawings, 210
    sample set of, 221–236
    Standard Filing System and Alphabetical Guide, 219
    Suggested Guide for Field Cost Accounting, 219
    supplementary general conditions, 210
    technical sections, 211
    test on, 245–246
Split entrance, 162
Split-level plans, 162–164
Square, 64
Stairs and stairwells
    balusters, 193
    bullnose, 193
    carriage, 193
    closed stringer, 194
    details, 193–195
    and floor plan, 157
    handrail, 194
    headroom, 194
    house stringer, 194
    landing, 194
    newel, 194
    nosing, 194
    open stringer, 194
    platform, 194
    railing, 194
    rise, 194

riser, 194
run, 194
stairwell, 194
step, 194
stringer, 194
tread, 194
winder, 194, 195
Stairwell, stairs and stairwells, 194; *See also* Stairs and stairwells
Standard Filing System and Alphabetical Guide, 219
Step, stairs and stairwells, 194
Stile, 64
Stone, symbols for, 53
Stool, 64
Story, 64
Strata Systems, 22
Stringer, stairs and stairwells, 194
Stucco, 64
Studs, 64, 164
Subfloor, 64
Subsoil, and construction, 8
Subtraction; *See* Construction mathematics
Suggested Guide for Field Cost Accounting, 219
Survey plats, 135–140; *See also* Site plans
Suspended ceiling, 64
Swale, 64
Switching arrangements, symbols for, 55
Switch outlets, symbols for, 55
Symbols
    architectural materials, 52–54
    auxiliary, 55
    concrete, 53
    convenience outlets, 55
    earthwork, 53
    electrical, 55
    elevation, 52
    exercises for, 107–108, 110, 121
    finishes, 54
    glass, 54
    general outlets, 55
    insulation, 54
    masonry, 53
    metals, 53
    partition, 52
    and plans, 7
    plumbing and heating, 56
    stone, 53
    switching arrangements, 55
    switch outlets, 55
    wood, 53

T

Tail joists, 64
Technology, and CADD, 17–22
Tensile strength, 64
Terminology
    abbreviations used, 65–73
    exercises for, 112–120
    glossary of construction, 58–65
    residential, 57
Terrazzo, 64
Thermal conductor, 64
Thermal resistance (R), 64
Threshold, 64
Throat, and fireplace and chimney, 191
Tick strip, 90
Toenail, 64
Topographic drawings, 139
Township, 138
T-post, 64
Transverse sections, 187
    and small church, 270
Trap, 64
Tread, stairs and stairwells, 194
Trim
    base, 195, 196
    casing, 195, 196
    details, 195, 196
    ogee, 195, 196
    plowed, 195, 196
Trimmer, 64
Truss, 64, 266, 268
Trussed rafter, 64s, 266
Typical
    concrete blocks, 143
    and residential terms, 57
    and working drawings, 30, 35

U

U.S. Geological Surveys, 139
Unexcavated, 149

V

Valley rafter, 65
Vanishing drawings, 38
Vapor barrier, 65
Vatican, working drawings of, 3
Veneer construction, 65
Vent stack, 65
Vestibule, 65
Visible lines, 39, 40
Visualization ability, and working drawings, 7

W

Wall details, 182–187
    brick-cavity, 183
    foundation, 182–183
    6" Norwegian brick, 184
Wainscot, 65
Wall section, 171
Wall tie, 65
Water table, 65
Weather stripping, 65
Weep holes, 65

Welded wire fabric (WWF), 65
Western frame construction, 164, 165
Winder, stairs and stairwells, 65, 194, 195
Windows and doors
    awning, 178
    casement, 178
    double-hung, 174, 188–189
    and elevation view, 174, 176–182
    fixed glass, 178
    and floor plan, 156
    hopper, 178
    horizontal slide, 178
    jalousie, 178
    skylights, 178
    typical residential doors, 179–181
    typical residential windows, 176–177
    window details, 189
Wood
    fire resistance of, 265–266
    symbols for, 53
Wood-frame walls, and floor plan, 156
Wood I beam, 65
Working drawings
    abbreviations used, 65–73
    auxiliary drawings, 36
    Babylonian civilizations, 1
    as communication, 5, 7
    and construction, 12, 14–15
    construction mathematics, 41–52
    details of, 5, 30–35
    diazo prints, 16
    elevation views, 28–30
    functions of, 12, 14–15
    and hidden lines, 30
    history of, 1, 4
    and lines on, 39–40
    making of prints, 15–17
    multicolor offset prints, 17
    pencil sketching, 86–99
    photographic reproduction, 17
    pictorial drawings, 36–39
    plain paper copies, 17
    reading of small church, 265–271
    regional variations of, 7–8
    residential, 135–207
    residential terms, 57
    scales and dimensions, 73–79
    sections of, 30–35
    sequence of, 15
    set of, 7
    statue of Gudea, 1
    symbols used, 51–57
    test on, 243–244, 250–251, 255–262
    Vatican, 3
    visualization ability, 7
Written specifications, set of, 7; *See also* Specifications
Wythe, 65

FIGURE 3-68

Perspective view of the Sycamore Walk home.

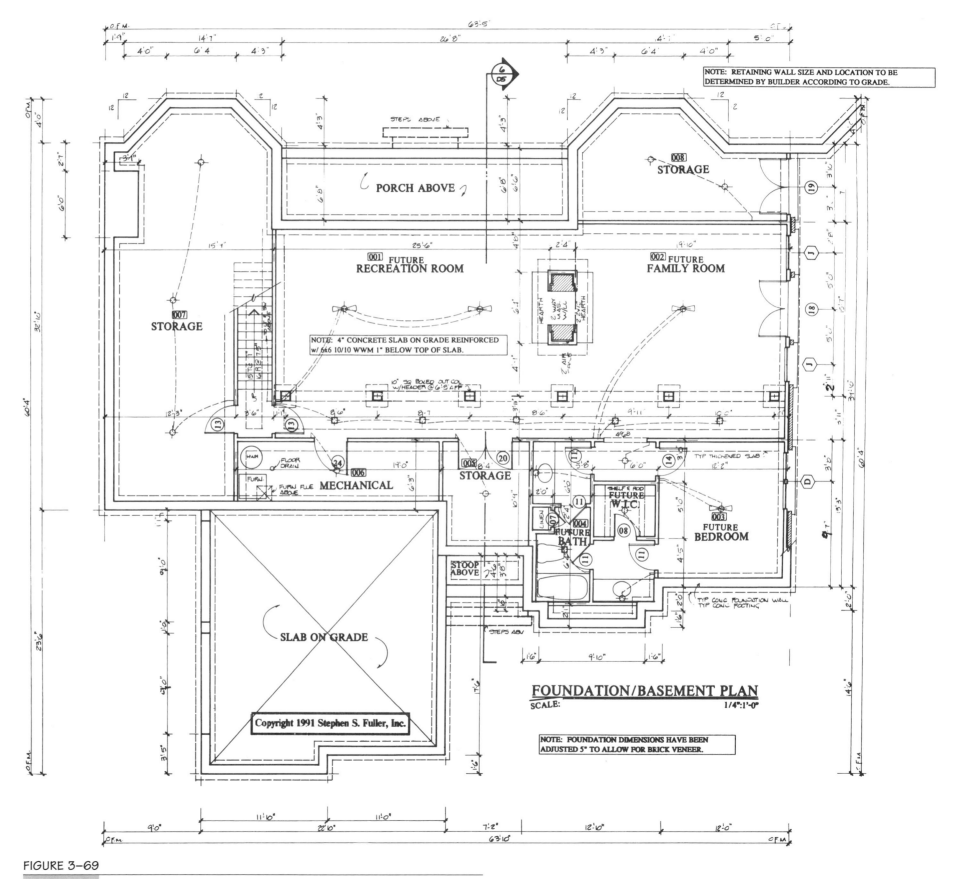

FIGURE 3-69

Foundation/basement plan.

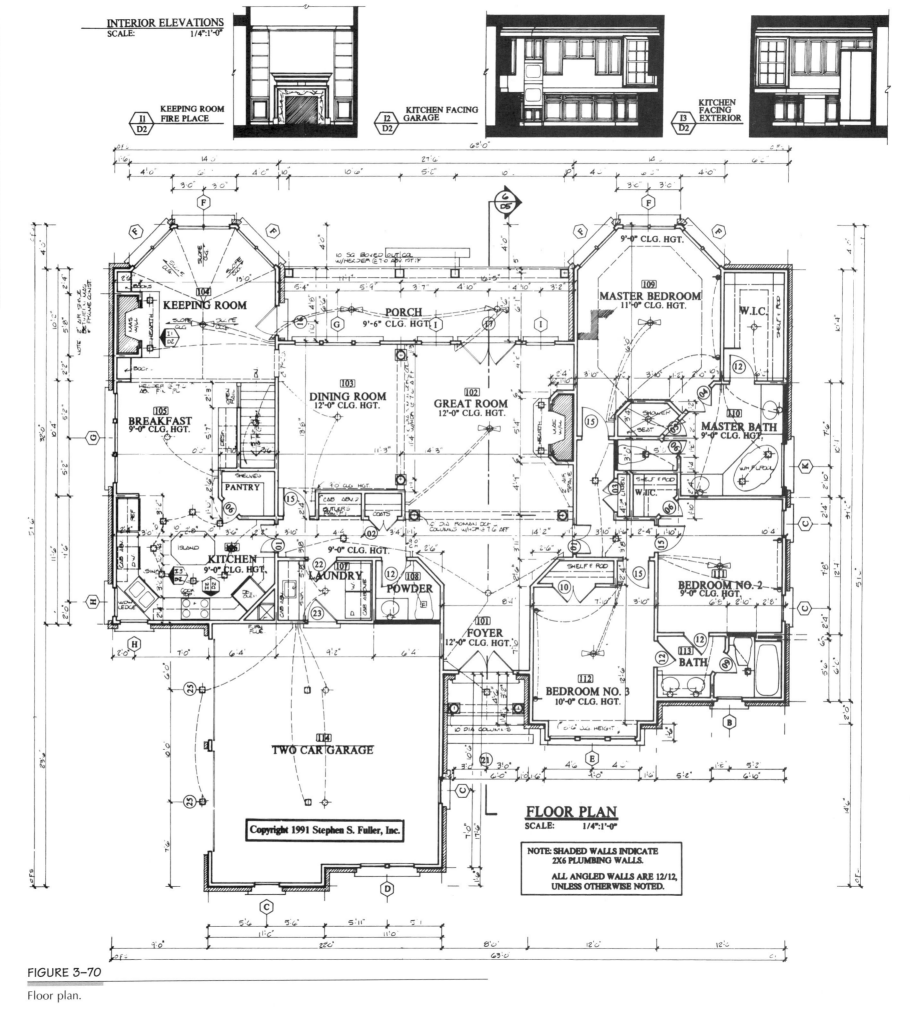

FIGURE 3-70

Floor plan.

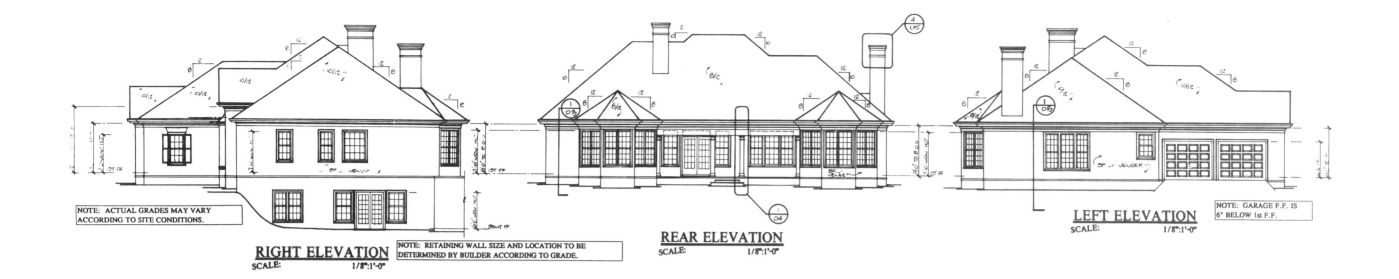
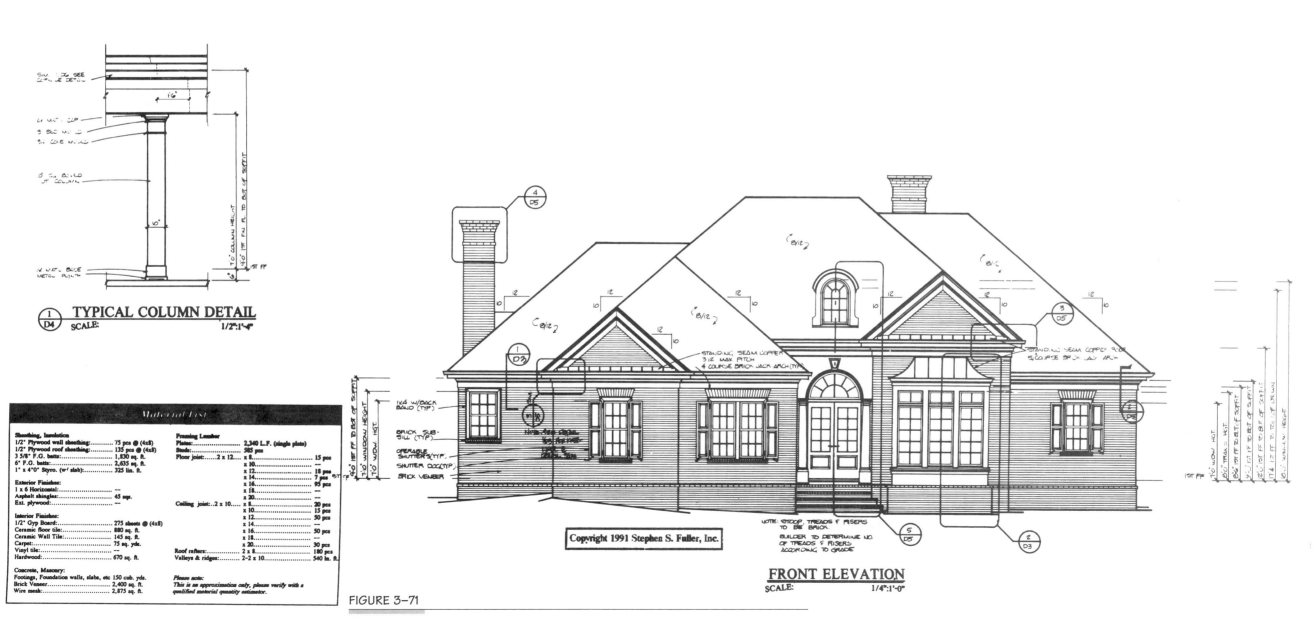

FIGURE 3-71

Elevation views.

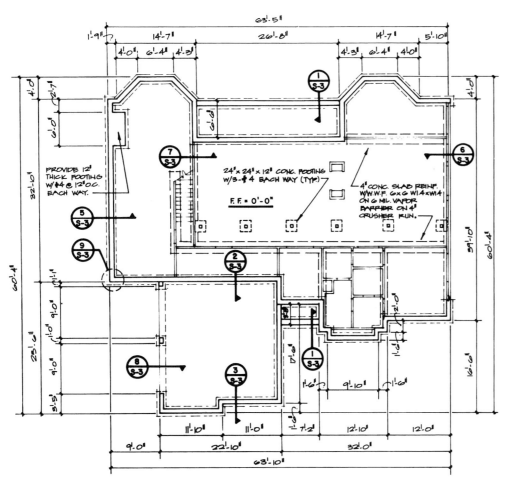

FOUNDATION PLAN
SCALE: 1/8" = 1'-0"

NOTE: 1. ALL PARTITIONS AT PERIMETER AND AT INTERIOR LOAD BEARING WALLS 2×6 @ 16" O.C. ON THIS LEVEL.
2. ALL OTHER PARTITIONS 2×4 @ 16" O.C. ON THIS LEVEL.
3. SEE ARCH. FOR LOCATION OF ALL INTERIOR PARTITIONS WITH FOOTINGS BELOW.

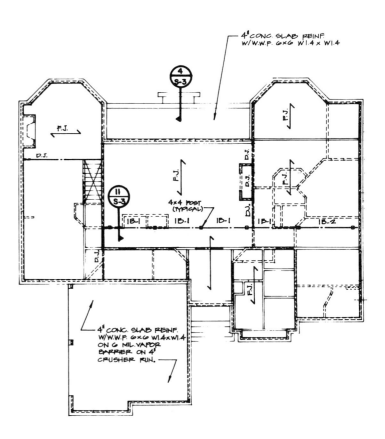

FIRST LEVEL FLOOR FRAMING PLAN

NOTE: 1. ALL JOISTS 2×12 @ 16" O.C. U.N.O.
2. ALL HEADERS IN BEARING WALLS 2-2×10 U.N.O.
3. FLOOR DECKED W/ 23/32" APA RATED STURDI-FLOOR 24" O.C. SPAN RATING. SEE FASTENER SCHEDULE FOR NAILING.

LEGEND

	BEARING WALL
	WALL ABOVE LEVEL BEING FRAMED
	WALL BELOW LEVEL BEING FRAMED
	STACKING WALL
●	POINT LOAD FROM ABOVE
○	SUPPORT BELOW
●—○	JACK TO WALL/BEAM BELOW
	FRAMING DIRECTION
	FRAMING DIRECTION BELOW
	BEAM/HEADER LINE

F.J.	FLOOR JOIST
C.J.	CEILING JOIST
1B-1	FIRST FLOOR BEAM Nº
2B-1	SECOND FLOOR BEAM Nº
CB-1	CEILING BEAM Nº
RB-1	ROOF BEAM Nº
U.N.O.	UNLESS NOTED OTHERWISE
W.W.F.	WIRE WELDED FABRIC
D.J.	DOUBLE JOIST
P.T.	PRESSURE TREATED
O.C.	ON CENTER
A.P.A.	AMERICAN PLYWOOD ASSOCIATION
F.P.	FLITCH PLATE
H-1	HEADER Nº

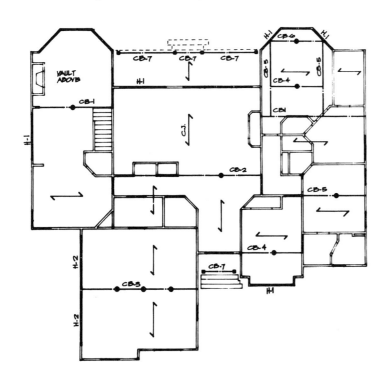

FIRST LEVEL CEILING FRAMING PLAN

1. ALL CEILING JOISTS 2×10 @ 16" O.C. U.N.O.
2. ALL HEADERS IN WALLS BEARING CEILING JOISTS 2-2×8 U.N.O.

BEAM SCHEDULE

MARK	BEAM	COMMENTS
1B-1	2-2×12 W/ 1/4" F.P.	
1B-2	2-2×12	
CB-1	2-2×12 W/ 1/4" F.P.	
CB-2	2-1¾" × 18" ML	
CB-3	5⅛" × 20⅝" GLULAM	
CB-4	2-2×12	
CB-5	3-2×12	
CB-6	2-2×10	
CB-7	2-2×10 P.T.	

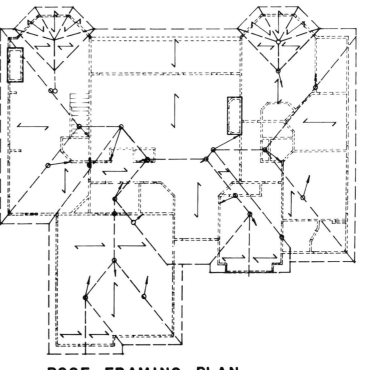

ROOF FRAMING PLAN

NOTE: 1. ALL RAFTERS 2×8 @ 16" O.C. U.N.O.
2. ROOF DECKED W/ 15/32" APA RATED 48/20 SHEATHING. SEE FASTENER SCHEDULE FOR NAILING.
3. ALL RIDGE BEAMS, HIP RIDGES AND VALLEY 2×12 U.N.O.

FIGURE 3-72

Framing plans and beam schedule.

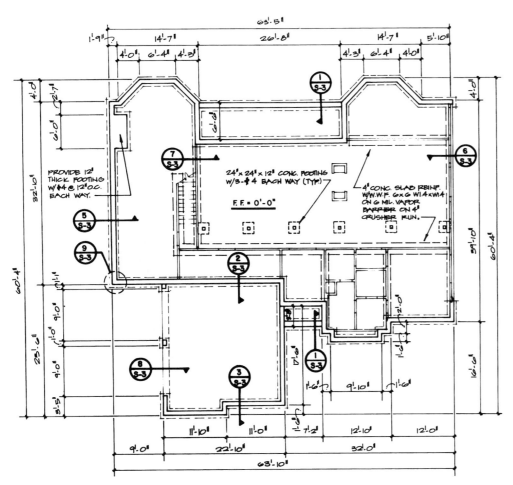

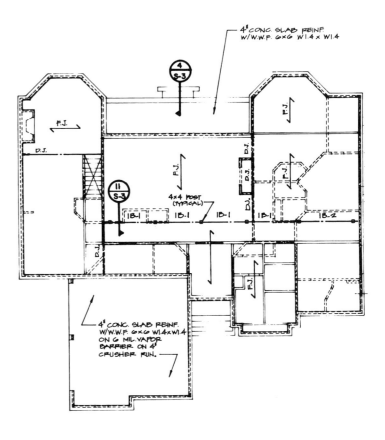

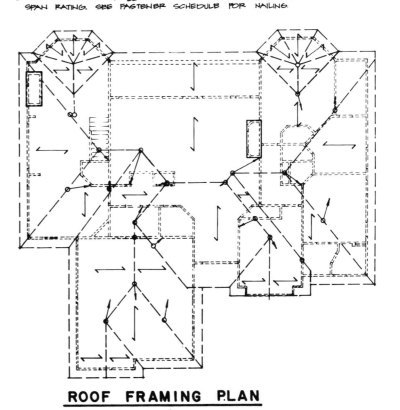

FIGURE 3-72

Framing plans and beam schedule.

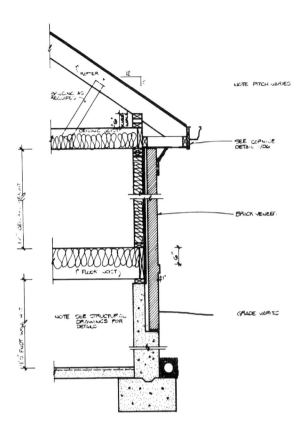
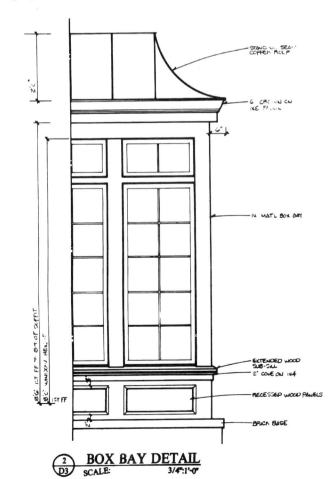
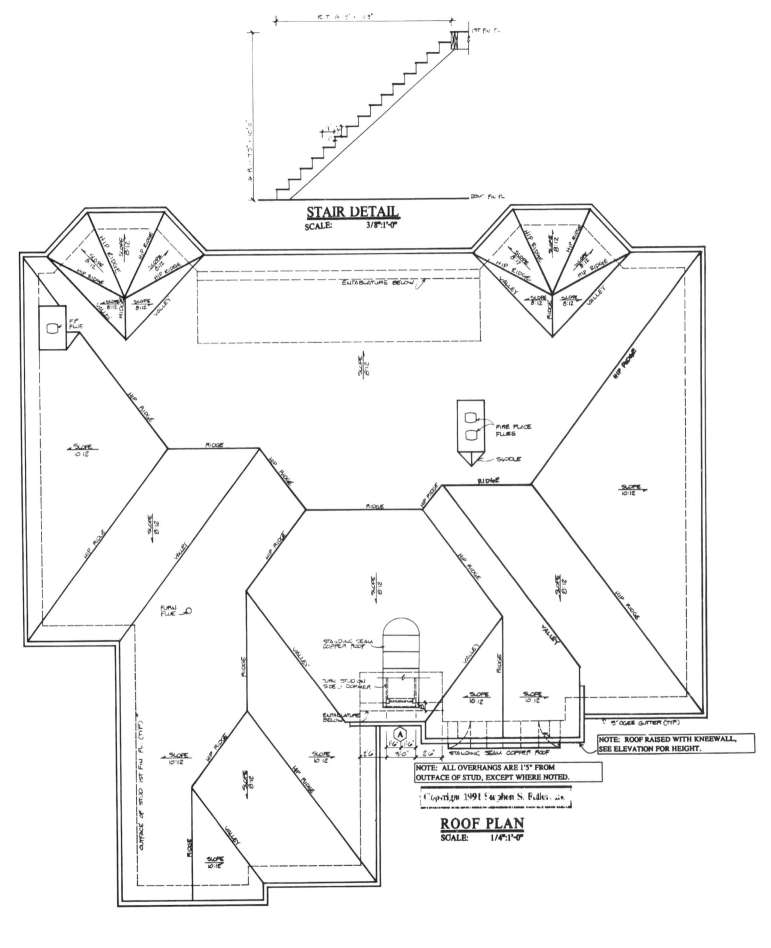

FIGURE 3–73

Roof plan and typical wall section.

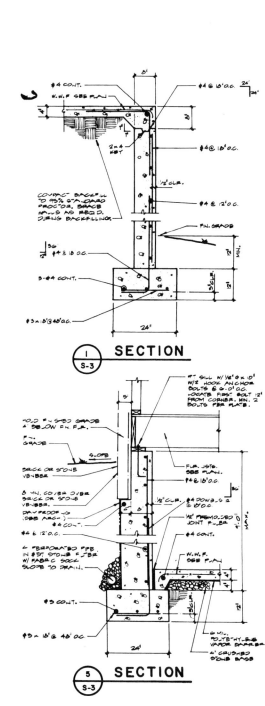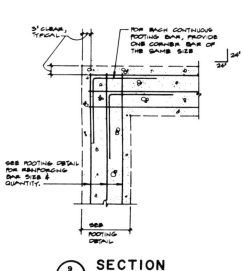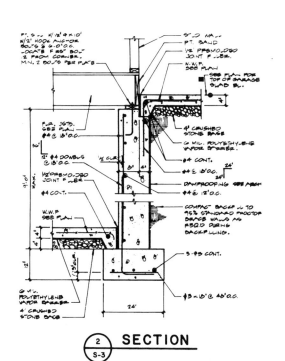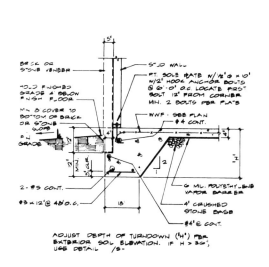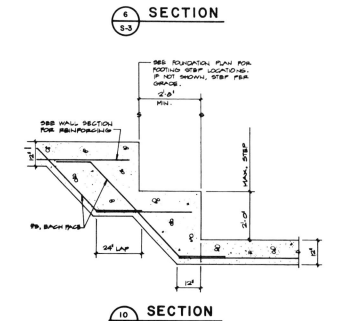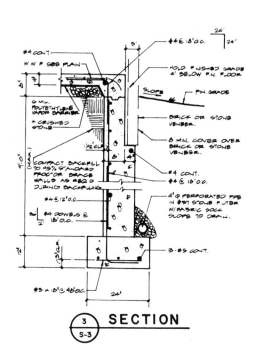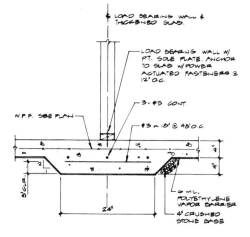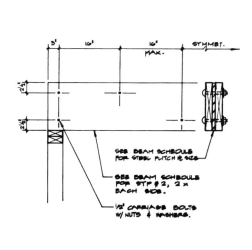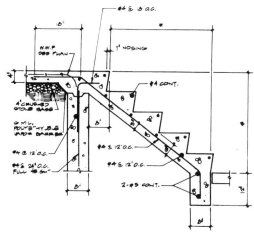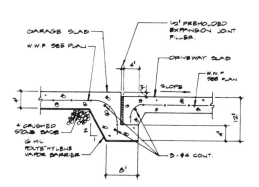

FIGURE 3-74

Section details.

NOTE: ALL DETAILS NOT TO SCALE

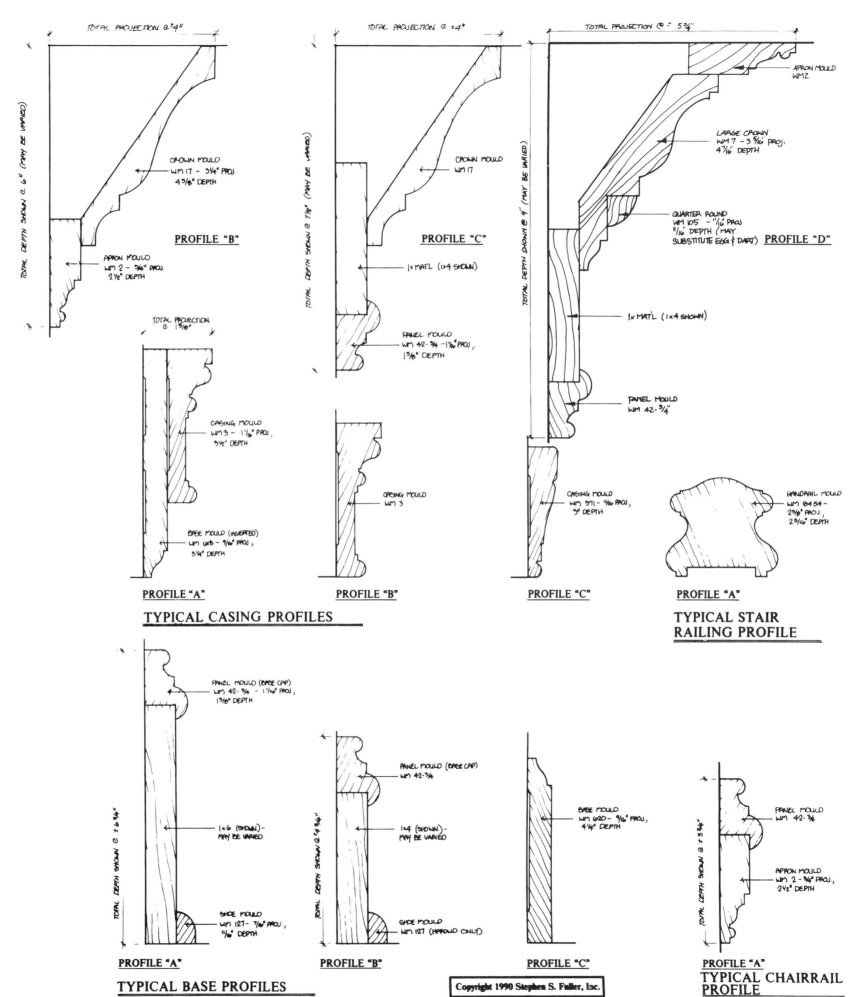

FIGURE 3-75

Molding details and room finish schedule.

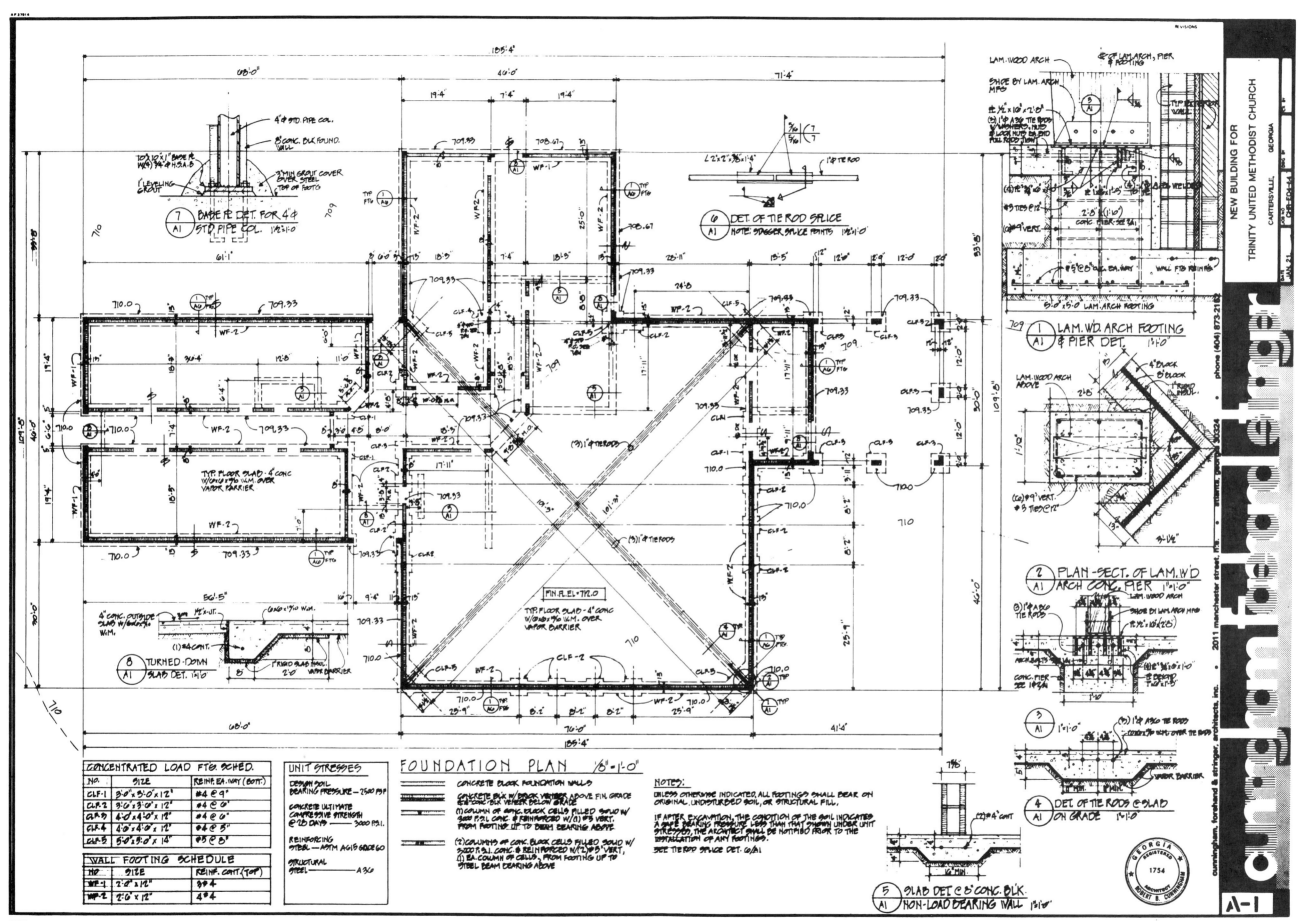

FIG. 5-7 Foundation plan.

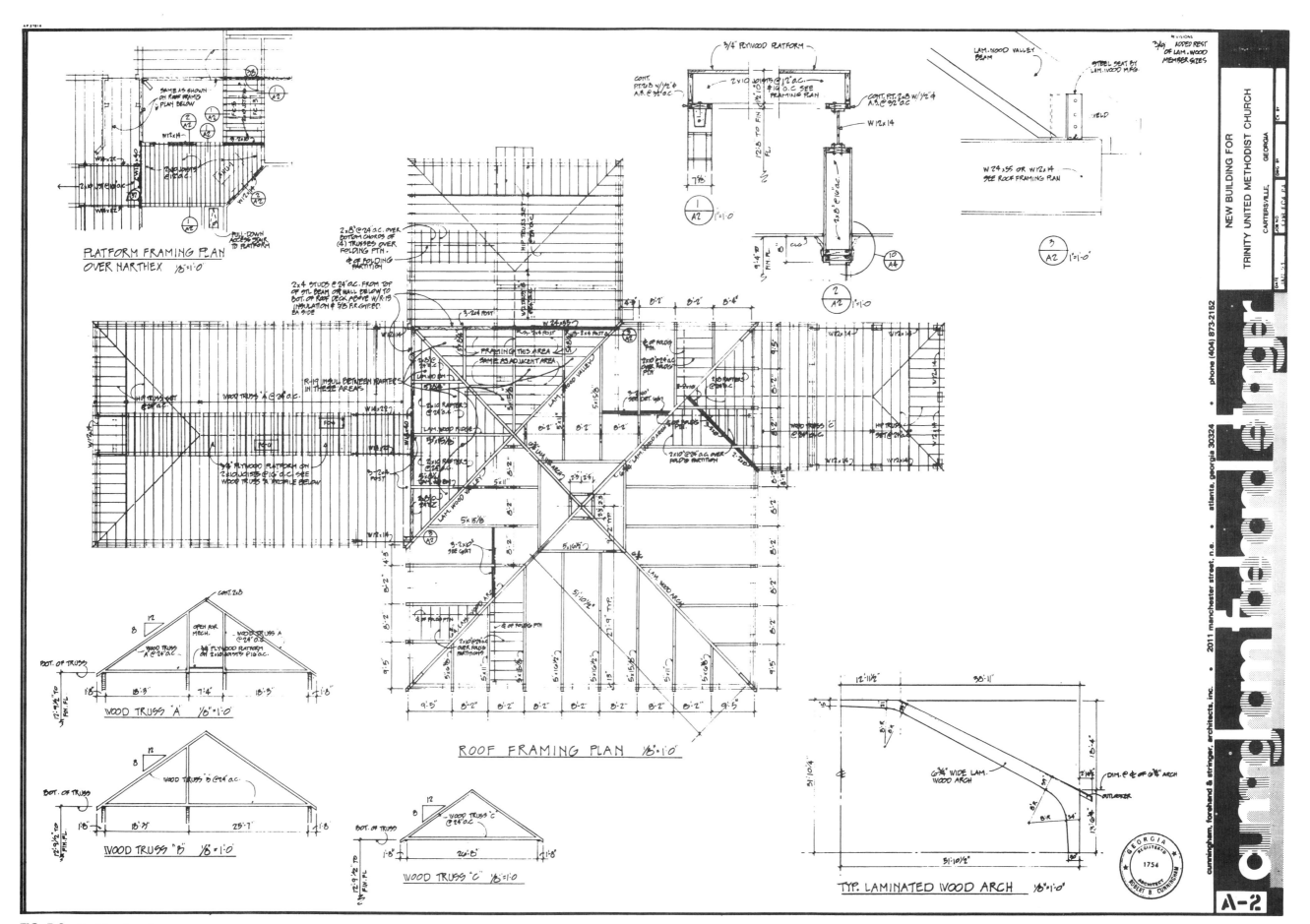

FIG. 5-8 Roof framing plan.

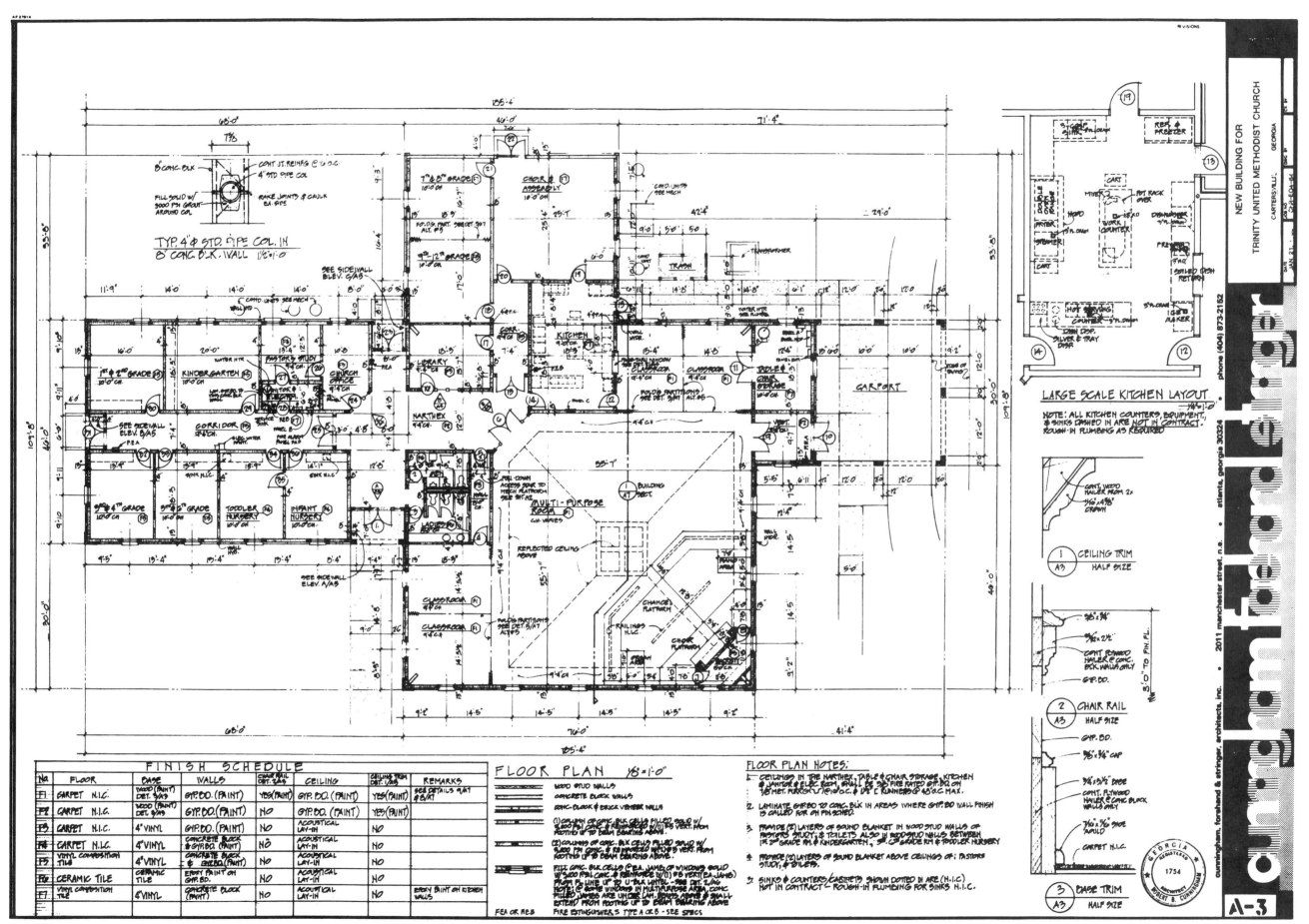

FIG. 5-9 Floor plan.

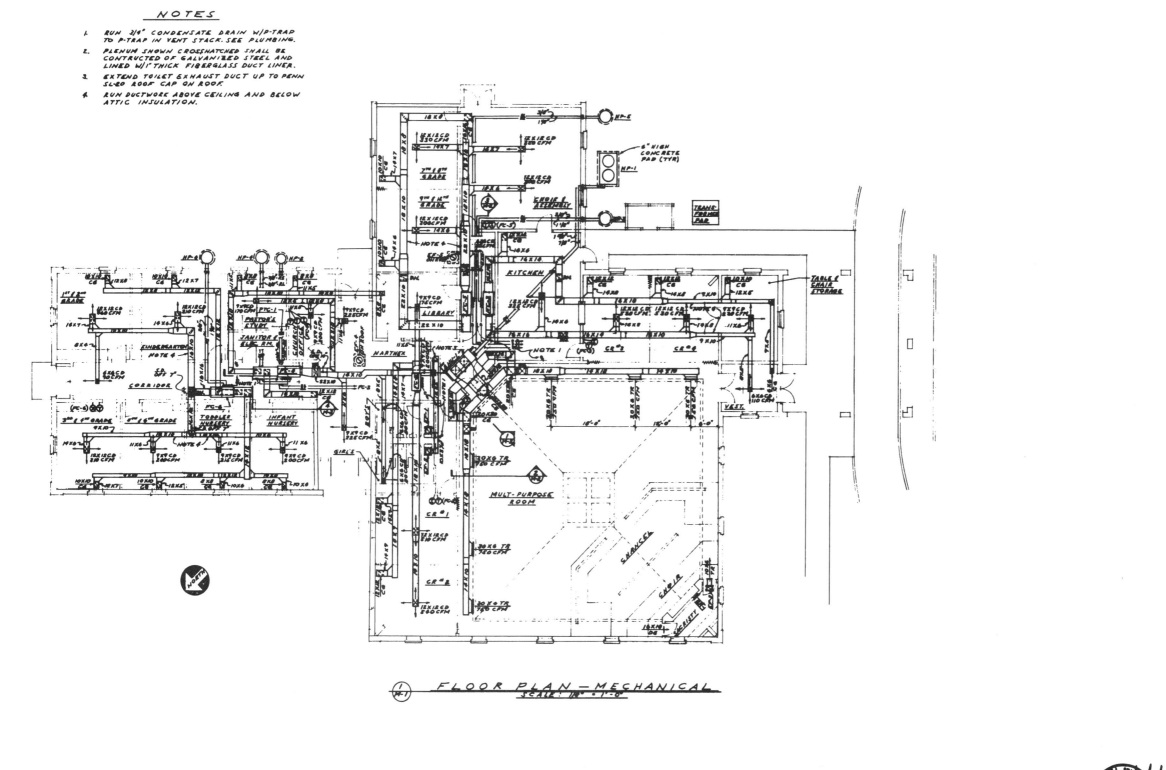

FIG. 5-15 Heating and cooling plan.

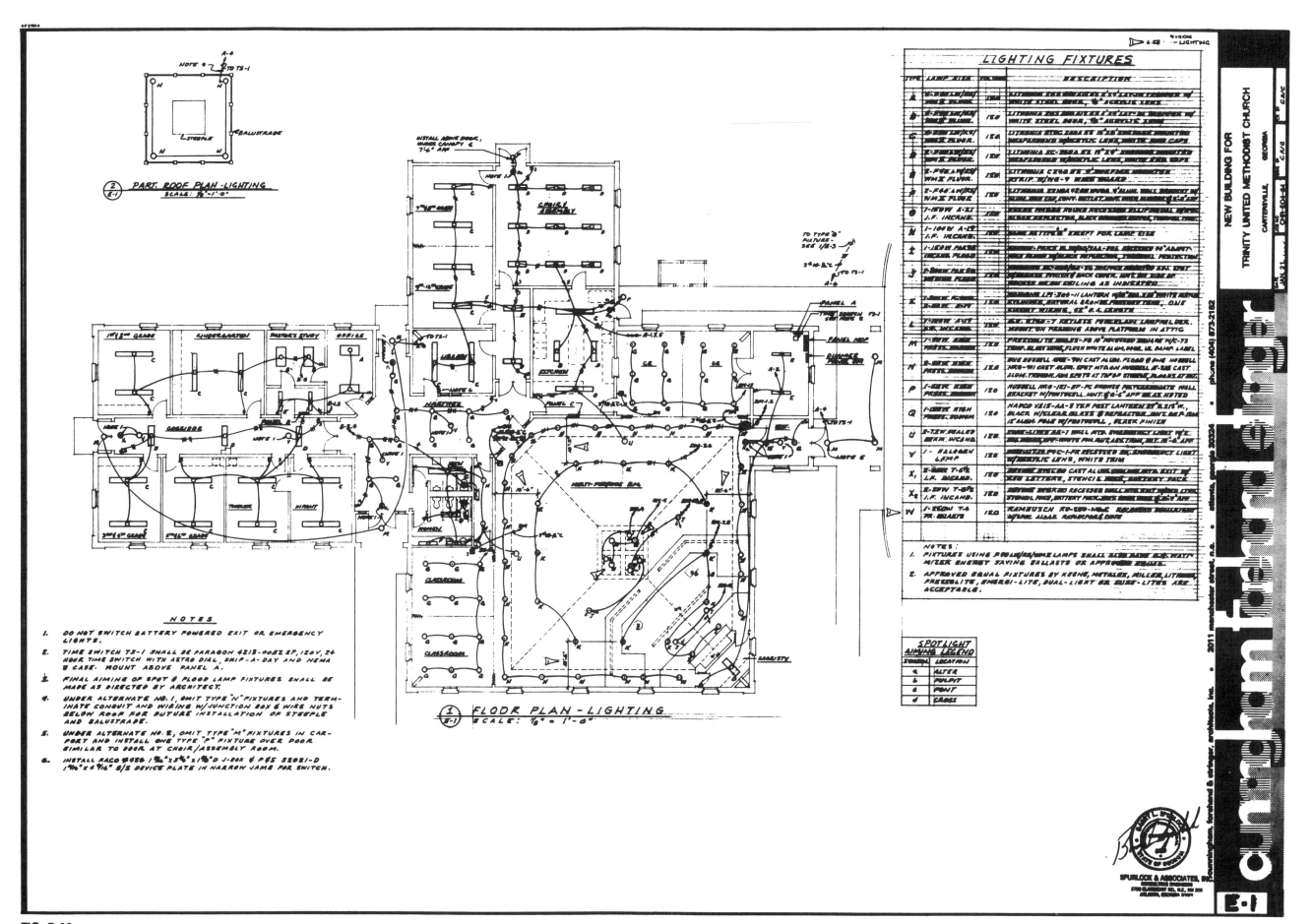

FIG. 5-16 Electrical lighting plan.

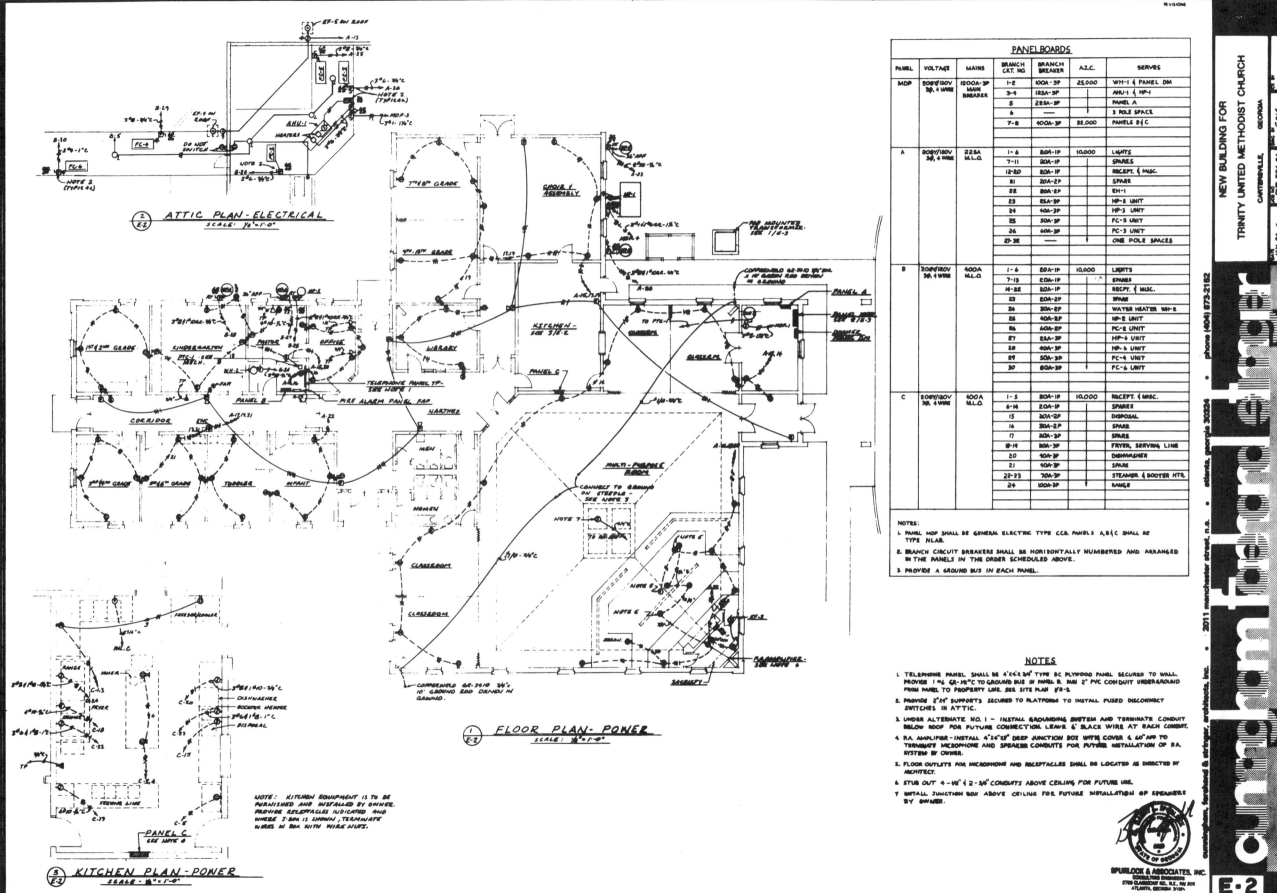

FIG. 5-17 Electrical power plan.

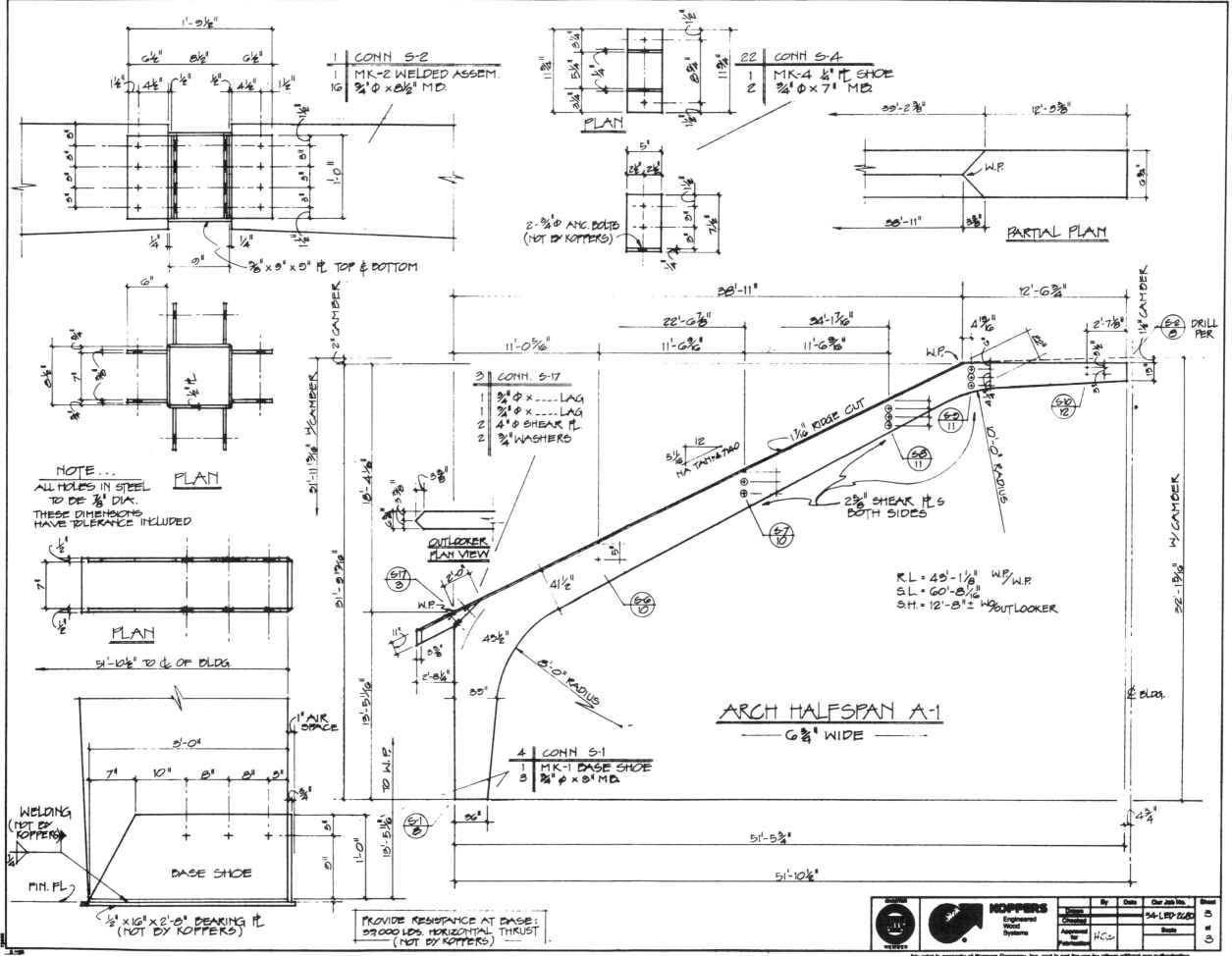

FIG. 5-18 Shop drawing of laminated roof structure.